The Theory and Application of Differential Games

NATO ADVANCED STUDY INSTITUTES SERIES

Proceedings of the Advanced Study Institute Programme, which aims at the dissemination of advanced knowledge and the formation of contacts among scientists from different countries

The series is published by an international board of publishers in conjunction with NATO Scientific Affairs Division

A	Life Sciences	Plenum Publishing Corporation
B	Physics	London and New York
C	Mathematical and Physical Sciences	D. Reidel Publishing Company Dordrecht and Boston
D	Behavioral and Social Sciences	Sijthoff International Publishing Company Leiden
E	Applied Sciences	Noordhoff International Publishing Leiden

Series C – Mathematical and Physical Sciences

Volume 13 – The Theory and Application of Differential Games

The Theory and Application of Differential Games

Proceedings of the NATO Advanced Study Institute held at the University of Warwick, Coventry, England, 27 August–6 September, 1974

edited by

J. D. GROTE

Control Theory Centre, University of Warwick, Coventry, England

D. Reidel Publishing Company

Dordrecht-Holland / Boston-U.S.A.

Published in cooperation with NATO Scientific Affairs Division

Library of Congress Cataloging in Publication Data

Nato Advanced Study Institute, University of Warwick, 1974.
The theory and application of differential games.

(NATO advanced study institutes series : Series C, mathematical and physical sciences ; v. 13)
Includes bibliographies.
1. Differential games—Congresses. I. Grote, J. D., ed.
II. Title. III. Series.

QA269.N37 1974 519.3 74–34041
ISBN-13: 978-94-010-1806-7 e-ISBN-13: 978-94-010-1804-3
DOI: 10.1007/978-94-010-1804-3

Published by D. Reidel Publishing Company
P.O. Box 17, Dordrecht, Holland

Sold and distributed in the U.S.A., Canada, and Mexico
by D. Reidel Publishing Company, Inc.
306 Dartmouth Street, Boston, Mass. 02116, U.S.A.

Softcover reprint of the hardcover 1st edition 1975

CONTENTS

PREFACE

The first international conference on differential games was held at Amherst, Massachusetts, in September 1969. A second meeting, partially supported by N.A.T.O., was held in Varenna, Italy, in June 1970. At these conferences many new theoretical results and applications, especially in economic problems, were presented.

The present volume consists of the lectures presented at a N.A.T.O. Advanced Study Institute on the "Theory and Applications of Differential Games" held at the University of Warwick, Coventry, England, from 27th August to 6th September, 1974.

The main contributions during the first week consisted of a survey of two person zero sum differential games by L. D. Berkovitz and four integrated lectures by R. J. Elliott and N. J. Kalton, who have made important contributions to the concept of "value" of a differential game.

Applications were featured during the second week and included tactical air games, pursuit and evasion problems, as well as computational aspects. A closing lecture with historical perspectives was given by Rufus Issacs, the recognised pioneer of differential games theory.

The time-table received much favourable comment and may be worth recording here as close to the optimum - there were no more than 4 lectures a day at 9.30, 11.30, 14.30 and 16.00, with a long break at 10.30 and a two hour lunch break. This left very ample time for formal discussions to run on if necessary, and also for much informal discussion outside the lecture theatre. Preprints were available for all participants on arrival. Evening social events were organised every other day and two weekend excursions were also arranged.

The lectures are presented here in the order of presentation, except for the Elliott-Kalton lectures which have been placed together. It is hoped that the present volume will provide an up-to-date picture of this active field of research, presented however in tutorial fashion and suitable for those who are beginners.

Much work goes into organising a meeting such as this one and I should like to thank particularly Dr. J. D. Grote who has acted as Editor of this volume and Mrs. Joan Carrington and Mrs. Caroline Howard-Williams who helped with the large volume of correspondence and other paper work that such a meeting generates.

Patrick C. Parks

Director of the NATO
Advanced Study Institute
on Differential Games,
27th August - 6th September 1974

Control Theory Centre,
University of Warwick,
Coventry, CV4 7AL,
England.

14th November, 1974.

LIST OF PARTICIPANTS

Abramatic, J.-F. C.	Centre d'Automatique, 35 Rue St. Honoré, 77305, Fontainbleau, France.
Allwright, J.C.	Department of Computing & Control, Electrical Engineering Building, Imperial College, London, S.W.7., England.
Baranger, J.R.	SANTI, Université de Lyon 1, 43 Boulevard 11 novembre 1918, 69 Villeurbane, France.
Basar, T.	Applied Mathematics Division, Marmara Research Institute, P.K. 141, Kadiköy, Istanbul, Turkey.
Battinelli, A.	Laboratorio Per Recerche di Dinamica dei Bistemi e di Elettronica Biomedica del C.N.R., Calleria Trieste 6, 35100 Padova, Italy.
Becker, H.E.	Universität Trier-Kaiserslautern, 675 Kaiserslautern, Pfaffenbergstrasse 95, W. Germany.
Berkovitz, L.D.	Department of Mathematics, Purdue University, West Lafayette, Indiana 47906, U.S.A.
Blaquière, A.	Laboratoire d'Automatique Theorique, Université de Paris 7, Tour 14-24 Se étage, 2 Place Jussieu, Paris 5, France.
Breakwell, J.V.	A/A Department, Durand 113, Stanford University, Stanford, California 94305, U.S.A.
Case, J.	Mathematical Sciences Department, John Hopkins University, Balitmore, Maryland 21218, U.S.A.

Chan, W.L.	Department of Mathematics, Chung Chi College, CUHK, Shatin, N.T., Hong Kong.
Crouch, P.E.	Control Theory Centre, Department of Engineering, University of Warwick, Coventry, CV4 7AL., England.
Cruz, J.B.	Coordinated Science Laboratory, University Illinois, Urbana, Illinois 61801, U.S.A.
Curtain, R.F.	Control Theory Centre, Department of Engineering, University of Warwick, Coventry, CV4 7AL., England.
Danskin, J.	Electronics Research Laboratory, University of California, Berkeley, California 94720, U.S.A.
Davies, M.J.	Department of Applied Mathematics, University College of Wales, Penglais, Aberystwyth, Wales.
Elliott, R.J.	Department of Pure Mathematics, University of Hull, Hull, HU5 2DW., England.
Ephremides, A.	Electrical Engineering Department, University of Maryland, College Park, MD 20742, U.S.A.
Forster, W.	Department of Mathematics, The University, Southampton, Hants., England.
Grote, J.D.	Control Theory Centre, Department of Engineering, University of Warwick, Coventry, CV4 7AL., England.
Hájek, O.	Department of Mathematics, Case Western Reserve University, Cleveland, Ohio 44106, U.S.A.
Hammond, J.K.	Department of Mathematics, Portsmouth Polytechnic, Hampshire Terrace, Portsmouth.
Holland, C.	Department of Mathematics, Brown University Providence, Rhode Island 02912, U.S.A.
Issacs, R.	Mathematical Sciences, John Hopkins University, Baltimore, Maryland 21218, U.S.A.

Kalton, N.J.	Department of Mathematics, University of Swansea, Swansea, S. Wales.
Kalver, L.C.	University of Pennsylvania, Graduate Group in Economics, Philadelphia, Pa. 19174, U.S.A.
Kelley, H.J.	Analytical Mechanics Associates Inc., 50 Jericho Turnpike, Jericho, N.Y. 11753, U.S.A.
Keppler, R.	Control Theory Centre, Department of Engineering, University of Warwick, Coventry, CV4 7AL., England.
Leitmann, G.	Department of Engineering, University of California, Berkeley, California 94720, U.S.A.
Lévine, P.	Laboratoire d'Econometrie, 4 Place Jussieu, Paris 5, France.
Lévine, J.	Laboratoire d'Econometrie, 4 Place Jussieu, Paris 5, France.
Lukes, D.L.	Department of Applied Mathematics, Thornton Hall, University of Virginia, Charlottesville, Va. 22901, U.S.A.
Marzollo, A.B.	Department of Electrical Engineering, University of Trieste, Trieste, Italy.
Matsumoto, T.	Department of Electrical Engineering, Waseda University, Shinjuku, Tokyo 160, Japan.
Markus, L.	University of Minnesota, School of Mathematics, 127 School Hall, Minneapolis, Minnesota 55455, U.S.A.
Olsder, G.J.	Twente University of Technology, P.O. Box 217, Enschede, The Netherlands.
Pachter, M.	Department of Applied Mathematics, Technion, Israel Institute of Technology, Haifa, Israel.
Parks, P.C.	Control Theory Centre, Department of Engineering, University of Warwick, Coventry, CV4 7AL. , England.

Pau, L.F.	Ecole Nationale Superieure des Telecommunications, 46 Rue Barrault, 75634 Paris, France.
Pierson, B.L.	304 Town Engineering Building, Iowa State University, Ames, Iowa 50010, U.S.A.
Pikkemaat, G.F.	Ecometric Institute, P.O. Box 800, Groningen, The Netherlands.
Plant, A.T.	Fluid Mechanics Research Institute, University of Essex, Colchester, CO4 3SQ. England.
Pritchard, A.J.	Control Theory Centre, Department of Engineering, University of Warwick, Coventry, CV4 7AL., England.
Roxin, E.O.	Department of Mathematics, University of Rhode Island, Kingston, R.I. 02881, U.S.A.
Rudge, J.	Statistics Laboratory, 16 Mill Lane, Cambridge, England.
Schmitendorf, W.E.	Mechanical Engineering Department, North Western University, Evanston, Illinois 60201, U.S.A.
Shell, K.	Department of Economics, University of Pennsylvania, 3718 Locust Walk CR, Philadelphia 19174, U.S.A.
Spielman, H.	Scientific Affairs Division, N.A.T.O., 1110 Brussels, Belgium.
Stalford, H.	Code 5308, Radar Division, Naval Research Laboratory, Washington D.C. 20375, U.S.A.
Stern, R.J.	Department of Applied Mathematics, Technion, Israel Institute of Technology, Haifa, Israel.
Szegö, G.P.	Numerical Optimisation Centre, Hatfield Polytechnic, 19 St. Albans Road, Hatfield, Herts., England.
Tabak, D.	Department of Industrial & Management Engineering, Ben Gurion University of the Negev, P.O. Box 2053, Beersheva 84 120, Israel.

Tweddell, E.T. — N.A.T.O. Channel Command, Northwood, Middlesex, England.

Van Swieten, A.C.M. — Mathematics Institute, P.O. Box 800, Groningen, The Netherlands.

Varaiya, P. — Department of Electrical Engineering, University of California, Berkeley, California 94720, U.S.A.

Vincent, T.L. — Department of Mathematics, University of Western Australia, Nedlands, West Australia 6009.

Witsenhausen, H.S. — Bell Laboratories, Murray Hill, New Jersey 07574, U.S.A.

Yu, P.L. — BEB 620, University of Texas, Austin, Texas 78712, U.S.A.

THE PAST AND SOME BITS OF THE FUTURE

Rufus Isaacs

The Johns Hopkins University

Since my book Differential Games [1] appeared in 1965, the subject has burgeoned vastly. I have not followed all its developments -- my research efforts have lately turned elsewhere -- and so there is some doubt as to my now qualifying as an expert. A second printing of [1] is due shortly and therein I describe some advances, particularly those that resolve quandaries appearing in the first. Here the material will be limited -- narrowly or broadly, as you deem -- with much of it essentially also in the introduction to the second printing.

How the subject started seems a good place to begin.

A PERSONAL HISTORY OF DIFFERENTIAL GAMES

In the mid 1940's the impact of game theory in the Rand Corporation was intense. Its source was the then new, and by now classic, book The Theory of Games and Economic Behaviour by von Neumann and Morgenstern. Then under the auspices of the U.S. Air Force, Rand was concerned largely with military problems and, to us there, this syllogism seemed incontrovertible:

Game Theory is the analysis of conflict.
Conflict analysis is the means of warfare planning.
Therefore game theory is the means of warfare planning.

But, despite some excellent output, the fruits proved meager. The initial ardor slowly abated. I pondered the reasons and sought a remedy.

J. D. Grote (ed.), The Theory and Application of Differential Games. 1-11.

One cause was that we adhered to the book's approach: matrices. The strategy matrix is the key device for proving the basic theorems and is indispensable for the birth of a meaningful theory. But how good was it for solving problems?

I thought of classical mathematical achievements. The calculation of a planetary orbit by Newtonian means, for example. This could be viewed as a one-player game, with the player navigating so to minimize the Hamiltonian which is here a payoff. Such a single player would mean that the game matrix was a single column with each row (or element) corresponding to an orbit candidate -- here a strategy -- with the entry being the ensuing Hamiltonian payoff. The game theory dictum "find the minimum" seems rather vacuous.

For success the vacuity is replaced by an effective tool, such as differential equations. Could something like this be done with two-player games? How? Revelation came with a specific class of problems.

We called it ess-ay-em then when it was an arcane and incipient term; the monosyllable SAM was still remote from popular jargon. The early Rand discussions proposed and discarded schemes for a pursuing missile to seem an enemy aircraft. Collision course was good? Yes. But if the targeted craft did this (or that), the missile won't hit. Constant bearing navigation? That seems easier for the missile. But what if the evader does that (or this)?

What optimal pursuit scheme for the missile? The crucial idea dawned: the question was unanswerable unless one also asked "What optimal evasive scheme for the aircraft?" The whole subject was game theory.

My attempt to forge this concept into a usable tool was inadvertently aided by a remark from a young colleague. He had devised a pursuit scheme, he said enthusiastically, in which the missile used the position,velocity, acceleration and, in fact, the first five time derivatives of the target's motion!

I mused on this. Should the missile act so? The evading pilot E controls something. The levers or knobs at his disposal must essentially regulate a time derivative of some order. This the missile P should not let affect his decisions, for E, having it under his instantaneous volition, could choose it to render them bad. Thus the key distinction between state and control variables!*

---

*I called them descriptive and navigation variables, but years later when I became aware of control theory, I adopted its better terms.

Once I had these concepts, it was not hard to find the principal governing their use. It is the

Tenet of Transition. If the play proceeds from one position to a second and V is thought of as known at the second, then it is determined at the first by demanding that the players optimize (that is make minimax) the increment of V during the transition.

It is mentioned often in my book (See Section 12), but was never quoted in the first printing, an omission I regret. The tenet was the key to a differential equation structure and such led to an actual means of solving simple pursuit games.

Some, to adumbrate reality, had kinematic limitations on the motion of the craft, such as the isotropic rocket and homicidal chauffeur games. The latter was a lucky choice; its at times baffling exploration brought to light basic concepts such as universal surfaces, barriers, and equivocal surfaces. The latter intrigued me as a phenomenon which belonged to competitive games alone and could not appear in anything like the calculus of variations.

The first paper, Games of Pursuit [2] appeared as a Rand report in 1951. It contained the tenet of transition but its presentation of continuous game techniques was crude and a bit erroneous. Understandably it was rejected by a mathematical journal. I was beginning to grasp the innovations inherent in my incipient theory. (See [5] or [4], first chapter, for amplification); its novel flavor could simply not be conveyed in a brief article.

Progress of pursuit games was steady, when a colleague, Arnold Mengel, brought me two air-war games, which later appeared in my book as the two versions of the war of attrition and attack. The first he had solved but the second baffled him. I was able to see that the cause was a universal surfact. New vistas opened; there was scope for my techniques beyond pursuit and evasion contests. I coined the name, differential games.

Through the Rand seminars, my colleagues were privy to my progress. Enthusiasm seemed ardent at first, but it dwindled. A proposed joint research project did not materialize. From my fellow mathematicians, I heard "Too hard," "Not regorous," or "No existence theorem." The latter two grounds were sound: I had obtained some solutions, but in what sense were they valid and how could they be proven? Although I did not know, I felt that answers were there and stuck to the subject. Finally, Samuel Karlin, then a Rand consultant, offered the splendid remedy of K-strategies, but apparently too late to recapture Rand's interest.

The late John von Neumann, then on appointment with the

Atomic Energy Commission, was on a evaluation tour of institutions serving the national interest. When he visited Rand, I spent an hour with him. His enthusiasm seemed great; his alacrity with new ideas astonishing. For several years thereafter I repeatedly heard from various reliable although informal sources that he deemed my work the best thing going at Rand.

If he said so apparently it was also too late. At the time I felt overwhelmed by the difficulties in making my new subject fulfill its rich promise. I wanted Rand to devote substantial resources to a joint effort, or, failing that, to support a long term investigation by myself alone. Neither was to be. Although I have never learned what went on behind administrative doors, my six-year stay at Rand ended in 1954.

Surely in this nation, then spending lavishly on research -- military, theoretical, and an opulence of other types -- there was support for differential games. I prepared a document on my plan for military applications in two versions: one detailed*; the other meant for a busy apocryphal executive or armed service officer.

I sent out applications for a job -- part - if not full-time -- or a grant. The country teemed with the latter then. But the floods of cash deluged orthodox channels only: research "yes" but new ideas inexorably "no."

When opportunities had long seemed nonexistent for simultaneously pursuing the subject and earning a living, I was offered a teaching fellowship abroad in 1963. The lecturing load was light; here was a chance to set down the ideas I had been carrying in my head for ten years. A book -- the only possible vehicle -- resulted.

During its galley proof stage, there was a national meeting on control theory. My friend, John Danskin, knew of it and kindly arranged for my attendance. The experience was unique and bizarre. The speakers were largely grappling with one-player versions of differential games such as I had solved years before. I had the eerie feeling of a bird who could fly with two wings watching fledlings attempt it with one. The eminent Pontryagin had come from Russia to present the featured address. It's title? Differential Games! How had this phrase, then published only in my Rand reports, reached the Soviet Union? His topic? A pursuit game, virtually the isotropic rocket, which he treated splendidly

---

*<u>A Proposal for Research in the Theory of Differential Games</u> (1955). I now possess the only copy extant, to my knowledge.

but plainly had just made a beginning.*

Since then the subject has bloomed. There have been national meetings exclusively on it and now an international one. It has even attained sufficient orthodoxy for the bestowal of research grants. Even the ears of the armed services, once so tenaciously sealed, are open a bit, for I have seen sponsored papers on the military use of differential games.

THE RELATION TO CONTROL THEORY

In the mid 1960's, at least, control theory was a widespread subject with an array of journals, national and international meetings. It was virtually identical to one-player differential games. Was there a common genesis to these two theories, on a special case of the other?

I do not fully know. I can state that my own work was done completely independently; I became aware of control theory only when my book was nearly complete; a few changes, such as the adoption of the names state and control variables and an appendix, were the sole effect. If there was any dependence at all, it was the other way around, for some auditors at my Rand seminars later produced papers on control theory.

At first at these seminars I stressed the tenet of transition as the pedestal of the theory. Later it came to seem more of an idle truism: the more ideas became familiar the more dispensable it grew. But it survived these seminars, although only in the one-player special case, and in this form was known as the Principle of Optimality.

But essentially I believe the two theories developed separately, for much of control theory came from the U.S.S.R. The history of mathematics has often shown parallel evolution when the times were ripe.

At times I have mused on this idle question. What would have been the impact on control theory if sponsorship for my book had been obtained with less than a ten year wait?

THEORY AND APPLICATIONS

Like many arenas of the mathematical sciences, differential

*In 1966, at the International Mathematics Congress in Moscow, I had the pleasure of a lengthy, congenial talk with Professor Pontryagin. He later wrote a foreword for the Russian translation of my book.

games has interest both as a theory and for its applications. The former dwells in the intellect alone, but here I think we find more than in the usual aesthetic appeal of pure mathematics. For there are contestents striving for opposing ends by strategic decisions. Purely abstract as we may wish to be, we cannot fully avoid endowing them with human characteristics. For example, we could try to envisage the outcome of a partie were one player acting optimally, but his opponent -- unaware of the exact solution -- can but apply obvious common sense.

The games of Sections 9.5 and 9.6 of my book offer some striking examples here. "Looping the blockader" (Example 9.6.2) seems to be of the stuff of pristine mathematics: could an application exist at all? In "Patrolling a Channel" (Example 9.6.3) I urge the reader -- when he has mastered the theory -- to envisage what would happen if P played optimally in a channel just a slight bit larger than the minimal width L_c. Here we have proved that E can pass, but, if he did not know the theory, he would have a rough time so doing, being driven further leftwards by each failed attempt. Is there a way (other than computer simulation) that we could experiment and see?

Thus I think differential games best lies in a realm somewhere between pure theory and models of the real world, a view expanded a bit in my chapter in [4]. Applications there are and the theory, as others, starts with idealized models of reality.

The history given earlier shows how the genesis of differential game was military. It was a time for patriotic ardor; the Cold and then Korean Wars spurred the nation; Viet Nam was not yet.

Even before this era ended, the emphasis on warfare of my work jabbed my conscience. Surely conflict and competition were human ubiquities. Should not differential games have a widespread scope? I sought it.

The trouble was that the most significant human contests are not zero-sum. There are recreational games of course but here theoretical solutions would likely annul the recreation. The only truly fertile domain I could find was warfare.*

---

*It can be soundly argued that wars are not zero-sum or else hardly any would be fought. This is true, but battles, campaigns, and such portions of the whole can usually be modelled with adequate versimilitude in the zero-sum format. Certain examples in the book -- even though necessarily very simplified -- indicate how.

Occasional zero-sum possibilities have come to light since. An election with a variety of campaign tactics available to the two parties could be modelled as a differential game in a variety of ways; Steve Alpern is considering the comparison of actual pursuit strategies of predatory animals with the optima as a facet of animal psychology.

But economics seems most promising. Among others, my competent colleague, Professor James Case, has built a non-zero-sum theory of differential games and treated thereby a number of promising and interesting economic situations of conflict; see his papers [3] and another in this volume.

There are difficulties here. Perhaps the prime one is this: with a distinct payoff for each player, there is not one main equation but several. There does not seem to exist an adequate technique for deriving the path (or characteristic) equations analogously to games with a single value. Case gets around this trouble in ingenious ways. No doubt further conquests will come and with them a rich, useful and important theory.

MILITARY APPLICATIONS

No one, no matter how ardent a pacifist, can deny the value to his country of improving military decision making. Can differential games be so used? As already stated, I have thought so for many years and once encountered only frustration thereby.

First, let us not underrate the difficulties in military (or pacific) applications. In [4] and [5] I have argued -- and shall sketch here by briefly -- the existence of a ladder of applied mathematics. The classical phases are the bottom rung and can be largely viewed as one-player games; the successive rungs are two-player zero-sum games with perfect information, then without it, then non-zero sum, etc. At each rung the difficulties increase enormously, not only in techniques but in the very nature of the problems and the existence and meaning of their answers. Most of us are handicapped by an education based on the bottom rung and there are times when we all blindly try to adopt our precepts to the higher levels.

Such is the philosophic basis of the difficulties of applying differential games to military problems. It translates into more mundane terms to anyone who seeks solutions without losing his sense of practical reality.

I believe the incubus can be surmounted. But not in the way tried until now: isolated research papers printed in journals or read at meetings -- although many are excellent -- is not the vehicle for the complex affair of applying differential games to warfare. A group must be devoted exclusively to it. It should consist of applied mathematicians under the advisory aegis of intelligent and knowing military men.

What kind of problems would such a group tackle? Those with moving craft are amply suggested by many extant examples, although, of course, often simplified.

There are battle games such as our two versions of the war of attrition and attack and the Battle of Bunker Hill, also similarly simplified. The crux of these is the basic question of distributing effort, means, supplies, and men between long and short range goals.

There are other broad and basic military questions. For example, under what conditions is a strategic retreat advisable -- that is, when is it better to yield now in the hope of gain later?

How should the group tackle such problems? The first step is clearly a choice of problem (in the broad sense as just discussed). The next is meeting of the group at which a variety of models is discussed, defined, and decided upon. These are to cover possible salient factors so as to span a large but reasonable spectrum of cases. As is true in all applied mathematics, the models must compromise between being reasonably solvable and being realistic. Thus, too much elaborate detail must be eschewed in any one case. But there will be facets whose effect we particularly wish to ascertain. Consequently one or some models will include such a facet, but will not embody too many others, so that the sought effect will stand out in bold relief.

The next step will be to allocate the ensuing assortment of problems to individuals or teams for solution. Each will be a substantial mathematical challenge.

What should we expect from a study and comparison of the results? Perhaps a certain phenomenon will be common to many solutions. It may be commonplace or it may be surprising. If the latter, possibly more models should be explored for better verification or simulated war games be played with an eye to see how our new phenomen withstands a wealth of realistic detail too involved for theoretical analysis. Out of this may come a new principle of warfare of a kind that could never have been reached by either armchair or empirical reasoning.

I say "may," for if the results of research could be foretold, the motive for research would be undone. Similarly, I must be vague about just what I mean by "principle." All I can say is that from my experience in solving problems, a glimmer of their existence -- dim but certain -- seems to be there. For example, in the war of attrition and attack, second version of [1], we see that under a large set of conditions, one player should continuously attack his enemy to just enough extent as to hold the latter's forces at strength $\sqrt{m_2/c_2}$. Why? Is there some genuine significance to this number? Or will it evaporate if the model is changed slightly? If the former, we would have uncovered a "principle" or "rule of warfare," -- possible a very advantageous one.

On the other side of the ledger, our comparison of the solved array of models may reveal that certain principles depend sharply on our assumptions (another way of saying model). Our conclusion would then be that these facets are not significant and the command should not waste valuable decision effort on them.

But all this, it seems, must await either a new generation of military minds or news that a possible hostile country has adopted such a program. Can we then expect a frenetic effort to close the "differential games gap" as once occurred in the U.S. with missiles?

GAMES OF INCOMPLETE INFORMATION

A formidable hiatus -- particularly in regard to military applications -- faces us here: when and how do we handle mixed strategies?

The latter generally appear in the solution only when the lack of information is really essential to good play of the game. Such means that it pays a player to move randomly so as to deter his opponent's prediction of future states; deception is vital and must be countered by outguessing.

There are cases when the cost of such randomness in terms of poor maneuvering is too high. For example, consider any simple pursuit game and suppose that P's knowledge of E's state is uncertain. Should E follow a simple course, such as a constant velocity path, P can offset his error by multiple observations and extrapolation (See [1] , Section 12.2). Then E can counter with a mixed strategy -- say, a randomized zig-zagging -- and so frustrate P's predictions. But the zig-zag will cut into E's effective speed in fleeing from P. Should E's ensuing loss exceed his gain from deception, we should expect his optimal strategy to be pure and not mixed. And this seems to happen often.

Hence one facet of the basic problem: when are optimal

strategies mixed?

When they are, what are they?

What I have learned, if anything, in the years since Chapter 12 of my book, is that the answers to these questions are generally more baffling than I had first thought. In that chapter I proposed some simple games of no information at all so that optimal play is pure outguessing. Would the spectrum be clearer from this light on one end? Here is the example which seems to have attracted most attention (including two of my students, Richard Worsham and Joe Foreman, whose doctoral dissertations were based on it):

Initially P and E are on the circle each with a uniform probability distribution. This fact is known by each player, but not his opponent's location initially or at any time preceding capture, which means both players occupy the same point. While P's speed is bounded by w, E may employ any speed. Payoff is (expected) time until capture.

Steve Alpern, as part of his dissertation at New York University (1973) found this remarkably simple and elegant solution:

The cohato (that is, coin-half-tour, as found independently and so named by Joe Foreman) strategy is: From his initial point a player traverses half the circle at a fixed speed, choosing each direction with probability 1/2. Once he reaches the diametrically opposite point, he again tosses a coin to decide the direction of a second half-tour and repeats such steps indefinitely.

The optimal strategies of both P and E are cohato with speed w. The Value is the time for either to traverse 3/4 of the circle.

At present such simplicity seems a rarity. There still appears much to be learned.

Obviously lack of information will be common in military problems. Not only can the players' ignorance pertain to the state, but often to the fixed parameters (which appear as coefficients in the formal equations). Can mixed strategies be optimal here too?

It would seem that here is a road we must learn to travel before warfare becomes a fruitful application of differential games.

REFERENCES

1. R. Isaacs, Differential Games, John Wiley and Sons, Inc., New York, 1956.
2. ———, Games of Pursuit, Rand Report P-257, 17 Nov. 1951.
3. James Case, On Recardo's Problem, J. of Econ. Theory, Vol. 3, No. 2, pp. 134-145(1971).
   ———, Differential Trading Games, In [4], pp. 377-400.
4. Topics in Differential Games, Edited by A. Blaquiere, North-Holland Publishing Company (1973).
5. R. Isaacs, Some Fundamentals of Differential Games, J. of Optimization Th. and App., Vol. 3, No. 5 (1969). This article is reprinted as part of my opening chapter of [4].

TWO PERSON ZERO SUM DIFFERENTIAL GAMES: AN OVERVIEW

Leonard D. Berkovitz

Department of Mathematics, Purdue University,
West Lafayette, Indiana U.S.A.

1. INTRODUCTION

In this lecture we shall acquaint the reader with two person zero sum differential games and with the difficulties involved in mathematically formulating the problem and solving it. We shall indicate some of the progress that has been made and some of the important work that remains to be done.

The study of differential games was started by Rufus Isaacs in a series of RAND Corporation memoranda that appeared in 1954 [1]. In these papers Isaacs also coined the term "differential game". Isaacs incorporated this work and his subsequent researches into a book [2] which stimulated much further work and interest in the subject. In [2] Isaacs studied many illuminating examples and stated some theorems about methods of solving differential games. Some of these theorems were justified heuristically and some were established under very stringent hypotheses. The examples in the book illustrated solution methods and presented phenomena that any general theory must take into account. The intrinsic interest of Isaacs's examples and the important mathematical questions that were implied in his work but were left unanswered helped generate interest in differential games. The relationship between differential games and optimal control theory and the publication of [2] at a time when interest in optimal control theory was very great served to further stimulate interest in differential games.

2. AN EXAMPLE

Originally, the study of differential games arose from the study of pursuit and evasion problems and from tactical problems. To

J. D. Grote (ed.), The Theory and Application of Differential Games. 13-22.

motivate the mathematical formulation of a zero sum two person differential game we consider the following simple pursuit game ([2], p.16). A pursuer P and an evader E move in the plane with constant speeds w_1 and w_2 respectively. Each can control the direction of his velocity vector. Thus if $\xi = (\xi^1,\xi^2)$ denotes the position of P, if $\eta = (\eta^1,\eta^2)$ denotes the position of E, if ϕ denotes the angle that P's velocity vector makes with the horizontal axis, and if ψ denotes the angle that E's velocity vector makes with the horizontal axis then the equations of motion of P and E are:

$$\begin{aligned} \dot{\xi}^1 &= w_1 \cos\phi & \dot{\eta}^1 &= w_2\cos\psi \\ \dot{\xi}^2 &= w_1 \sin\phi & \dot{\eta}^2 &= w_2\sin\psi, \end{aligned} \tag{1}$$

where the dot denotes differentiation with respect to time. The pursuer P wishes to steer, i.e. choose ϕ, at each instant of time so as to "capture" E in as short a time as possible. The evader E wishes to steer, i.e. choose ψ, at each instant of time so as to avoid "capture" or to maximize the time of "capture", if "capture" is unavoidable. The definition of "capture" is quite arbitrary. We can either take it to be a coincidence of the positions of P and E or we can take it to be a given proximity. Both are included in the following scheme. Let $\varepsilon \geq 0$ be given. We say that capture occurs if

$$(\xi^1 - \eta^1)^2 + (\xi^2 - \eta^2)^2 = \varepsilon.$$

If $\varepsilon = 0$ then capture means coincidence. If t_o denotes the initial time and t_f the time of capture then

$$t_f - t_o = \int_{t_o}^{t_f} dt.$$

The pursuer wishes to choose ϕ at each instant so as to minimize the integral and the evader wishes to choose ψ at each instant so as to maximize this integral.

3. A PRELIMINARY FORMULATION

The pursuit problem just stated is an example of a differential game, which is a two person zero sum game that can be described somewhat imprecisely as follows.

Let t denote time. Let $x = (x^1,\ldots,x^n)$ be a vector in real Euclidean space E^n and let $\mathscr{R}$ be a fixed region in (t,x) space. The position, or state, of the game at time t is given by a vector $x(t)$ and is determined by a system of first order differential equations

$$\frac{dx^i}{dt} = g^i(t,x,u,v) \qquad x^i(t_o) = x^i_o\,, \tag{2}$$

where $u = (u^1,\ldots,u^\sigma)$ is chosen at each instant of time by Player I and $v = (v^1,\ldots,v^s)$ is chosen at each instant of time by Player II. We call the system (2) the state equations. In vector notation they are written as:

$$\frac{dx}{dt} = g(t,x,u,v) \qquad x(t_o) = x_o,$$

where $g = (g^1,\ldots,g^n)$.

In the pursuit-evasion example $x = (\xi,\eta) = (\xi^1,\xi^2,\eta^1,\eta^2)$, $u = \phi$, $v = \psi$ and the state equations are given by (1).

It is usually assumed that both players know the present state of the game and that they know how the game proceeds; that is they know the state equations (2). At time t when Player I chooses $u(t)$ he does not know the choice $v(t)$ of Player II. Similarly when Player II chooses $v(t)$ at time t he does not know the choice $u(t)$ of Player I at time t. Since both players know the state $x(t)$ at time t it is reasonable to permit each player to take the state of the game into account in making his choice. Thus Player I can let his choice be governed by a vector valued function U of position and time

$$U(t,x) = (U^1(t,x),\ldots,U^\sigma(t,x))$$

defined on $\mathscr{R}$. Similarly, Player II can let his choice of v be governed by a vector valued function V of position and time

$$V(t,x) = (V^1(t,x),\ldots,V^s(t,x))$$

defined on $\mathscr{R}$. The functions U and V are called pure strategies and the variables u and v are called strategic variables. Player I selects his strategy from a class $\mathscr{U}$ of permissible strategies and Player II selects his strategy from a class $\mathscr{V}$ of permissible strategies. The classes $\mathscr{U}$ and $\mathscr{V}$ are prescribed in advance and the choice of strategy is made by each player prior to the start of play. Therefore, in selecting a pure strategy a player selects a set of instructions for choosing his strategic variable in all possible situations (t,x).

Play begins at some initial time and position (t_o,x_o) in $\mathscr{R}$ and the state $x(t)$ at time t must be such that $(t,x(t))$ is in $\mathscr{R}$. Play terminates whenever t and the position $x(t)$ are such that $(t,x(t))$ is a point of a previously specified set $\mathscr{T}$ in $\bar{\mathscr{R}}$, the closure of $\mathscr{R}$. If (t_1,x_1) denotes the point of termination of play of the game starting at (t_o,x_o), then the payoff to Player I is given by

$$P(t_o,x_o,U,V) = h(t_1,x_1) + \int_{t_o}^{t_o} f(t,x(t),u(t),v(t))dt, \qquad (3)$$

where h is a real valued function on $\mathscr{T}$, f is a real valued function on (t,x,u,v)-space, $u(t) = U(t,x(t))$ and $v(t) = V(t,x(t))$. The objective of Player I is to select a strategy that maximizes P; the objective of Player II is to select a strategy that minimizes P.

In the pursuit-evasion example if we relabel the variables as follows: $(\xi^1,\xi^2,\eta^1,\eta^2) = (x^1,x^2,x^3,x^4)$, then $\mathscr{R} = \{(t,x):t \geq 0,\ x \in E^4\}$ and $\mathscr{T} = \{(t,x):(x^1 - x^3)^2 + (x^2 - x^4)^2 = \varepsilon\}$. The payoff can be written either in the form $h = t_1$, $f \equiv 0$, or $h \equiv 0$, $f \equiv 1$.

Usually certain constraints are placed on the choice of strategies as follows. Let Ω_1 be a mapping that assigns subsets of the u-space E^σ to points of the (t,x)-space and let Ω_2 be a mapping that assigns subsets of the v-space E^s to points of (t,x)-space. Thus:

$$\Omega_1:(t,x) \to \Omega_1(t,x) \quad \text{a subset of } E^\sigma$$

$$\Omega_2:(t,x) \to \Omega_2(t,x) \quad \text{a subset of } E^s.$$

In applications the mappings Ω_1 and Ω_2 are usually determined by systems of inequalities

$$K^i(t,x,u) \geq 0 \qquad i = 1,\ldots,\pi$$
$$R^i(t,x,v) \geq 0 \qquad i = 1,\ldots,p.$$

We restrict the choice of strategies by requiring

$$U(t,x) \in \Omega_1(t,x) \qquad V(t,x) \in \Omega_2(t,x). \qquad (4)$$

The assumptions made concerning the knowledge that a player takes into account when he makes his choice of strategic variable is sometimes referred to as the "information pattern" of the game. Some writers use information patterns different from the one described above. Some assume that the players do not know the state of the game, or if they do, they make their choice without reference to the state of the game. In this situation the players choose functions U and V that are functions of time alone. Also, the following information pattern is used by some authors. Player I chooses a strategy U that is a function of (t,x) and Player II chooses a strategy V that is a function of U. Finally, we note that some authors use the term "differential game" to describe problems that are "min-max" or "max-min", but are not games in the sense used in the mathematical theory of games.

4. PROBLEMS OF FORMULATION

Although the discussion in the preceding paragraphs should give the reader a reasonable idea as to the nature of a differential game it is not adequate for a mathematical investigation. We have not really set up the problem as a game in the sense used in the mathematical theory of games.

A two person zero sum game is defined by specifying a set $\mathscr{A}$, called the strategy space for Player I, a set $\mathscr{B}$, called the strategy space for Player II, and a real valued function Π defined on $\mathscr{A}\times\mathscr{B}$. If Player I chooses a in $\mathscr{A}$ and Player II chooses b in $\mathscr{B}$, then the payoff to Player I is $\Pi(a,b)$. The payoff to Player II is $-\Pi(a,b)$. Player I wishes to maximize $\Pi(a,b)$ and Player II wishes to minimize $\Pi(a,b)$. The game is defined to be the triple $(\mathscr{A}, \mathscr{B}, \Pi)$. The game is said to have a value π^* if

$$\pi^* = \sup_a \inf_b \Pi(a,b) = \inf_b \sup_a \Pi(a,b),$$

where a ranges over $\mathscr{A}$ and b ranges over $\mathscr{B}$. If the game has a value π^* and if there exists an (a^*,b^*) such that $\pi^* = \Pi(a^*,b^*)$ and

$$\Pi(a,b^*) \leq \Pi(a^*,b^*) \leq \Pi(a^*,b) \qquad a \in \mathscr{A},\ b \in \mathscr{B}, \tag{5}$$

then we say that a^* is an optimal strategy for Player I and that b^* is an optimal strategy for Player II. The pair (a^*,b^*) is also called a saddle point. Since we always have

$$\sup_a \inf_b \Pi(a,b) \leq \inf_b \sup_a \Pi(a,b),$$

it readily follows that if (5) holds then the game has a value π^* and $\pi^* = \Pi(a^*,b^*)$.

The following theorem, which is a slightly less general but easier to state version of a theorem of Ky Fan [3], gives a sufficient condition for the existence of a saddle point. For further generalizations see Sion [4].

THEOREM 1. Let $\mathscr{A}$ and $\mathscr{B}$ be compact convex subsets of linear topological spaces. Let Π be upper semicontinuous and concave on $\mathscr{A}$ for each b in $\mathscr{B}$ and convex and lower semicontinuous on $\mathscr{B}$ for each a in $\mathscr{A}$. Then there exists a point (a^*,b^*) in $\mathscr{A}\times\mathscr{B}$ such that (5) holds.

If the game $(\mathscr{A}, \mathscr{B}, \Pi)$ does not have a value and a saddle point it is sometimes possible to define a game $(\mathscr{A}_1, \mathscr{B}_1, \tilde{\Pi})$ such that $(\mathscr{A}_1, \mathscr{B}_1, \tilde{\Pi})$ has a value and a saddle point and such that $\mathscr{A}_1 \supseteq \mathscr{A}$, $\mathscr{B}_1 \supseteq \mathscr{B}$ and $\tilde{\Pi}(a,b) = \Pi(a,b)$ for (a,b) in $\mathscr{A}\times\mathscr{B}$. This is usually accomplished by the introduction of mixed strategies.

In order to define a differential game as a two person zero sum game we must specify the strategy spaces $\mathscr{A}$ and $\mathscr{B}$ and the payoff function Π. It is clear that we will want to take $\mathscr{A} = \mathscr{U}$, to take $\mathscr{B} = \mathscr{V}$, and to take $\Pi = P(t_o,x_o,U,V)$ where P is given by (3). A major difficulty arises when we try to specify the classes $\mathscr{U}$ and $\mathscr{V}$. These classes must have the property that for every U in $\mathscr{U}$ and V in $\mathscr{V}$, $P(t_o,x_o,U,V)$ is defined. This requires that the differential equation

$$\frac{dx}{dt} = g(t,x,U(t,x),V(t,x)) \qquad x(t_o) = x_o$$

have a solution ϕ defined on an interval $[t_o,t_1]$ such that for $t \in (t_o,t_1)$ the points $(t,\phi(t))$ are in $\mathscr{R}$ and $(t_1,\phi(t_1))$ is in $\mathscr{T}$. If g is continuously differentiable with respect to all of its arguments and the same is true of U and V, then standard existence and uniqueness theorems for differential equations guarantee the existence of a unique solution in a neighborhood of (t_o,x_o). The resulting solution need not reach the set $\mathscr{T}$. In that event $P(t_o,x_o,U,V)$ is not defined.

From examples of differential games and from other considerations we know that the functions in $\mathscr{U}$ and $\mathscr{V}$ cannot be restricted to the class of $C^{(1)}$ functions. One is led to the consideration of piecewise $C^{(1)}$ functions or even larger classes of functions such as bounded measurable functions. The right hand side of (6) then need not be Lipschitzian in x, even though g is $C^{(1)}$. In fact if U and V are not continuous in x, the right hand side of (6) need not be continuous in x. This leads to even more severe complications concerning the solutions of (6).

Let us suppose that somehow we can find strategy spaces $\mathscr{U}$ and $\mathscr{V}$ and we can define the game $(\mathscr{U}, \mathscr{V}, P)$. Functions U in $\mathscr{U}$ and V in $\mathscr{V}$ will be called pure strategies. The following questions now arise. Does the game $(\mathscr{U}, \mathscr{V}, P)$ have a value? If the game has a value do optimal pure strategies exist? If the game has a pure strategy solution, can one characterize it in a reasonable way?

If we attempt to answer these questions by applying Theorem 1, or some variant of it, we see that a precise specification of the strategy spaces $\mathscr{U}$ and $\mathscr{V}$ is required. They must be topologized so that they are compact. Moreover, f, h and g must be such that P has the requisite semicontinuity, concavity, and convexity properties. This is readily carried out only in very special cases such as f, g, h linear and $\mathscr{T}$ given by $t_1 = T$. (Linear games of fixed duration).

5. SADDLE POINTS

There have been two main thrusts in the attack on the problems posed in the last paragraph. In this section we discuss the more pragmatic one in which the focus is on solving the game and in which the formulation problem is to some extent "assumed away". We shall discuss the other thrust, which focuses on the formulation, in the next section.

Let P be given by (2). For each initial condition (t,x) in $\mathscr{R}$ we assume that we can define a game $(\mathscr{U}, \mathscr{V}, P)$, where $\mathscr{U}$ and $\mathscr{V}$ are appropriate classes of functions whose elements U and V satisfy the constraints (4). We note that because we are unable to assert that the solutions of (6) are unique it follows that for a given U in $\mathscr{U}$, V in $\mathscr{V}$, and initial point (t,x), the payoff P(t,x,U,V) will be multivalued. The different values of P(t,x,U,V) correspond to different trajectories emanating from (t,x).

The game $(\mathscr{U}, \mathscr{V}, P)$ starting at (t,x) is said to have a pure strategy solution (U*,V*) if P(t,x,U*,V*) is single valued and if for all U in $\mathscr{U}$ and V in $\mathscr{V}$

$$P(t,x,U,V^*) \le P(t,x,U^*,V^*) \le P(t,x,U^*,V). \tag{7}$$

We note that if P(t,x,U*,V) is multivalued then (7) must hold for all values of P(t,x,U*,V). A similar statement holds concerning P(t,x,U,V*). We also note that the requirement that P(t,x,U*,V*) be single valued does not preclude the possibility of more than one trajectory corresponding to (U*,V*) emanating from (t,x); it requires that P take on the same value along all such trajectories.

The functions (U*,V*) of the preceding paragraph are called an <u>optimal pair</u> or <u>saddle point</u>. Let

$$W(t,x) = P(t,x,U^*,V^*).$$

Thus, W(t,x) is the value of the game $(\mathscr{U}, \mathscr{V}, P)$ with initial point (t,x). The function W is called the <u>value function</u> or simply the <u>value</u> of the game.

We now focus our attention upon each of the two inequalities in (7). We relabel the initial point (t,x) as (τ,ξ). The right hand inequality suggests the following control problem.

PROBLEM. Find a control v that minimizes

$$h(t_1,x_1) + \int_{\tau}^{t_1} f(t,x(t),U^*(t,x(t)),v(t))dt \tag{8}$$

subject to

$$\frac{dx}{dt} = g(t,x,U^*(t,x),v) \qquad x(\tau) = \xi \tag{9}$$

$$v(t) \in \Omega_2(t,x(t)) \tag{10}$$

$$(t_1,x(t_1)) \in \mathscr{T}. \tag{11}$$

If in (9) we take $v = V(t,x)$, where V belongs to $\mathcal{V}$ then we obtain a solution ϕ that reaches $\mathcal{T}$ at $(t_1,x_1) = (t_1,\phi(t_1))$. Corresponding to this solution we obtain a control function of time

$$v(t) = V(t,\phi(t)),$$

such that if we take $v = v(t)$ in (9) we again get ϕ. Moreover, since $V(t,x) \in \Omega_2(t,x)$, it follows that $v(t)$ satisfies (10). Let $\mathcal{V}'(\tau,\xi)$ denote the class of control functions $v(t)$ obtained this way. Let $v^*(t)$ denote the control function corresponding to $V^*(t,x)$. It follows from (7) that v^* minimizes (8) in the class $\mathcal{V}'(\tau,\xi)$.

An analogous discussion can be carried out for the left hand inequality. Thus, corresponding to each of the inequalities in (7) we obtain a control problem. This suggests that we should be able to obtain necessary conditions that hold along a trajectory corresponding to a saddle point (U^*,V^*) by considering the two associated variational problems and using results and techniques from the calculus of variations and optimal control theory.

At this point a price must be paid for not knowing the classes $\mathcal{U}$ and $\mathcal{V}$ precisely. All we know is that $v^*(t)$ minimizes (8) in the class $\mathcal{V}'(\tau,\xi)$. We do not know the nature of the class $\mathcal{V}'(\tau,\xi)$. The necessary conditions in the calculus of variations and in optimal control theory are derived on the assumption that $v^*(t)$ minimizes in the class of all admissible controls. Therefore, since $\mathcal{V}'(\tau,\xi)$ may be too small a subclass of the class of all admissible controls we cannot apply the maximum principle and other necessary conditions from the calculus of variations and control theory directly to our problem. We must devise new arguments based on whatever information we have. We shall not go into the details of these arguments here. We only point out that to derive useful necessary conditions certain assumptions about the regularity of the functions U^* and V^* must be made, together with assumptions as to the nature of the field of trajectories resulting from the use of (U^*,V^*). Although many known examples fall within these assumptions, many other known examples do not.

The necessary condition that is obtained is a two-sided maximum principle. A partial differential equation that W satsifies is also obtained. These results agree with the results obtained by Isaacs in a formal manner or in a rigorous manner under more restrictive hypotheses. The program referred to above was first applied to differential games with two dimensional state vectors in [5]. It was generalized to higher dimensions in [6] and an improved treatment was given in [7]. The reader is referred to [7] for precise and detailed statements of results.

The fact that a pair of functions (U^*,V^*) are such that the necessary conditions hold along trajectories resulting from the use of U^* and V^* in (6) does not guarantee that U^* and V^* are optimal. One can, however, by introducing an extension of the notion of a field in the calculus of variations obtain sufficiency theorems of the following type. If the necessary conditions hold along all the trajectories corresponding to (U^*,V^*) and if the field of trajectories has certain regularity properties then (7) holds in the following sense. If V is any function satisfying (4) and is such that $P(t,x,U^*,V)$ is defined for all (t,x) in $\mathscr{R}$, then the right-most inequality in (7) holds, provided the trajectories corresponding to (U^*,V) lie in the region covered by the trajectories corresponding to (U^*,V^*) and are not tangent to manifolds of discontinuity of U^* or V. A similar statement holds for functions U. For details and precise theorems, see [6].

We emphasize that functions (U^*,V^*) that satisfy the sufficiency theorems are not necessarily saddle points. To assert that (U^*,V^*) is a saddle point we must know precisely what the strategy classes $\mathscr{U}$ and $\mathscr{V}$ are and we must know that U^* is in $\mathscr{U}$ and V^* is in $\mathscr{V}$. Otherwise, all we can assert as a consequence of our sufficiency theorem is that (U^*,V^*) is an equilibrium point. By that we mean that (7) holds for all U and V as described in the preceding paragraph.

6. EXISTENCE OF VALUE

The second major effort, which in a sense is the more logical one, is directed toward first giving an adequate definition of the game and showing that the defined game has a value. The pertinent references here are Fleming [8], [9], Varaiya and Lin [10], and Roxin [11] for the preliminary work in this direction and Friedman [12] and Elliott and Kalton [13] for the full development of the theory.

The approaches of the various authors are different, but they are the same in spirit. The direct definition of a game as attempted in the previous sections is abandoned because of the many difficulties. Instead, somewhat indirect definitions are given. The games so defined are well defined mathematically and they do possess values. Friedman also gives conditions under which a saddle point exists and shows how to compute saddle points under his definition. The method is more complicated than the method of the previous section, but on the other hand, one is sure of what one is computing. Friedman also identifies his value with the value defined by the methods of the previous section, but the differentiability assumptions needed to make this identification are such that most interesting examples are ruled out. Elliott and Kalton establish the existence of a value for a much wider class of games than does Friedman. Their definition of value, however, does not permit an answer to the question as to whether a saddle point exists. No techniques for computing saddle points are suggested, although approximate saddle points are discussed.

An adequate summary of the work mentioned in this section would be beyond the scope of this paper. Since other lecturers are scheduled to discuss this work, our omission will not be too serious.

7. CONCLUSION

In the last two sections of this paper we have indicated the achievements and the shortcomings of two major efforts in the area of two person zero sum games. The achievements in each area appear to complement the gaps in the other area. The full connection between these two bodies of results is yet to be made, and presents us with a challenge.

REFERENCES

1. R. Isaacs, Differential Games I, II, III, IV, The RAND Corporation, Research Memoranda RM-1391, RM-1399, RM-1411, RM-1486 (1964).
2. R. Isaacs, Differential Games, John Wiley and Sons, New York, London, Sydney, 1965.
3. K. Fan, Minimax theorems, Proc. Natl. Acad. Sci. 39 (1953), 42-47.
4. M. Sion, On general minimax theorems, Pacific J. Math. 8 (1958), 171-176.
5. L. D. Berkovitz and W. H. Fleming, On differential games with integral payoff, Contribution to the Theory of Games Vol. III, Annals of Math Study No. 39, Princeton University Press, Princeton, N.J. (1957), 413-435.
6. L. D. Berkovitz, A variational approach to differential games, Annals of Math Study No. 52, Princeton University Press, Princeton, N.J., 1964, 127-174.
7. W. H. Fleming, The convergence problem for differential games, J. Math. Anal. Appl. 3(1961), 102-116.
8. W. H. Fleming, The convergence problem for differential games II, Annals of Mathematics Study No. 52, Princeton University Press, Princeton, N.J., 1964, 195-210.
9. P. Varaiya and J. Lin, Existence of saddle points in differential games, SIAM J. Control 7(1969), 141-157.
10. E. Roxin, The axiomatic approach in differential games, J. Optimization Theory and Applications, 3(1969), 153-163.
11. A. Friedman, Differential Games, Wiley-Interscience, New York, 1971.
12. R. J. Elliott and N. J. Kalton, The existence of value in differential games. Mem. Amer. Math. Soc. 126 (1972).

INTRODUCTION TO DIFFERENTIAL GAMES

I

COMPETITIVE DYNAMIC SITUATIONS, STRATEGIES, AND VALUES

Robert J. Elliott

Department of Pure Mathematics, University of Hull, Hull, Yorkshire, England

1. TWO PERSON GAMES

Consider a static competitive situation where there are two players J_1 and J_2. Suppose that J_1 can do any one of m things and J_2 can do any one of n things, and that the outcome of J_1 choosing his i^{th} course of action or play and J_2 choosing his j^{th} course of action or play is that J_1 receives an amount a_{ij} (of money, say), and J_2 receives an amount b_{ij}. The game is completely described, therefore, by two matrices with real entries $A = (a_{ij})$ and $B = (b_{ij})$, $1 \le i \le m$, $1 \le j \le n$. It is supposed that J_1 and J_2, who are both trying to maximize their outcome, know A and B but they do not know the choice of play made by the other player.

2. ZERO SUM GAMES

In the above situation there is no general agreement as to what is the joint best that J_1 and J_2 can force, so we now consider the case where, if J_1 receives an amount a_{ij} it must come from J_2: i.e. $a_{ij} = - b_{ij}$ for all i and j. Such a situation is called a zero sum game and because $A = -B$ only the "payoff matrix" A for J_1 need be considered.

In terms of A, J_1 is trying to maximize the outcome. For

J. D. Grote (ed.), The Theory and Application of Differential Games. 23-34. All Rights Reserved.

each of his i actions he will look at $\min_j a_{ij}$ and he will then choose the i that maximizes $\min_j a_{ij}$. The best that J_1 can force is therefore $\max_i \min_j a_{ij}$. Similarly, in terms of A, J_2 is trying to minimize the outcome and the best that J_2 can force is $\min_j \max_i a_{ij}$. These quantities are called the lower and upper values of the game and it is trivially the case that:

$$\max_i \min_j a_{ij} \leq \min_j \max_i a_{ij}$$

If there are actions i^* for J_1 and j^* for J_2 such that

$$\max \min a_{ij} = \min \max a_{ij} = a_{i^*j^*} ,$$

then $a_{i^*j^*}$ is a saddle point and is a natural quantity to call the value of the game. This is because $a_{i^*j} \geq a_{i^*j^*} \geq a_{ij^*}$, so that if either J_1 or J_2 digresses from the saddle point play they will not do so well.

In general, however, a matrix does not have a saddle value. To discuss such cases von Neumann [15] introduced the idea of mixed strategies. A mixed strategy for J_1 is a probability

$$x = (x_1, \ldots , x_m) , x_i \geq 0 , \Sigma x_i = 1 ,$$

over his possible plays. Similarly a mixed strategy for J_2 is a probability $y = (y_1, \ldots , y_n)$, $y_j \geq 0$, $\Sigma y_j = 1$. The outcome of a pair of mixed strategies is defined to be the expected value

$$x A y .$$

Mixed strategies can be thought of in terms of taking weighted averages of repeated plays of the original game and von Neumann's celebrated "minimax theorem" states that in terms of mixed strategies a two person zero-sum matrix game always has a saddle value. That is, there are mixed strategies x^*, y^* such that

$$\min_y \max_x x A y = \max_x \min_y x A y = x^*Ay^*$$

3. OPTIMAL CONTROL

Let us move now from a static competitive situation and

consider a dynamical system in R^m of the following form:

$$\dot{x}(t) = f(t,x,y)$$

$$x(0) = 0 \in R^m , \quad x \in R^m , \quad t \in [0,1] .$$

f satisfies continuity, Lipschitz and growth conditions which ensure the integrability of the system. Here the variable y is to be chosen at each time $t \in [0,1]$ from a compact metric space Y with the object of minimizing a cost function:

$$P(y) = g(x(1)) + \int_0^1 h(t,x,y)dt.$$

$g : R^m \to R$ and $h : [0,1] \times R^m \times Y \to R$ are both supposed continuous. This model is of the kind discussed in optimal control theory.

4. TWO PERSON ZERO SUM DIFFERENTIAL GAMES

Consider now a dynamical system

$$x(t) = f(t,x,y,z) , \quad x(0) = 0 \in R^m , \quad x \in R^m , \quad t \in [0,1] , \tag{1}$$

where f satisfies the usual continuing Lipschitz and growth conditions, and a cost, or payoff, of the form

$$P(y,z) = g(x(1)) + \int_0^1 h(t,x,y,z)dt \quad . \tag{2}$$

We now suppose that a controller, or player, J_1 chooses y from a compact metric space Y at each time $t \in [0,1]$ with the object of maximizing the final payoff P, and a second player J_2 chooses z from a compact metric space Z at each time t with the object of minimizing P. This model incorporates both the competitive element of von Neumann's zero sum games and the dynamic setting of optimal control theory, so a dynamical system (1) and payoff (2) is called a two person zero-sum differential game.

5. STRATEGIES AND VALUE

Differential games were first studied in a pioneering way in the early 1950's by R. Isaacs. His results were not published until his book [11] appeared in 1965. Isaacs' approach centred around looking for solutions of the so-called Issacs equation (see section 6 below).

In setting up a mathematical model of how a differential game develops a basic problem is to try and describe the information available to each player as the game proceeds; in particular, it is not realistic to suppose that either player has advance knowledge. To model the information flow both Fleming ([5], [6]) and Friedman

([9], [10]) discussed approximating differential games obtained by considering for any integer n, a partition of the time interval $[0,1]$. Write $t_j = j/n$ for $j = 0,1,\dots n$ and $I_1 = [0,t_1)$, $I_k = [t_{k-1},t_k]$, $k = 2,\dots n$.

For each such partition Fleming considered an upper game and a lower game. In the upper game J_2, the minimizing player, chooses a constant control value z_j at each time t_j, with the knowledge of $z_0, \dots, z_{j-1}, y_0, \dots, y_{j-1}$. Then, with the knowledge of $z_0, \dots, z_{j-1}, y_0, \dots, y_{j-1}$ and of z_j, J_1 chooses a control value y_j at time t_j. Therefore, in the upper game J_1 has an information advantage at each step.

Fleming replaces the differential equations by difference equations so that, for example,

$$x(t_{j+1}) = x(t_j) + \delta f(t_j,x(t_j),y_j,z_j) \text{ where } \delta = 1/n,$$

and the final payoff is approximated by

$$P(y,z) = g(x(1)) + \delta \sum_{j=0}^{n-1} h(t_j,x(t_j),y_j,z_j) .$$

The best the two players can obtain with the above information pattern, is the so-called upper value W_n^+ in this approximating game, and

$$W_n^+ = \min_{z_0} \max_{y_0} \dots \min_{z_{n-1}} \max_{y_{n-1}} P(y,z) .$$

If J_1 is thought to play first at each time t_j, by disclosing his choice of control value y_j, then the best the two players can obtain is the so-called lower value W_n^-, and then:

$$W_n^- = \max_{y_0} \min_{z_0} \dots \max_{y_{n-1}} \min_{z_{n-1}} P(y,z) .$$

For the case when f and h have the forms $f_1(x,y) + f_2(x,z)$ and $h_1(x,y) + h_2(x,z)$ respectively Fleming shows in [5] that

$$\lim_{n\to\infty} W_n^+ = W^+ \text{ and } \lim_{n\to\infty} W_n^- = W^- \text{ exist.}$$

The existence of the limits W^+ and W^- for the general case is implicitly established in [7] by considering related stochastic differential games (see below).

Write $M_1([a,b])$ (resp. $M_2([a,b])$) for the space of measurable functions on $[a,b]$ with values in Y(resp. Z). (A function y from $[a,b]$ to Y is measurable if $\phi \circ y$ is measurable for every continuous real valued function ϕ on Y).

Friedman [9] also considers an approximating upper and lower game for every partition of the time interval $[0,1]$ into n equal sub-intervals. However, instead of a constant control, at time t_{j-1} each player can choose a measurable control function $y_j(t)$, $z_j(t)$ resp. over I_j. The trajectory then evolves according to

$$x(t_{j+1}) = x(t_j) + \int_{t_j}^{t_{j+1}} f(t,x(t),y_j(t),z_j(t))dt.$$

In the upper game J_2 plays first at each step and the best the two players can obtain is an upper value

$$V_n^+ = \inf_{z_0(t)} \sup_{y_0(t)} \dots \inf_{z_{n-1}(t)} \sup_{y_{n-1}(t)} (g(x(1)) + \int_0^1 h(t,x(t),y(t)z(t))dt) .$$

The lower value, obtained if J_1 plays first at each step, is

$$V_n^- = \sup_{y_0(t)} \inf_{z_0(t)} \dots \sup_{y_{n-1}(t)} \inf_{z_{n-1}(t)} (P(y,z)) .$$

Because control functions are now used it is easier to establish that $\lim_{n\to\infty} V_n^+ = V^+$ and $\lim_{n\to\infty} V_n^- = V^-$ exist.

Further, if f and h have the form $f_1(t,x,y) + f_2(t,x,z)$ and $h_1(t,x,y) + h_2(t,x,z)$ respectively then Friedman shows that $V^+ = V^-$.

Instead of the approximating methods obtained by considering partitions, another method of modelling the information flow was described by Roxin [14] and Varaiya and Lin [13] and developed by N. Kalton and the author [2]. As defined above, $M_1 = M_1([0,1])$ and $M_2 = M_2([0,1])$ are the measurable functions with values in Y and Z respectively, that is the control functions for J_1 and J_2.

A pseudo-strategy is a map $\alpha : M_2 \to M_1$, that is, a rule telling J_1 what to do, and each such α has a value

$$u(\alpha) = \inf_{z \in M_2} P(\alpha z, z)$$

giving the worst that can happen to J_1 if he uses α.

Similarly a pseudo-strategy for J_2 is a map $\beta : M_1 \to M_2$ and each such β has a value

$$v(\beta) = \sup_{y \in M_1} P(y, \beta y) .$$

What we are trying to model is the information flow and for α to be a genuine strategy it is not realistic that it should anticipate any future change in the action of J_2. Consequently we say α is a strategy if for any intermediate time T, $0 \le T \le 1$ if $z_1(t) = z_2(t)$ a.e. $0 \le t \le T$ then $(\alpha z_1)(t) = (\alpha z_2)(t)$ a.e. $0 \le t \le T$. A pseudo-strategy β for J_2 is said to be a strategy if it satisfies a similar non-anticipatory condition.

Write Γ for the set of strategies for J_1 and Δ for the set of strategies for J_2. Then the value to J_1 of the game, the upper value, is the best J_1 can ensure using strategies: i.e.

$$U = \sup_{\alpha \in \Gamma} u(\alpha) .$$

Similarly, the lower value V is $\inf_{\beta \in \Delta} v(\beta)$

U and V are not immediately comparable but it is shown in [2] that both are less than V^+ which in turn is less than W^+. Similarly, U and V are both greater than V^- which is greater than W^-. By introducing related stochastic games, however, we shall show in our second lecture that all upper values are equal, that is $U = V^+ = W^+$, similarly $V = V^- = W^-$.

6. DYNAMIC PROGRAMMING

Instead of a differential game starting at $t = 0$, $x(0) = 0$, a game can be considered which starts at an intermediate time and position (t,x), however, it is still supposed the game ends at time 1. By considering $M_1([t,1])$ and $M_2([t,1])$ appropriate strategies can be introduced and upper and lower values $U(t,x)$ and $V(t,x)$ defined. For a small time interval δ, by dynamic programming considerations one can see that approximately

$$V(t,x) \simeq \max_{z \in Z} \min_{y \in Y} \left[\int_t^{t+\delta} h(t,x,y,z)dt + V(t+\delta, x(t+\delta))\right],$$

where $x(t+\delta) = x(t) + \int_t^{t+\delta} f(t,x,y,z)dt.$

Arguing formally and expanding V in a Taylor series (unfortunately V is not necessarily differentiable) we see that V might satisfy the (lower) Isaacs-Bellman equation:

$$L^{-}V = \frac{\partial V}{\partial t} + \max \min (\nabla V.f + h) = 0,$$

together with the boundary condition

$$V(1,x) = g(x).$$

In fact at points of differentiability it can be shown (see [3]) that V does satisfy this equation. Similarly the upper value satisfies the upper Isaacs-Bellman equation

$$L^{+}U = \frac{\partial U}{\partial t} + \min_z \max_y (\nabla U.f + h) = 0 .$$

Unfortunately these Isaacs-Bellman equations are non-linear and degenerate, and there are no theorems guaranteeing either the existence or uniqueness of solutions. In his papers [6] , [7] Fleming circumvented this problem by considering a related stochastic difference game as follows.

Let us suppose that f and h satisfy uniform Lipschitz conditions in both t and x, and that g is twice continuously differentiable and that g, $\frac{\partial g}{\partial x_i}$ and $\frac{\partial g}{\partial x_i \partial x_j}$ all satisfy uniform Lipschitz conditions in x. Fleming then considers the parabolic equation

$$\frac{\lambda^2}{2} \nabla^2 R + \frac{\partial R}{\partial t} + \min_z \max_y (\nabla R.f + h) = 0$$

subject to the boundary condition $R(1,x) = g(x)$. From results of Friedman [8] , or Oleinik and Kruzkov or [12] , it is known that the above equation has a unique solution $W_\lambda^+(t,x)$ for $\lambda > 0$ and W_λ^+ is continuously differentiable in x. Further W_λ^+ ,

$\frac{\partial W_\lambda^+}{\partial t}$, $\frac{\partial W^+}{\partial x_i}$, $\frac{\partial^2 W_\lambda^+}{\partial x_i}$, $\frac{\partial^2 W_\lambda^+}{\partial x_i \partial x_j}$ each satisfy Hölder conditions of the form

$$|\psi\,(t,x) - \psi(t',x')| \leq K(|t - t'|^{\gamma/2} + \|x - x'\|^{\gamma}) .$$

For $\lambda > 0$ and $\delta = 2^{-N}$ with N an integer, Fleming considers a stochastic difference equation related to the above parabolic equation:

$$W^+_{N,\lambda}(t_j,x) = \min_z \max_y \{E(W^+_{N,\lambda}(t_{j+1},x') + \delta h(t_j,x,y,z)\}$$

where $\quad x' = x + \delta f(t_j,x,y,z) + \delta^{\frac{1}{2}}\lambda\eta_j$,

$(\eta_o, \ldots , \eta_{2^N-1})$ is a sequence of normalized mutually independent Gaussian random variables and E denotes expectation. $W^+_{N,\lambda}$ is determined for $t_j = j$, $j = 0,1,2,\ldots,2^N$ by the boundary condition $W^+_{N,\lambda}(1,x) = g(x)$. In his papers [6] , [7] Fleming obtains the following results:

THEOREM 1. $\lim_{N\to\infty} W^+_{N,\lambda}(t,x) = W^+_\lambda(t,x)$ uniformly on compacta for $\lambda > 0$ and dyadically rational t.

THEOREM 2. $\lim_{\lambda\to 0} W^+_{N,\lambda}(t,x) = W^+_N(t,x)$ uniformly in N for each dyadically rational t and N such that $t = p.\ 2^{-N}$ with p an integer.

From these two theorems he deduces

THEOREM 3. $\lim_{\lambda\to 0} W^+_\lambda(t,x) = \lim_{N\to\infty} W^+_N(t,x)$ for all dyadically rational t. In particular, $W^+ = \lim_{N\to\infty} W^+_N$ exists and $\lim_{\lambda\to 0} W^+_\lambda(t,x) = W^+(t,x)$ exists for all $t \in [0,1]$ and $x \in R^m$.

Fleming shows that the function W^+ is a "generalized solution" of the Isaacs-Bellman equation in the sense that it is Lipschitz in t and x and satisfies this equation almost everywhere. Further we notice that $W^+(t,x)$ depends only on the function $\min_z \max_y (p.f + h)$, (and the boundary condition $g(x)$), because it is the limit of the unique solutions $W^+_\lambda(t,x)$.

By considering the unique solution $W^-_\lambda(t,x)$ of the parabolic equation

$$\frac{\lambda^2}{2}\nabla^2 R + \frac{\partial R}{\partial t} + \max_y \min_z (\nabla R.\ f + h) = 0$$

and the related stochastic difference equation it can be shown similarly that

$$W^- = \lim_{N\to\infty} W_N^- \quad \text{exists and} \quad \lim_{\lambda\to 0} W_\lambda^-(t,x) = W^-(t,x) \quad \text{exists}$$

for all $t \in [0,1]$ and $x \in R^m$. Consequently, we have that, if f,h,g satisfy the Lipschitz conditions described above and also the Isaacs condition

$$\min_z \max_y (p.f + h) = \max_y \min_z (p.f + h)$$

for $t \in [0,1]$, $x \in R^m$, $p \in R^m$, then $W^+ = W^-$.

The Lipschitz requirements can be removed by approximation arguments, so by the inequalities described at the end of section 5 the following result can be proved.

THEOREM. If the game satisfies the Isaacs condition then all upper and lower values are equal: $W^- = V^- = V = U = V^+ = W^+$. For details see the Memoir [2] .

Note that if f and h satisfy Friedman's separation conditions then the Isaacs condition certainly holds.

7. RELAXED CONTROLS

We have seen how von Neumann introduced probabalistic ideas using mixed strategies into matrix games; the use of relaxed controls in differential games is in some ways analogous. Relaxed controls were first introduced into control theory in [17] and into differential games in [1], where a more detailed discussion may be found.

Write $\Lambda(Y)$ and $\Lambda(Z)$ for the sets of regular probability measures on Y and Z. By the Riesz representation theorem we may consider $\Lambda(Y)$ as a subset of $C(Y)^*$ and, because Y is metrizable, with the weak-* topology $\sigma(C(Y)^*, C(Y))$ $\Lambda(Y)$ is a compact metrizable space. Similarly with the topology $\Lambda(C(Z)^*, C(Z))$ $\Lambda(Z)$ is also a compact metrizable space. The definitions of f and h can be extended by stating:

$$f : I \times R^m \times \Lambda(Y) \times \Lambda(Z) \to R^m$$

$$f_i(t,x,\sigma,\tau) = \iint_{ZY} f_i(t,x,y,z)d\sigma(y)d\tau(z) \qquad i = 1,2 \dots m,$$

and $h : I \times R^m \times \Lambda(Y) \times \Lambda(Z) \to R$

$$h(t,x,\sigma,\tau) = \int_Z\int_Y h(t,x,y,z)d\sigma(y)d\tau(z)$$

It is easily checked that f and h still satisfy the same Lipschitz and continuity conditions and we may now consider four versions of the differential game which we denote by G, G_1, G_2, G_{12}.

G is the original differential game, whilst in G_1 J_1 may use relaxed controls, that is he may choose a probability measure σ from the compact metric space $\Lambda(Y)$ at each time $t \in [0,1]$. Similarly in G_2 J_2 may use relaxed controls whilst in G_{12} both players may use relaxed controls. Below a subscript 1, 2 or 12 indicates that the quantity refers to the game G_1, G_2 or G_{12}. Similarly write

$$F^+(t,x,p) = \min_z \max_y (p.f + h)$$

$$F^-(t,x,p) = \max_y \min_z (p.f + h)$$

$$F_1^+(t,x,p) = \min_z \max_{\sigma\in\Lambda(Y)} (p.f + h)$$

$$F_{12}^+(t,x,p) = \min_{\tau\in\Lambda(Z)} \max_{\sigma\in\Lambda(Y)} (p.f + h)$$

$$F_{12}^-(t,x,p) = \max_{\sigma\in\Lambda(Y)} \min_{\tau\in\Lambda(Z)} (p.f + h)$$

It is easy to check that

$$F^+(t,x,p) \equiv F_1^+(t,x,p).$$

Also $p.f + h$ is linear (or rather affine) and separately continuous on $\Lambda(Y) \times \Lambda(z)$ so we may quote Wald's extension [16] of von Neumann's minimax theorem to obtain

$$F_{12}^+(t,x,p) \equiv F_{12}^-(t,x,p).$$

Consequently, if f, h and g satisfy the Lipschitz conditions described in section 6 above when discussing Fleming's work, we may, by considering the related parabolic equations and stochastic games deduce the following results:

THEOREM. (i) $W_1^+ = W^+$

(ii) $W_2^- = W^-$

(iii) $W^+_{12} = W^-_{12} = W^+_2 = W^-_1$.

By approximation arguments the need for Lipschitz conditions can again be removed and we can deduce from (iii) that in terms of relaxed controls all upper and lower values are equal. i.e.:

COROLLARY. $W^-_{12} = V^-_{12} = V_{12} = U_{12} = V^+_{12} = W^+_{12}$.

Consequently the introduction of relaxed controls gives an analogous result to that obtained by using mixed strategies in matrix game theory.

Full details of the work described above can be found in the Memoir [2] of N. Kalton and the author.

REFERENCES

1. Elliott, R. J., Kalton, N. J. and Markus, L., Saddle points for linear differential games, J. S.I.A.M. Series A. Control 11. 1973, 100-112.
2. Elliott, R. J. and Kalton, N. J., The existence of value in differential games, Memoir of the American Math. Society. 126 Providence, R.I., 1972.
3. Elliott, R. J. and Kalton, N. J., Cauchy problems for certain Issacs-Bellman equations and games of survival, Trans. Amer. Math. Soc. To appear.
4. Elliott, R. J. and Kalton, N. J., Upper values of differential games, J. Diff. Equations 14, 1973, 89-100.
5. Fleming, W. H., The convergence problem for differential games, J. Math. Analysis and App. 3, 1961, 102-116.
6. Fleming, W. H., The convergence problem for differential games II, Ann. of Math. Study 52, 195-210, Princeton N. J., 1964.
7. Fleming, W. H., The Cauchy problem for degenerate parabolic equations, J. Math. and Mech. 13, 1964, 987-1008.
8. Friedman, A., On quasi-linear parabolic equations of the second order. II., J. Math. and Mech. 9, 1960, 539-556.
9. Friedman, A., Differential Games, Pure and Applied Maths., Vol. 25, Wiley-Interscience, New York, 1971.
10. Friedman, A., Differential Games, Regional Conference Series in Mathematics, No. 18, Amer. Math. Soc. Providence R.I., 1973.
11. Isaacs, R., Differential Games, John Wiley and Sons, New York, 1965.
12. Oleinik, O. A. and Kruzkov, S. N., Quasilinear parabolic equations of second order with many independent variables, Usp. Mat. Nauk., 16, 1961, 115-155.
13. Varaiya, P. and Lin, J., Existence of saddle points in differential games, J. S.I.A.M. Series A., Control 7, 1969, 141-157.

14. Roxin, E., The Axiomatic approach in differential games, J. Optimization Theory and App. 3, 1969, 153-163.

15. von Neumann, J. and Morgenstern, O., Theory of Games and Economic Behaviour, Princeton Univ. Press, Princeton 2nd Ed. 1947.

16. Wald, A., Statistical Decision Functions, John Wiley and Sons, New York, London 1950.

17. Warga, J., Functions of relaxed controls, J. S.I.A.M., Series A. Control 5, 1967, 628-641.

INTRODUCTION TO DIFFERENTIAL GAMES

II

STOCHASTIC GAMES AND PARABOLIC EQUATIONS

Robert J. Elliott

Department of Pure Mathematics, University of Hull,
Hull, Yorkshire, England

1. INTRODUCTION

In section 6 of the first lecture we stated that the non-linear parabolic equation related to the Isaacs-Bellman equation has a unique solution, and we mentioned how Fleming [5] , [6], by considering a related stochastic difference equation, established the convergence of the upper and lower values of his approximating games. However, we did not explicitly introduce any stochastic differential games; this was because in a stochastic situation certain measurability problems enter and care has to be taken when defining strategies and value.

In this lecture we consider a stochastic situation in which noise enters the dynamics in a discrete way. Controls and strategies which depend on the noise do not then involve measurability difficulties. We show that the upper value U defined in section 5 of the first lecture is approximated by the solution ϕ^{ε} of the parabolic equation. Because, as stated in section 6 of the first lecture, the upper value W^{+} introduced by Fleming [6] is the limit of ϕ^{ε}, and because $U \leq V^{+} \leq W^{+}$ we can conclude that $U = V^{+} = W^{+}$, where V^{+} is the upper value introduced by Friedman [8].

Having defined our model of a stochastic differential game our proofs,for details of which we refer to [3], are adaptations of Fleming [5], [6]. However, we have presented in detail the proof of Lemma B below, which we consider to be fundamental to the theory of differential games.

This result, whilst intuitively obvious, is not trivial to prove even in our case of discrete noise, and in Fleming's paper

J. D. Grote (ed.), The Theory and Application of Differential Games. 35-43.

[6, p.992] the equivalent result is hidden in the statement that "A proof by induction on the number of moves shows that $V_N(s,x)$ is the value of the game with initial data (s,x)".

2. STOCHASTIC DIFFERENTIAL GAMES

Consider first the deterministic differential game G of prescribed duration $[0,1]$ with control sets Y and Z which are compact metric spaces. The dynamics of G are given by ($x \in R^m$)

$$\frac{dx}{dt} = f(t,x(t),y(t),z(t)),$$
$$x(0) = 0 \in R^m ,$$

where f is continuous and satisfies a constant Lipschitz condition in t and x. To simplify the exposition we suppose that the payoff has the form

$$P = g(x(1)),$$

where g is twice differentiable and its derivatives $\frac{\partial g}{\partial t}$, $\frac{\partial g}{\partial x_i}$, $\frac{\partial^2 g}{\partial x_i \partial x_j}$ are all Lipschitz continuous in (t,x). K denotes the Lipschitz constant in all cases. Further, from [2, Section 9] we can suppose that f and g vanish outside some bounded set.

For $(t,x,p) \in [0,1] \times R^m \times R^m$ write

$$H(t,x,p) = \min_z \max_y (p.f(t,x,y,z)) .$$

Then from results of Friedman [7] or Oleinik-Kruzkov [9] we can quote:

<u>THEOREM A.</u> For $\varepsilon > 0$ there is a unique solution of the equation

$$\left(\frac{\varepsilon^2}{2}\right)\nabla^2\phi + \left(\frac{\partial\phi}{\partial t}\right) + H(t,x,\nabla\phi) = 0$$

subject to

$$\phi(1,\xi) = g(\xi),$$

and ϕ has the property that $\frac{\partial\phi}{\partial t}$ and $\frac{\partial^2\phi}{\partial x_i \partial x_j}$ are bounded and satisfy Hölder inequalities of the form $(0 < \gamma < 1)$

$$|\psi(t,x) - \psi(t',x')| \le Q \{|t - t'|^{\gamma/2} + \|x - x'\|^{\gamma}\} .$$

Let N be an integer and $\delta = 1/N$; for $0 \le j \le N$ write $t_j = j\delta$. Suppose $\{\eta_{ij}$; $i = 1,2,\ldots,N$, $j = 1,2,\ldots,m\}$ is a collection of independent random variables each taking the values ± 1 with probability $1/2$. Write: $\eta_i = (\eta_{i1}, \ldots, \eta_{im}) \in R^m$,

$$\eta = (\eta_1, \ldots, \eta_N),$$

and L for the lattice in $(R^m)^N$ of values of η.

Again we denote by $M_1(j)$ and $M_2(j)$ the spaces of control functions for J_1 and J_2 defined on $[t_j, 1]$, and we denote by Γ_j the set of strategies on $[t_j, 1]$ for J_1, i.e. maps $\alpha : M_2(j) \to M_1(j)$ such that if

$$z_1(t) = z_2(t) \qquad \text{a.e. } t_j \le t \le \tau.$$

then $j_1(t) = z_2(t) \qquad$ a.e. $t_j \le t \le \tau$

We now wish to define controls which are functions of the noise, so a *stochastic control* θ of order j for J_2 is a map

$$\theta : L \to M_2(j)$$

$$\eta \to \theta_\eta$$

such that:

(i) $\theta_\eta(t)$ is independent of η for $t_j \le t \le t_{j+1}$

(ii) If $\eta_k = \eta_k^*$ $k = j + 1, \ldots, l$,

$(l < N)$, then $\theta_{\eta_k}(t) = \theta_{\eta_k^*}(t)$ a.e. for $t_j \le t \le t_{l+1}$.

Note that θ is independent of $\eta_1, \ldots, \eta_j$ and write Θ_j for the set of stochastic controls of order j.

A *stochastic strategy* A for J_1 is a map $A : L \to \Gamma_j$

$$\eta \to A_\eta$$

such that (i) if $z_1(t) = z_2(t)$ a.e. $t_j \le t \le \tau < t_{j+1}$ and $\eta, \eta^* \in L$ then $A_\eta(z_1)(t) = A_{\eta^*}(z_2)(t)$ a.e. for $t_j \le t \le \tau$

and (ii) if $\eta_k = \eta_k^*$, $k = j + 1, \ldots, l$

$(l < N)$ $\quad z_1(t) = z_2(t)$ a.e. $t_j \le t \le \tau$

where $\tau \le t_{l+1}$ then

$$A_\eta(z_1)(t) = A_{\eta^*}(z_2)(t) \text{ a.e. for } t_j \le t \le \tau .$$

Note again that A is independent of $\eta_1, \ldots, \eta_j$, and write

A_j for the set of stochastic strategies of order j.

For $\xi \in R^m$ and $\varepsilon > 0$ we can now define the game $G^\delta_\varepsilon(t_j,\xi)$ which starts at position ξ at time t_j.

Given any $y \in M_1(j)$, $z \in M_2(j)$ and $\eta \in L$ we define the trajectory $x_\eta(t)$ as the discontinuous solution of the equation:

$$x_\eta(t) = \xi + \int_{t_j}^{t} f(s,x_\eta(s), y(s), z(s))ds + \varepsilon\delta^{\frac{1}{2}} \sum_{t_j<t_k\leq t} \eta_k .$$

The payoff in $G^\delta_\varepsilon(t_j,\xi)$ is

$P^j_\eta(\xi,y,z) = g(x_\eta(1))$. For $A \in A_j$ and $\theta \in \Theta_j$ we have a payoff

$P^j(\xi;A,\theta) = E(P^j_\eta(\xi;A_\eta\theta_\eta\theta_\eta))$,

and the value of A to J_1 is

$$u_j(\xi;A) = \inf_{\theta\in\Theta_j} P^j(\xi;A,\theta) .$$

The value of the game to J_1 is defined to be

$$U^\delta_\varepsilon(t_j,\xi) = \sup_{A\in A_j} u_j(\xi;A) .$$

Note that $U^\delta_\varepsilon(1,\xi) = g(\xi)$.

3. A BASIC LEMMA

As mentioned in the introduction the following dynamic programming result is fundamental to the theory of stochastic differential games. It allows U^δ_ε to be estimated step by step and is used in the approximation argument of lemma F below.

LEMMA B. For $j < N$

$$U^\delta_\varepsilon(t_j,\xi) = \sup_{\alpha\in\Gamma_j} \inf_{z\in M_2(j)} E(U^\delta_\varepsilon(t_{j+1}, x_\eta(t_{j+1})))$$

where x is the trajectory corresponding to $(\alpha z,z)$.

Proof. Let $A \in A_j$; then the behaviour of A on (t_j, t_{j+1}) is that of a fixed strategy $\alpha \in \Gamma_j$ (condition (i)). Suppose $z_0(t) \in M_2(j)$; then $(\alpha z_0,z_0)$ induce a trajectory $x_\eta(t)$ for $\eta \in L$. In fact $x_\eta(t_{j+1})$ depends on η_{j+1} only. We denote the possible values in R^m of η_{j+1} by S; to $\sigma \in S$ there corresponds a value ξ_σ of $x_\eta(t_{j+1})$. We define $A^\sigma \in \mathsf{A}_{j+1}$ by

$$A^\sigma_\eta(z)(t) = A_{\eta^\sigma}(\hat{z})(t) \qquad t_{j+1} \le t \le 1$$

where

$$\text{(a)} \quad \eta^\sigma_k = \eta_k , \qquad k \neq j+1$$
$$\eta^\sigma_{j+1} = \sigma ,$$

(b) $\hat{z} \in M_2(j)$ is defined by

$$\hat{z}(t) = z_0(t) \qquad t_j \le t \le t_{j+1}$$
$$= z(t) \qquad t_{j+1} \le t \le 1 .$$

Clearly for each $\sigma \in S, A^\sigma$ is independent of $\eta_1, \ldots, \eta_{j+1}$ and belongs to A_{j+1}. Now

$$u_{j+1}(\xi_\sigma ; A^\sigma) \le U^\delta_\varepsilon(t_{j+1}, \xi_\sigma)$$

and hence, given $v > 0$, there exists $\theta^\sigma \in \Theta_{j+1}$ such that

$$P^{j+1}(\xi_\sigma ; A^\sigma, \theta^\sigma) \le U^\delta_\varepsilon(t_{j+1}, \xi_\sigma) + v.$$

Now define $\theta \in \Theta_j$ by

$$\theta_\eta(t) = z_0(t), \qquad t_j < t \le t_{j+1}$$
$$= \theta^\sigma_\eta(t), \qquad t_{j+1} < t \le 1 \qquad \text{and} \qquad \eta_{j+1} = \sigma$$

Then

$$P_j(\xi;A,\theta) = \frac{1}{2^m} \sum_{\sigma \in S} P^{j+1}(\xi_\sigma;A^\sigma,\theta^\sigma)$$
$$\le E(U^\delta_\varepsilon(t_{j+1}, x_\eta(t_{j+1})) + v.$$

Hence $u_j(\xi;A) \le E(U^\delta_\varepsilon(t_{j+1}, x_\eta(t_{j+1})))$ and so as $z_0 \in M_2(j)$ is arbitrary $u_i(\xi;A) \le \inf_{z_0 \in M_2(j)} E(U^\delta_\varepsilon(t_{j+1}, x_\eta(t_{j+1})))$

and hence,

$$U^\delta_\varepsilon(t_j, \xi) \le \sup_{\alpha \in \Gamma_j} \inf_{z \in M_2(j)} E(U^\delta_\varepsilon(t_{j+1}, x_\eta(t_{j+1}))).$$

Conversely fix $\alpha \in \Gamma_j$; then for $z \in M_2(j)$ there is a trajectory $x_\eta^{(z)}(t)$ corresponding to $(\alpha z, z)$ and $\eta \in L$. Now let $\xi_\sigma^{(z)} = x_\eta^{(z)}(t_{j+1})$ when $\eta_{j+1} = \sigma$ (note $x_\eta^{(z)}(t_{j+1})$ depends only on η_{j+1}). For $\xi \in R^m$ and $v > 0$ there exists $A(\xi) \in A_{j+1}$ such that

$$u_{j+1}(\xi, A(\xi)) \geq U_\varepsilon^\delta(t_{j+1}, \xi) - v.$$

Define $A \in A_j$

$$\begin{aligned} A_\eta z(t) &= \alpha z(t), & t_j \leq t \leq t_{j+1}, \\ &= A(\xi_\sigma^{(z)})\, \hat{z}(t), & t_{j+1} \leq t \leq 1,\ \eta_{j+1} = \sigma. \end{aligned}$$

where $\hat{z}$ is the restriction of z to $M_2(j+1)$. It is easy (but tedious!) to check that $A \in A_j$. For $\theta \in \Theta_j$ define $\theta^\sigma \in \Theta_{j+1}$ by

$$\theta_\eta^\sigma(t) = \theta_{\eta^\sigma}(t), \qquad t_{j+1} \leq t \leq 1.$$

As $\theta \in \Theta_j$, $\theta_\eta(t)$ is independent of η on $[t_j, t_{j+1}]$ e.g. $\theta_\eta(t) = z(t)$.

If we condition $\eta_{j+1} = \sigma$ we have

$$P_\eta^j(\xi;\ A_\eta\theta_\eta, \theta_\eta) = P_\eta^{j+1}(\xi_\sigma^{(z)}\ ;\ A_\eta(\xi_\eta^{(z)})\theta_\eta^\sigma, \theta_\eta^\sigma)$$

and since $A(\xi_\sigma^{(z)})$ and θ^σ are independent of η_{j+1} we obtain

$$\begin{aligned} E(P_\eta^j(\xi; A_\eta\theta_\eta, \theta_\eta) \,|\, \eta_{j+1} = \sigma) &= E(P_\eta^{j+1}(\xi^{(z)}); A_\eta(\xi_\sigma^{(z)})\theta_\eta^\sigma, \theta^\sigma) \\ &= P^{j+1}(\xi_\sigma^{(z)}; A(\xi_\sigma^{(z)}), \theta^\sigma) \\ &\geq u_{j+1}(\xi_\sigma^{(z)}; A(\xi_\sigma^{(z)})) \\ &\geq U_\varepsilon^\delta(t_{j+1}, \xi_\sigma^{(z)}) - v. \end{aligned}$$

Hence,

$$P^j(\xi;A,\theta) \geq E_\sigma(U_\varepsilon^\delta(t_{j+1},\xi_\sigma^{(z)})) - v$$
$$= E(U_\varepsilon^\delta(t_{j+1},x_\eta^{(z)}(t_{j+1}))) - v.$$

Therefore

$$U_\varepsilon^\delta(t_j,\xi) \geq \sup_{\alpha\in\Gamma_j} \inf_{z\in M_2(j)} E(U_\varepsilon^\delta(t_{j+1}),\ x_\eta(t_{j+1}))$$

as required. The lemma is thus proved.

4. APPROXIMATION RESULTS

Having presented in detail Lemma B we now briefly give the more straightforward results needed.

An application of the dynamic programming idea of Theorem 3.1 of [3] enables us to state:

Lemma C. For any $\delta > 0$

$$U_0^\delta(t_j,\xi) = U(t_j,\xi)$$

If $y \in M_1(j)$, $z \in M_2(j)$ and $\eta \in L$ write x_η^ε for the trajectory in $G_\varepsilon^\delta(t_j,\xi)$ and x for the trajectory in $G_0^\delta(t_j,\xi)$. An application of Gronwall's lemma enables us to show

Lemma D. $\|x_\eta^\varepsilon(1) - x(1)\| \leq \varepsilon\delta^{1/2}e^K\|w\|$

where $\|w\| = \sup_{j+1\leq k\leq N} \| \sum_{j+1} \eta_i\|$.

Applying a result in Doob [1, p.106] we can show that $E(\|w\|) \leq 2mN^{\frac{1}{2}}$ and so because g is Lipschitz we have the following estimate:

Lemma E. $|U^{\delta}_{\varepsilon}(t_j,\xi) - U(t_j,\xi)| \leq 2m\varepsilon Ke^{K}$.

Suppose ϕ^{ε} is the solution of the parabolic equation of Theorem A. Using lemma B and a step-by-step approxmiation agrument we can prove:

Lemma F. $\lim_{\delta\to 0} U^{\delta}_{\varepsilon}(0,0) = \phi^{\varepsilon}(0,0)$.

Finally, as mentioned in the first lecture, we know from the work of Fleming [5], [6] that

$$\lim_{\varepsilon\to 0} \phi^{\varepsilon}(0,0) = W^{+}$$

But from lemmas E and F

$$|\phi^{\varepsilon}(0,0) - U| \leq 2mKe^{K}\varepsilon$$

so $$\lim_{\varepsilon\to 0} \phi^{\varepsilon}(0,0) = U$$

Consequently, as $U \leq V^{+} \leq W^{+}$ we can conclude

THEOREM G. $U = V^{+} = W^{+}$.

i.e. all upper values are equal for the kind of differential game described in section 2. More general payoffs can be treated by approximation arguments as in [2, section 10], and details of the above lemmas can be found in the paper [3] by N. Kalton and the author.

Finally, it is perhaps of interest to point out that, as defined in the first lecture, U is a supremum (over strategies) of an infimum (over controls), whereas V^{+}, as defined by Friedman [8], is the limit as $\delta \to 0$ of an infimum (over his lower δ-strategies) of a supremum, so at first sight it is suprising the two are equal.

REFERENCES

1. J. L. Doob, Stochastic Processes, Wiley, New York, 1953.
2. R. J. Elliott, N. J. Kalton, Upper Values of differential games. J. Diff. Eqns. 14 (1973), 89-100.
3. R. J. Elliott, N. J. Kalton, The existence of value in differential games, Mem. Amer. Math. Soc. 126, (1972).
4. R. J. Elliott, N. J. Kalton, Cauchy problems for certain Isaacs-Bellman Equations and games of survival, Trans. Amer Math. Soc. To appear.
5. W. H. Fleming, The convergence problem for differential games II, Advances in game theory, Ann.Math. Studies, 52 (1964), 195-210.
6. W. H. Fleming, The Cauchy problem for degenerate parabolic equations, J. Math. & Mech. 13 (1964), 987-1008.
7. A. Friedman, Partial Differential Equations of Parabolic Type, Prentice-Hall, Englewood Cliffs., N. J., 1964.
8. A. Friedman, Differential Games, Wiley-Interscience, New York, 1971.
9. O. A. Oleinik and S. N. Kruzkov, Quasi-linear parabolic equations of second order with many independent variables, Uspeh Mat. Nauk. 16 (1961), 115-155.

DIFFERENTIAL GAMES OF SURVIVAL

N.J. Kalton

Department of Pure Mathematics, University College of Swansea, Singleton Park, Swansea.

1. INTRODUCTION

In these lectures we study differential games which are not of fixed duration. For such games the value does not depend in a nice "continuous" way on the control variables selected by the players, and our results are therefore less precise than those for games of fixed duration. Nevertheless it is important for practical applications to develop a theory which goes beyond this basic type of game. We shall start by giving examples of the main types of differential game.

Suppose Y and Z are compact metric spaces and $f: \mathbb{R} \times \mathbb{R}^m \times Y \times Z \to \mathbb{R}^m$ is a continuous function which is Lipschitz in the first two variables (these conditions can be weakened). We suppose that two competing players J_Y (the maximizer) and J_Z (the minimizer) select control functions $y: [t_0, \infty) \to Y$ and $z: [t_0, \infty) \to Z$ which are Lebesgue measurable. The precise method of selection is discussed in the next section. The resulting control functions $y(t)$, $z(t)$ determine a trajectory given by

$$\dot{x} = f(t,x,y(t),z(t)) \quad \text{a.e.} \quad t_0 \leq t$$

subject to an initial condition,

$$x(t_0) = x_0.$$

Corresponding to the trajectory $x(t)$, and the control functions $y(t)$ and $z(t)$, we calculate a pay-off $P[y(\cdot),z(\cdot)]$

J. D. Grote (ed.), The Theory and Application of Differential Games. 45-61.

which J_Y aims to maximize and J_Z aims to minimize. In a differential game of *survival* (Friedman [6] Ch.5) the pay-off takes the form

$$P = \int_{t_0}^{t_F} h(t,x(t),y(t),z(t))dt + g(t_F,x(t_F)) \qquad (1)$$

where $h: R \times R^m \times Y \times Z \to R$, and $g: R \times R^m \to R$ are continuous and t_F is the first time for which $(t,x(t)) \in F$, where F is a given closed subset of $R \times R^m$.

If $g \equiv 0$ and $h \equiv 1$, then $P = t_F - t_0$. In this case we call the game, a *pursuit-evasion* game ([6] Ch.3). If $g \equiv 0$ and $h \geq 0$ then the game is called a *generalized pursuit-evasion* game.

The other possible type of pay-off which we consider is a differential game of *optional stopping* ([9]) where

$$P = \min_{t_0 \leq t \leq t_F} g(t,x(t))$$

In all these examples we shall impose a finiteness condition on the game, that for some $T > t_0$, $F \supset [T,\infty) \times R^m$. Thus we will always have $t_F \leq T$.

Two important types of game will not be treated. In [6] Ch.6, Friedman studies games with restricted phase co-ordinates, while in [1] Bensoussan and Friedman have studied a generalization of optional stopping games in which both players (rather than just the minimizer as above) have the option of halting the game at any time $t_0 \leq t \leq t_F$.

2. THE DEFINITION OF VALUE

We now turn to the problem of defining the upper and lower value of a differential game. There is an abundance of different definitions in the literature; in this section we shall discuss some of the ideas involved. Let $\mathcal{M}_Y$ and $\mathcal{M}_Z$ denote the spaces of equivalence classes of Lebesgue measurable functions $y: [t_0,T] \to Y$ and $z: [t_0,T] \to Z$ where functions equal almost everywhere are identified. The pay-off P is defined on $\mathcal{M}_Y \times \mathcal{M}_Z$ (this abstract setting, without reference to the underlying differential equation, we call an *evolutionary* game [9]).

The most mathematically attractive definition of value was proposed by Varaiya (see Roxin [10] or Varaiya-Lin [11]). A *strategy* for J_Z is a map $\beta: \mathcal{M}_Y \to \mathcal{M}_Z$ satisfying the condition that if

$$y_1(t) = y_2(t) \quad \text{a.e.} \quad t_0 \le t \le \tau,$$

then

$$\beta y_1(t) = \beta y_2(t) \qquad t_0 \le t \le \tau.$$

The value of β to J_Z will then be

$$v(\beta) = \sup_y P(y, \beta y)$$

and clearly the value of the game to J_Z is $\inf_\beta v(\beta)$. Similarly a strategy for J_Y is a map $\alpha: \mathcal{M}_Z \to \mathcal{M}_Y$ satisfying the same condition, and the value $v(\alpha)$ of that strategy to J_Y is

$$u(\alpha) = \inf_z P(\alpha z, z).$$

The value of the game to J_Y is $\sup_\alpha u(\alpha)$.

However there are problems with this definition. The use of a strategy β allows the minimizer to instantaneously anticipate his opponent's choice of control function; it allows J_Z too much of an advantage. The appropriate definition of the upper value or the value of the game to the minimizer, in which we suppose the minimizer instantaneously at a disadvantage is

$$U^+ = \sup_\alpha u(\alpha)$$

and conversely

$$U^- = \inf_\beta v(\beta).$$

Even with this definition it is, however, unknown whether $U^+ \ge U^-$ in general (for fixed time differential games this is the case).

These difficulties are circumvented by Varaiya and Lin [11] and Friedman [6], [7], by a partition method. Let $\mathcal{P} = \{t_0, t_1, \ldots, t_n\}$ be a partition of $[t_0, T]$ where

$t_0 < t_1 < \ldots < t_n = T$. Corresponding to the partition $\mathcal{P}$ they consider a game $G_{\mathcal{P}}$ in which the players J_Y and J_Z alternately choose their control functions on the intervals $[t_0, t_1), [t_1, t_2), \ldots, [t_{n-1}, t_n]$. If at each step, J_Z chooses his control function first then the game G has value denoted by $V_{\mathcal{P}}^+$; conversely if J_Y plays first the game has value $V_{\mathcal{P}}^-$ where $V_{\mathcal{P}}^+ \geq V_{\mathcal{P}}^-$. It is then easy to show that the nets $(V_{\mathcal{P}}^+ : \mathcal{P}$ a partition of $[t_0, \tau))$ and $(V_{\mathcal{P}}^-)$ are monotonic and we define the "Friedman upper value" and the "Friedman lower value" of the game by

$$V_F^+ = \lim_{\mathcal{P}} V_{\mathcal{P}}^+$$

$$V_F^- = \lim_{\mathcal{P}} V_{\mathcal{P}}^- .$$

It then follows that $V_F^+ \geq V_F^-$. We note that we may express $V_{\mathcal{P}}^+$ and $V_{\mathcal{P}}^-$ in the language used above, thus

$$V_{\mathcal{P}}^+ = \inf v(\beta)$$

where β runs over all strategies for J_Z which satisfy that if

$$y_1(t) = y_2(t) \quad \text{a.e.} \quad t \leq t_j$$

then

$$\beta y_1(t) = \beta y_2(t) \quad \text{a.e.} \quad t \leq t_{j+1}$$

for $j < n$.

This definition has the added attraction that the approximating process mirrors the physical constraints on the players. It is a reasonable restriction to impose as in $G_{\mathcal{P}}$ that J_Z may only make decisions at a finite number of times $t_0, t_1, \ldots, t_n$. However, it seems to the author that if we are attempting to make the definition fit a physical model, we should also consider reaction times. In $G_{\mathcal{P}}$, with J_Z playing first at each step, J_Z is still allowed at the times $t_1, t_2, \ldots, t_n$ to be aware of the full history of the game and to act instantaneously on that knowledge. In a real situation there would be a slight delay before J_Z could act, i.e. he would respond to the position at time t_j at a time $t_j + \sigma$ where σ is a reaction time. If we attempt to incorporate this idea then we reach a definition of value which can be made without reference to a partition ([3]).

We shall say that a strategy β for J_Z is an s-delay strategy (where $s > 0$) if whenever

$$y_1(t) = y_2(t) \quad \text{a.e.} \quad t_0 \leq t \leq \tau$$

then

$$\beta y_1(t) = \beta y_2(t) \quad \text{a.e.} \quad t_0 \leq t \leq \min(T, \tau + s).$$

We define

$$V_s^+ = \inf v(\beta)$$

where β runs over s-delay strategies for J_Z. Then the upper value V^+ is given by $V^+ = \lim_{s \to 0} V_s^+$.

We can similarly define V^- and we have $V^+ \geq V^-$.

There is one question however which arises in this definition. In selecting an s-delay strategy J_Z chooses a fixed control function $z(t)$ to operate on the interval $[t_0, t_0 + s]$. The choice of this control depends on exact knowledge of the starting conditions (t_0, x_0). There may be situations in which it is unrealistic to suppose that J_Z is able to act on a precise knowledge of the starting conditions until after a delay s. This also turns out to have mathematical importance in the next section.

For $s > 0$ and $z_0(\cdot)$ a measurable Z-valued function on $[t_0, t_0 + s]$ we define

$$Q_s^+[z_0] = \inf v(\beta)$$

where β runs over all s-delay strategies such that $\beta y(t) = z_0(t)$ for $t_0 \leq t \leq t_0 + s$. Then

$$Q_s^+ = \sup(Q_s^+[z_0]) \quad (\text{over all } z_0(\cdot))$$

and $Q^+ = \lim_{s \to 0} Q_s^+$. We similarly define Q_s^- and Q^- ([5]).

We have thus defined four different notions of upper value which are related by the inequalities $Q^+ \geq V^+ \geq V_F^+ \geq U^+$. Similarly for the lower values we have $Q^- \leq V^- \leq V_F^- \leq U^-$; and we have $V_F^- \leq V_F^+$ although it is not known whether $U^- \leq U^+$. Of course in fixed time games all these concepts reduce to two; however in games of survival it is by no means clear that they are equivalent.

We know of no example to distinguish between V^+, V_F^+ and U^+, although there seems no reason to believe that such an example does not exist. However it is very easy to illustrate the difference between Q^+ and V^+ by a control problem [5] (i.e. a trivial differential game). Let $Z \subset R^2$ be the set of all $z = (\xi_1, \xi_2)$ where $\xi_1^2 + \xi_2^2 \leq 1$ and $\xi_1 \geq 0$. Consider the game in R^2 with dynamics

$$\dot{x} = z$$

initial condition $$x(0) = (0, \tfrac{1}{4})$$

and terminal set $F = \{(t,\xi_1,\xi_2) : \xi_1 \leq 0 \text{ and } \xi_2 \leq 0\} \cup \{(t,\xi_1,\xi_2)\ t \geq 1\}$ Suppose the pay-off is given by $P = t_F$.

J_Z's optional strategy is to adopt $z(t) \equiv (0,-1)$ and so $V^+ = \frac{1}{4}$. However it is easy to see that $Q_s^+ = 1$ for all $s > 0$ and so $Q^+ = 1$.

3. DYNAMIC PROGRAMMING

We now specialize to differential games of survival, although the same techniques can be applied equally to optional stopping games ([9]); all the results of §§3-5 are taken from [5]. Consider a game then with pay-off defined by (1). As we allow the initial conditions to vary we obtain functions $U^+(t,x)$, $V^+(t,x)$, $Q_s^+(t,x)$, etc. Suppose now that τ is any function $\tau : \mathcal{M}_Y \times \mathcal{M}_Z \to [t_0, T]$ such that if

$$y_1(t) = y_2(t) \quad \text{a.e.} \quad t \leq \tau(y_1,z_1)$$

$$z_1(t) = z_2(t) \quad \text{a.e.} \quad t \leq \tau(y_1,z_1)$$

then $\tau(y_1,z_1) = \tau(y_2,z_2)$. We call such a map τ a non-anticipating stopping time; an example is $\tau \equiv t_F$, for the trajectory corresponding to y_1, z_1. Consider now the game $G(\tau;\varphi)$, where φ is some real-valued function on $R \times R^m$, in which the pay-off corresponding to a pair of controls (y,z) is given by

$$P = \int_{t_0}^{\tau} h(t,x(t),y(t),z(t))dt + \varphi(\tau,x(\tau)).$$

$G(\tau;\varphi)$ is an evolutionary game, and therefore has values as defined in the preceding section $U^+_{\tau,\varphi}$, etc. By a dynamic programming theorem we shall mean a result relating the values of $G(\tau;\varphi)$ where φ is itself a value function to the values of the original game.

Theorem 3.1

$$U^+_{\tau,U^+} = U^+(t_0,x_0) \quad \text{and} \quad U^-_{\tau,U^-} = U^-(t_0,x_0).$$

This is a very nice result whose only defect is the difficulty in treating U^+ and U^- as discussed in the preceding section. For V^+ and V^- the results are less fortunate.

Theorem 3.2

$$V^+_{\tau,V^+} \leq V^+(t_0,x_0) \quad \text{and} \quad V^-_{\tau,V^-} \geq V^-(t_0,x_0).$$

However we can now justify the need for the Q-values by providing the reverse inequalities

Theorem 3.3 If $0 < s \leq \tau$ everywhere then

$$Q_{s,\tau,Q^+_s} \geq Q^+_s(t_0,x_0) \quad \text{and} \quad Q_{s,\tau,Q^-_s} \leq Q^-_s(t_0,x_0).$$

Furthermore

$$Q_{s,\tau,U^+} \leq Q^+_s(t_0,x_0) \quad \text{and} \quad Q_{s,\tau,U^-} \geq Q^-_s(t_0,x_0).$$

These dynamic programming results are the basic tools for the study of differential games. We do not know a result of the type of Theorem 3.3 for the Q-value.

One immediate application of Theorem 3.1 can be given

Theorem 3.4 Let (t_0,x_0) be a point of differentiability of the function U^+. Then at (t_0,x_0)

$$\frac{\partial U^+}{\partial t} + \min_z \max_y (\nabla U^+ . f + h) = 0.$$

i.e. the Isaacs-Bellman equation is satisfied.

4. REGULARITY OF THE BOUNDARY

We now introduce some ideas inspired by similar concepts in potential theory. We denote by $\overline{Q}_s^+$ the upper semi-continuous regularization of Q_s^+ given by

$$\overline{Q}_s^+(t,x) = \limsup_{(\tau,\xi)\to(t,x)} Q_s^+(\tau,\xi).$$

We shall consider Q_s^+ and $\overline{Q}_s^+$ as defined on all $R \times R^m$; of course if $(t,x) \in F$ then $Q_s^+(t,x) = g(t,x)$. Suppose now $(t,x) \in \partial F$; we shall say that (t,x) is *Q^+-regular* if

$$\left.\begin{aligned} &\text{(i)} \quad \lim_{s\to 0} \overline{Q}_s^+(t,x) = g(t,x) \\ \text{and} \\ &\text{(ii)} \quad \lim_{(\tau,\xi)\to(t,x)} U^+(\tau,\xi) = g(t,x) \end{aligned}\right\} \qquad (2)$$

Similarly we shall say that (t,x) is *Q^--regular* if

$$\left.\begin{aligned} &\text{(i)} \quad \lim_{s\to 0} \tilde{Q}_s^-(t,x) = g(t,x) \\ &\text{(ii)} \quad \lim_{(\tau,\xi)\to(t,x)} U^-(t,x) = g(t,x) \end{aligned}\right\} \qquad (3)$$

where

$$\tilde{Q}_s^-(t,x) = \liminf_{(\tau,\xi)\to(t,x)} Q_s^-(\tau,\xi),$$

is the lower semi-continuous regularization of Q_s^-.

These conditions seem rather complicated, but in the next section we give some sufficient criteria for regularity. In this section, however, we shall simply list some applications.

Theorem 4.1 Suppose every point of ∂F is Q^+-regular. Then for any $(t_0,x_0) \in R \times R^m$, $Q^+(t_0,x_0) = V^+(t_0,x_0) = U^+(t_0,x_0)$. Furthermore the function Q^+ is continuous in (t,x), and

$$\lim_{s\to 0} Q_s^+(t,x) = Q^+(t,x)$$

uniformly on compact sets.

[Remark: this result depends on the Lipschitz conditions assumed on the function $f(t,x,y,z)$; see §1.]

The proof of Theorem 4.1 uses the dynamic programming theorems of §3. For example to establish continuity of Q^+ we consider two initial positions (t_1,x_1) and (t_2,x_2) and a stopping time $\tau(y,z)$ which is the first t at which one of the trajectories $(t,x_1(t))$, $(t+t_2-t_1,x_2(t+t_2-t_1))$ enters F where

$$\dot{x}_1(t) = f(t,x_1(t),y(t),z(t))$$

and

$$x_1(t_1) = x_1$$

while

$$\dot{x}_2(t) = f(t,x_2(t),y(t+t_1-t_2),z(t+t_1-t_2))$$

with

$$x_2(t_2) = x_2.$$

In the case of fixed time games, the Isaacs condition

$$\min_z \max_y (p.f+h) = \max_y \min_z (p.f+h) \tag{4}$$

for all $p \in R^m$, guarantees that the upper and lower values of the game coincide. For general games of survival this result may be extended in the presence of regularity conditions.

Theorem 4.2 Suppose every point of ∂F is both Q^+- and Q^--regular and that the Isaacs condition (4) is fulfilled. Then for any (t_0,x_0), $Q^+(t_0,x_0) = Q^-(t_0,x_0)$.

Again the proof uses dynamic programming arguments.

5. CRITERIA FOR REGULARITY

For a differentiable function u defined on an open subset of $R \times R^m$ we define

$$L^+u = \frac{\partial u}{\partial t} + \min_{z\in Z} \max_{y\in Y} (\nabla u.f+h)$$

$$L^-u = \frac{\partial u}{\partial t} + \max_{y\in Y} \min_{z\in Z} (\nabla u.f+h)$$

$$L_O^+ u = \frac{\partial u}{\partial t} + \min_{z\in Z} \max_{y\in Y} \nabla u.f$$

$$L_O^- u = \frac{\partial u}{\partial t} + \max_{y\in Y} \min_{z\in Z} \nabla u.f \ .$$

Thus L^+ and L^- are the differential operators associated with the Isaacs-Bellman equations, while L_O^+ and L_O^- are the corresponding operators for the game with h replaced by O. If the Isaacs condition is satisfied then $L^+ \equiv L^-$ and $L_O^+ \equiv L_O^-$. We shall show in this section how regularity of the boundary is related to the operators L^+, L^-, L_O^+ and L_O^-.

Theorem 5.1 Suppose θ_1, θ_2 are two functions on $R \times R^m$ which are C^1 on $R \times R^m$-int F and such that $\theta_1 = \theta_2 = g$ on ∂F with

$$L^+\theta_1 \leq O \leq L^+\theta_2 \quad \text{on} \quad R \times R^m - \text{int} F.$$

Then every point of ∂F is Q^+-regular. Furthermore for any $(t_O, x_O) \in R \times R^m$- int F

$$\theta_1(t_O,x_O) \geq Q^+(t_O,x_O) \geq \theta_2(t_O,x_O) .$$

Corollary 5.2 If θ is a C^1-solution of the Isaacs-Bellman equation $L^+\theta = O$ subject to $\theta(t,x) = g(t,x)$ for $(t,x) \in \partial F$, then $\theta = Q^+ = U^+$.

Corollary 5.3 Suppose the Isaacs condition (4) is fulfilled and there exist C^1-functions θ_1, θ_2 on $R \times R^m$-int F such that $\theta_1 = \theta_2 = g$ on ∂F and

$$L\theta_1 \leq O \leq L\theta_2 \quad \text{on} \quad R \times R^m - \text{int} F$$

(where $L \equiv L^+ \equiv L^-$). Then for any (t_O,x_O) $Q^+(t_O,x_O) = Q^-(t_O,x_O)$.

Corollary 5.4 In general suppose there exist gunctions θ_1, θ_2 which are C^1 on $R \times R^m$- int F and such that $\theta_1 = \theta_2 = g$ on ∂F with

$$L^+\theta_1 \leq O \leq L^-\theta_2 .$$

Then every point of ∂F is both Q^+-regular and Q^--regular.

We now turn to criteria involving L_0^+ and L_0^-. Here we are able to give a condition under which a single point of ∂F is Q^+-regular.

Theorem 5.5 Let $(t_0,x_0) \in \partial F$ and suppose N is a neighbourhood of (t_0,x_0) such that there exist C^1-functions θ_1, θ_2 on N such that

$$\text{(i)} \quad \theta_1(t_0,x_0) = \theta_2(t_0,x_0) = 0,$$

$$\text{(ii)} \quad \theta_2(t,x) < 0 < \theta_1(t,x)$$

$$\text{for } (t,x) \in N\text{-int}F \text{ with } (t,x) \neq (t_0,x_0),$$

$$\text{(iii)} \quad L_0^+\theta_1 \leq 0 \leq L_0^+\theta_2 \text{ on } N\text{-int}F.$$

Then (t_0,x_0) is Q^+-regular.

To interpret this result suppose for convenience that the Isaacs condition is satisfied so that $L_0^+ \equiv L_0^-$. Then the conditions express the ability of both players to steer the trajectory towards (t_0,x_0) from nearby points. We can derive a global result from Theorem 5.5.

Theorem 5.6 Suppose there exist C^1-functions θ_1, θ_2 such that $\theta_1 = \theta_2 = 0$ on ∂F, $\theta_1 \geq 0 \geq \theta_2$ on $R \times R^m - F$ and $L_0^+\theta_1 < 0 < L_0^+\theta_2$ on ∂F. Then every point of ∂F is Q^+-regular.

A special case of the conditions of Theorem 5.6 is used by Friedman [6] p.81 and [8]. This illustrates the meaning of the Theorem. Suppose F has a C^2-boundary, i.e. ∂F can be expressed by the equation

$$x_i = \psi(t,x_1 \ldots x_{i-1},x_{i+1} \ldots x_m)$$

where ψ is twice differentiable. Suppose also that any point $(t,x) \in \partial F$ we have

$$\nu_0 + \min_{z\in Z} \max_{y\in Y} \sum_{i=1}^{m} \nu_i f_i < 0 \tag{3}$$

where $\nu = (\nu_0, \nu_1, \ldots, \nu_m)$ is the normal to ∂F at (t,x). This is called condition (F) by Friedman. Then consider $\rho(t,x)$ the distance of (t,x) from ∂F. It follows from (3) that $L_0^+\rho < 0$

on ∂F and so ρ will serve for θ_1.

Equally suppose we also have

$$\nu_0 + \min_{y\in Y} \max_{z\in Z} \sum_{i=1}^{m} \nu_i f_i < 0 \tag{4}$$

on ∂F (condition $(\tilde{F})$ of Friedman [8]). Then

$$-\nu_0 + \max_{y\in Y} \min_{z\in Z} (- \sum_{i=1}^{m} \nu_i f_i) > 0$$

from which it follows that $L_0^-(-\rho) > 0$ on ∂F. Then since $L_0^+ u \geq L_0^- u$ for any function u, it follows that every point of ∂F is both Q^+- and Q^--regular. Then Theorem 4.2 will apply under conditions (F) and $(\tilde{F})$.

The criteria of Theorems 5.1 and 5.6 can be mixed with θ_1 satisfying Theorem 5.1 and θ_2 satisfying Theorem 5.6 or vice versa.

In one special case, if the game is a generalized pursuit-evasion game, θ_2 may be taken identically zero in Theorem 5.1. Thus for example in the discussion above, condition (3) alone suffices for regularity in a generalized pursuit-evasion game.

6. OPTIONAL STOPPING GAMES

Many of the results of §§3-5 extend with suitable interpretation to optional stopping games. We shall not give detailed results here, but some typical results (see [9]). We recall that the pay-off takes the form

$$P = \min_{t_0\leq t\leq t_F} g(t,x(t))$$

where g is a continuous function on $\mathbb{R} \times \mathbb{R}^m$. The definitions of Q^+- and Q^--regularity are as before ((2) and (3)). However it is easy to see that 2(i) is trivially satisfied since

$$Q_s^+(t,x) \leq g(t,x)$$

for any (t,x). It follows easily from this observation that if $(t,x) \in \partial F$ is Q^--regular then it must also be Q^+-regular. From this we can deduce

Theorem 6.1 Suppose the Isaacs condition

$$\min_{z\in Z}\max_{y\in Y} p.f = \max_{y\in Y}\min_{z\in Z} p.f, \quad p\in R^m \quad (t,x)\in R\times R^m\text{-int } F$$

is satisfied, and that every point of ∂F is Q^--regular. Then for any (t_0,x_0), $Q^+(t_0,x_0) = Q^-(t_0,x_0)$.

Theorem 6.1 above can be deduced from a generalization of Lemma 7.1 and Theorem 6.2 of [9] (taking $h \equiv 0$).

A special type of optional stopping game is a *restricted fuel* game in which we introduce an extra space co-ordinate ξ governed by

$$\dot{\xi} = h(t,x,y,z)$$

where $h: R\times R^m \times Y \times Z$ is continuous (note that h does not depend on ξ). We suppose that the pay-off takes the form

$$P = \min_{t_0\le t\le t_F} g(t,x(t),\xi(t))$$

where g is monotonically increasing in ξ, and F is of the form $F = \{(t,x,\xi) : t \ge T \text{ or } \xi \ge \Lambda\}$. This describes a game in which J_Z seeks to reach a point of minimum g without exhausting his resources either of time t or of energy ξ. If $h \ge 0$ everywhere we can think of ξ as measuring fuel consumption, and Λ as a fuel constraint. In this case we obtain

Theorem 6.2 Suppose every point of ∂F is Q^--regular and the Isaacs condition

$$\min_{z\in Z}\max_{y\in Y}(p.f+h) = \max_{y\in Y}\min_{z\in Z}(p.f+h) \quad p\in R^m \quad (t,x,\xi)\in R\times R^m\times R - \text{int } F$$

holds; then for every (t_0,x_0,ξ_0),

$$Q^+(t_0,x_0,\xi_0) = Q^-(t_0,x_0,\xi_0).$$

Note that this is stronger than simply Theorem 6.1 applied to restricted fuel games in that the Isaacs condition is not homogeneous. We remark that in [9] Theorem 7.2 the hypothesis of regularity should be included.

Theorem 6.3 Suppose in a restricted fuel game we also have

(i) $g(t,x,\xi) \equiv \xi$

(ii) $\max_{y\in Y}\min_{z\in Z} h(t,x,y,z) = \varepsilon > 0$ for all $(t,x)\in R\times R^m - F$.

Then every point of ∂F is Q^--regular and so the conclusion of Theorem 6.2 is valid.

7. EXTENDED VALUES

An alternative definition of value has been suggested particularly with reference to games of pursuit and evasion. (Friedman [6] p.78.) We shall describe here a definition of extended value in the context of games of survival which differs from Friedman's definition, but has the same underlying idea; however we do not know whether the two definitions coincide in general.

We shall suppose that the pay-off is of the form

$$P = \int_{t_0}^{t_F} h(t,x(t),y(t),z(t))dt + g(t_F,x(t_F))$$

as before. We then introduce a 'dummy' space-variable ξ governed by

$$\xi = \int_{t_0}^{t} h(s,x(s),y(s),z(s))ds$$

and write the pay-off P as

$$P = g(t_F,x(t_F)) + \xi(t_F).$$

Now an *approximate strategy* **A** for J_Y is a sequence (α_n) where each α_n is a delay-strategy. Similarly an approximate strategy $B = (\beta_n)$ for J_Z is a sequence of delay-strategies. Each pair (α_n, β_n) induces a unique pair of controls $(y_n(\cdot),z_n(\cdot))$ satisfying

$$\alpha_n z_n = y_n,$$

$$\beta_n y_n = z_n,$$

and the controls induce trajectories $(x_n(t),\xi_n(t))$ in $\mathbb{R}^{m+1}$. This sequence of trajectories is relatively compact in the topology of uniform convergence on $[t_0,T]$. Let $(\bar{x}(t),\bar{\xi}(t))$ be any limit point; then let

$$P_{\bar{x},\bar{\xi}} = g(t_F,\bar{x}(t_F)) + \bar{\xi}(t_F)$$

where t_F is the time of entry of $(t,x(t))$ into F. Then $P[A,B]$ is the set of such $P_{\bar{x},\bar{\xi}}$ and we define the extended upper and lower values by

$$V_e^+ = \inf_B \sup_A \sup P[A,B]$$

$$V_e^- = \sup_A \inf_B \inf P[A,B] .$$

It is clear that $V_e^+ \geq V_e^-$. We say the game has extended value if $V_e^+ = V_e^-$.

Let us now restrict to the special case when $g \equiv 0$. We can consider an associated restricted fuel game G_Λ^* with dynamics

$$\dot{x} = f(t,x,y,z)$$

$$\dot{\xi} = h(t,x,y,z)$$

subject to $x(t_0) = x_0$ and $\xi(t_0) = 0$, and pay-off

$$P = \min_{t_0 \leq t \leq t_{F_\Lambda^*}} \rho(t,x(t))$$

where $\rho(t,x)$ is the distance of (t,x) from F and $F_\Lambda^* = \{(t,x,\xi): t \geq T \text{ or } \xi \geq \Lambda\}$. Obviously if this restricted fuel game can be shown to have value zero then this means that, in some sense, the original game of survival reaches the terminal set F with $\xi \leq \Lambda$. This argument can be made precise to give the following result using Theorem 6.3.

Theorem 7.1 Suppose in a game of survival $g \equiv 0$, and

$$\max_{y \in Y} \min_{z \in Z} h(t,x,y,z) \geq 0 \quad \text{for} \quad (t,x) \in R \times R^m - F.$$

Then if the Isaacs condition (4) is satisfied we have

$$V_e^+(t_0,x_0) = V_e^-(t_0,x_0) .$$

This result is given in [9] (Theorem 8.3); the corresponding result for Friedman's definition of extended value is obtained by Elliott and Friedman in [2]. We call a game satisfying the conditions of Theorem 7.1 a *quasi-pursuit-evasion* game. Obviously the theorem applies to generalized pursuit-evasion games (see [4]).

Finally we can extend the result to general games of survival, and our result is interesting to compare with Corollary 5.3.

Theorem 7.2 Suppose in a game of survival the Isaacs condition (4) holds and there is a C^1-function θ on $R \times R^m$ such that

$$L\theta = \frac{\partial\theta}{\partial t} + \min_{z\in Z}\max_{y\in Y}(f.\nabla\theta + h) \geq 0$$

for $(t,x) \in R\times R^m - F$, and $\theta = g$ on ∂F. Then for any (t_0,x_0), $V_e^+(t_0,x_0) = V_e^-(t_0,x_0)$.

To prove Theorem 7.2 we simply consider the game of survival with pay-off

$$P = \int_{t_0}^{t_F} h_1(t,x(t),y(t),z(t))dt$$

where $h_1(t,x,y,z) = h(t,x,y,z) + \frac{\partial\theta}{\partial t} + f.\nabla\theta$.

The new 'dummy' variable η then satisfies

$$\eta(t) = \int_{t_0}^{t} h_1(t,x,y(t),z(t))dt$$

$$= \xi(t) + \theta(t,x(t)) - \theta(t_0,x_0)$$

along any trajectory.

It follows that in the definition of $P[A,B]$, to any limit trajectory $(\bar{x}(\cdot),\bar{\xi}(\cdot))$ there corresponds a limit trajectory $(\bar{x}(\cdot),\bar{\eta}(\cdot))$ where

$$\bar{\eta}(t) = \bar{\xi}(t) + \theta(t,\bar{x}(t)) - \theta(t_0,x_0)$$

and so

$$\bar{\eta}(t_F) = \bar{\xi}(t_F) + g(t_F,\bar{x}(t_F)) - \theta(t_0,x_0).$$

From this it follows that if $V_{e,1}^+$ and $V_{e,1}^-$ denote the extended values of the new game then

$$V_{e,1}^+ = V_e^+ - \theta(t_0,x_0),$$

$$V_{e,1}^- = V_e^- - \theta(t_0,x_0)$$

and then we can apply Theorem 7.1.

REFERENCES

1. A. BENSOUSSAN and A. FRIEDMAN, Nonlinear variational inequalities and differential games with stopping games, to appear.

2. R.J. ELLIOTT and A. FRIEDMAN, A note on generalized pursuit-evasion games S.I.A.M. J. Control, to appear.

3. R.J. ELLIOTT and N.J. KALTON, The existence of value in differential games, Mem. Amer. Math. Soc. No.126 (1972).

4. ——— ———, The existence of value in differential games of pursuit and evasion, J. Diff. Eqns. 12 (1972) 504-523.

5. ——— ———, Cauchy problems for certain Isaacs-Bellman equations and games of survival, Trans. Amer. Math. Soc. to appear.

6. A. FRIEDMAN, Differential Games, Wiley-Interscience, New York 1971.

7. ———, On the definition of differential games and the existence of value and saddle points, J. Diff. Eqns. 7 (1970) 69-91.

8. ———, Remarks on differential games of survival, J. Diff. Eqns. 14 (1973) 121-128.

9. N.J. KALTON, Differential games with optional stopping, Proc. Cambridge Philos. Soc. to appear.

10. E. ROXIN, Axiomatic approach in differential games, J. Optim. Theory Appl. 3 (1969) 153-163.

11. P. VARAIYA and J. LIN, Existence of saddle points in differential games, SIAM J. Control. 7 (1969) 142-157.

SOLUTION SETS FOR N-PERSON GAMES

J. D. Grote

Control Theory Centre
University of Warwick

<u>Introduction</u> The phrase 'solution sets' of the title does not refer to the standard concepts of solution for N-person games, a survey of which can be found in (6, Chs. VIII, IX). It refers rather, to a new concept arising from a dynamic approach to game theory, introduced by Smale, (7). This latter paper can be viewed as a dynamic view of cooperative N person games without side payments. The author has then developed this approach to non cooperative and partially cooperative games, (2, 3).

This dynamic approach presents the game as a simple differential game with terminal payoff, the terminal surface being the solution set of the game. The main tool in the method is the differential calculus and we naturally take the most general setting for this; i.e. we assume that the set of pure strategies of each player is a differentiable manifold. There are no restrictions of convexity etc. placed on the sets of pure strategies or the payoff functions and indeed Ekeland, (1), has given a topological classification of the equilibria of such games in the spirit of Thom's catastrophe classification (8).

<u>Definitions</u> We define an <u>N-person game</u> as follows :-

a) N smooth (i.e. infinitely differentiable) manifolds $M_1, \ldots M_N$. M_i is the set of pure strategies for the i th player. Let $W = M_1 \times M_2 \times \ldots \times M_N$.

b) N smooth functions $f_i : W \to R$, the payoff functions for each player.

J. D. Grote (ed.), The Theory and Application of Differential Games. 63-75.

c) Two rules.

rule I The players are constrained to change their strategies continuously and piecewise smoothly. Player i changes his strategy in order to maximise f_i subject to any constraints in rule 2.

rule 2 Constraints on the ability of the players to cooperate and compete.

A generalised control system, g.c.s., is a set valued section $G : W \to TW$, of the tangent bundle of W. (i.e. at each point w of W is given a set of directions at w which represent the 'admissible' directions at w for the game.)

An admissible path, ϕ , for a g.c.s. G is a continuous piecewise smooth path $\phi: R \to W$, such that

$$\frac{d\phi}{dt}(t) \in G(\phi(t) \text{ wherever } \frac{d\phi}{dt} \text{ exists.}$$

The set of points w in W for which $G(w) = \{0\}$ for a g.c.s. G is called stat G.

w_o in stat G is called asymptotically stable if given any open neighbourhood V of w_o in stat G there exists an open neighbourhood U of w_o in W such that for all w in U and any admissible path ϕ such that $\phi(0) = w$, then $\lim_{t\to\infty} \phi(t)$ is in V.

Throughout we restrict attention to the generic case. Thus we assume all the f_i to be non-degenerate. For simplicity we assume the M_i are Riemannian and thus W is Riemannian and we can introduce the gradient vector fields, grad f_i .

1. The Cooperative Game

The interesting case, considered by Smale (7), is that of cooperation without side payments. (The introduction of side payments merely acts to reduce the effective number of players in the game. We shall throughout this paper assume any cooperation is without side payments.) Thus from any given initial strategies the players change their strategies along any path $\phi : R \to W$ such that

$$\frac{d}{dt}(f_i \circ \phi)(t) > 0 \qquad \forall\ i, t \qquad (I)$$

i.e. No move is allowed which is to the disadvantage of any player.

Smale constructs a g.c.s. $H : W \to TW$ such that the associated admissible paths satisfy (I).

Define $H_i : W \to TW$ as follows :-

$$H_i(w) = \{ V \in T_w W \mid \text{grad } f_i \,.\, V \geq V.V\}$$

where . denotes the Riemannian metric on W. Define $H : W \to TW$ by

$$H(w) = \bigcap_i H_i(w) ,$$

then the admissible paths of H satisfy (I).

Definition. Stat H is called the critical Pareto set, θ

Smale denotes the asymptotically stable part of θ by θ_s. This is the classical "Pareto Optimum" for such cooperative games, representing the set of best positions for each player given the cooperative rule. However, in this setting there is no guaranteed Pareto optimum. We have, (7)

Theorem 1.1 If W is compact then generically θ_s is not empty.

However, θ_s, may have many components and the initial strategies determine towards which component of θ_s the game evolves.

Example 1 $N = 2$, $M_1 = M_2 = R$, $f_1 : R^2 \to R$ is defined by

$$f_1(x_1, x_2) = x_1^2 - x_2^2 \quad , \quad f_2 : R^2 \to R \text{ is defined by}$$

$$f_2(x_1, x_2) = (x_1 - \alpha)(x_2 - \beta)$$

If we let $F : R^2 \to R^2$ be the map

$F(x_1, x_2) = (f_1(x_1, x_2), f_2(x_1, x_2))$ and let

$$S'(F) = \{(x_1, x_2) \in R^2 \mid dF(x_1, x_2) \text{ has rank } < 2\}$$

then we see that $S'(F)$ is the circle

$$(x_1 - \tfrac{\alpha}{2})^2 + (x_2 - \tfrac{\beta}{2})^2 = \frac{\alpha^2+\beta^2}{4} \quad , \text{ see Fig. 1.}$$

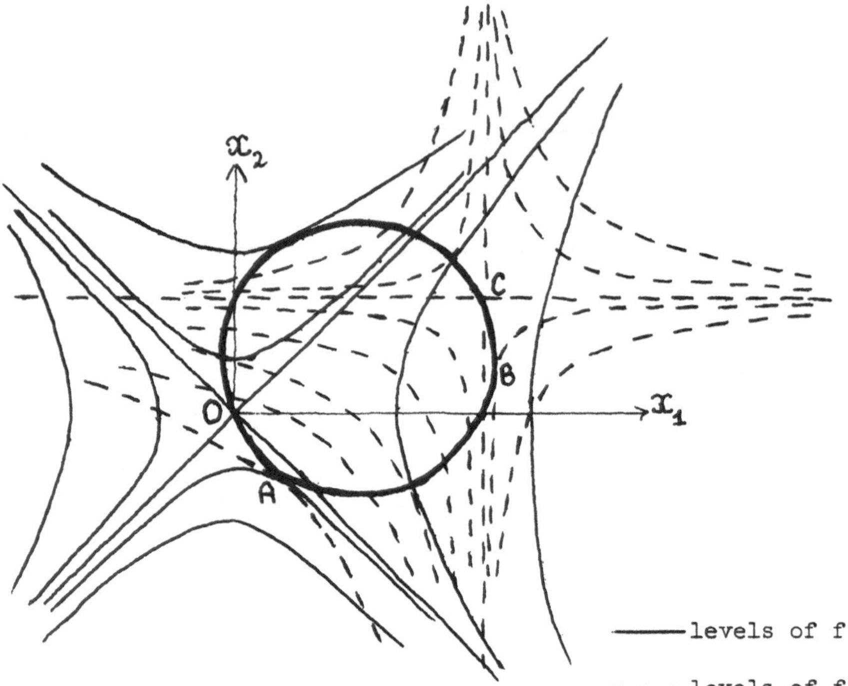

Fig. 1

S'(F) is the set of points at which the levels of f_1 and f_2 are mutually tangential. Denote by A and B the points at which S'(F) is itself tangential to the levels of f_1 and f_2. (i.e. A and B are in S'''(F) (5)) Clearly $\theta \subset S'(F)$. In fact if we denote the point (o,o) by O and the point (α, β) by C, then θ is the arc OABC. θ_s is the arc AB and the arcs OA and BC are asymptotically unstable and may be regarded as Pareto minima in some sense. Clearly if we played a game in higher dimensions we could encounter "Pareto saddles" and this suggests a Morse theory of cooperative games (c.f. (9)).

In (7), Smale does indeed develop such a theory introducing a generalised second differential of F at a point S'(F) to determine the "type" of the point (i.e. Pareto min., max. or saddle) and giving Morse Inequalities relating the topology of W to the existence of the different types of critical Pareto set.

2. The Non Cooperative Game

In this section we adapt the preceeding ideas to fit the

non cooperative N person game. We show that the Nash equilibrium points of the N person game, when they exist, do not necessarily coincide with the asymptotically stable points of the given g.c.s.

Let $\pi_i : W \to M_i$ be the canonical projection, and $(\pi_i)_* : T_w W \to T_{\pi_i(w)} M_i$ the induced projection on the tangent spaces.

Then define $C_i : W \to TW$ by

$$C_i(w) = \{V \in T_w W \mid V.(\pi_i)_* \operatorname{grad} f_i > (\pi_i)_* V.(\pi_i)_* V\}$$

Define $C : W \to TW$ by $C(w) = \bigcap_i C_i(w)$, then we have immediately

Theorem 2.1 If ϕ is an admissible path of the g.c.s., C such that $\phi(t_o) = (x_1, \ldots, x_N) = w$ and $(\pi_i)_* C_i(w) \neq \{0\}$ then there exists $\varepsilon > 0$ such that

$$f(x_1, \ldots \pi_i(\phi(t)), \ldots, x_N) \geqslant f(w) \text{ for all } t_o < t < t_o + \varepsilon .$$

Moreover, if ω is a Nash equilibrium point of the game then we have w is in stat C.

However, the following example shows that a Nash equilibrium point is not necessarily an asymptotically stable point of stat C.

Example 2 $N = 2$, $M_1 = M_2 = R$, $f_1 : R^2 \to R$ defined by

$$f_1(x_1, x_2) = -x_1^2 + 4x_1 x_2 + x_2 \quad , \; f_2 : R^2 \to R \text{ defined by}$$

$$f_2(x_1, x_2) = 4x_1x_2 - x_2^2 + x_1$$

The point (0, 0) is seen to be a Nash equilibrium point.

$$f_1(x_1, 0) = -x_1^2 \leqslant f_1(0, 0)$$

$$f_2(0, x_2) = -x_2^2 \leqslant f_2(0, 0) .$$

However it is readily seen that at any point (x_1, x_2) such that

$$0 < \frac{x_2}{2} < x_1 < 2x_2$$

or $2x_2 < x_1 < \frac{x_2}{2} < 0$, then any admissible path associated with the g.c.s. C through (x_1, x_2) moves away from (0, 0).

It is also easy to construct games with no Nash equilibrium point.

Example 3 Take any zero sum game with a unique Nash equilibrium point, O, defined for $N = 2$, $M_1 = M_2 = R$. Assume that in a neighbourhood of the Nash equilibrium point the levels of the payoff function f are as in Fig. 2a. Perturb this game to a new game for which the payoff function f_1, agrees with f outside a neighbourhood of O. Assume that inside the neighbourhood of O the levels of f_1 are as in Fig. 2b. Then we see that O is no longer a Nash equilibrium and no new Nash equilibrium has been introduced, since stat C remains unchanged.

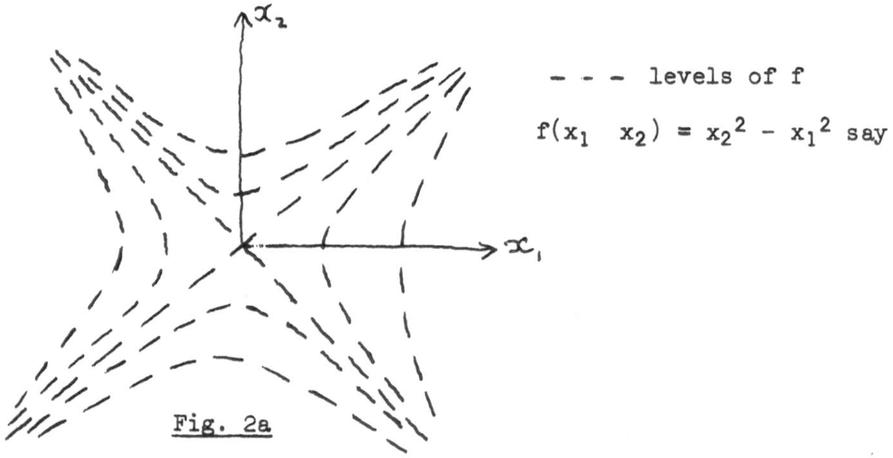

Fig. 2a

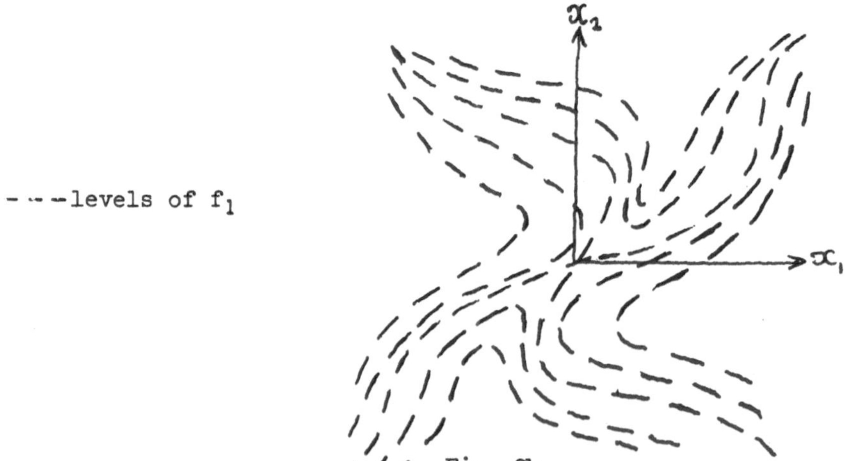

Fig. 2b

Thus we see that the Nash equilibrium point has little relevance to such games with no existence theorem and without the asymptotic stability properties necessary for a dynamic description.

3. Solution Sets for Non Cooperative Games

In this section we define an alternative concept of solution arising naturally from the dynamic description introduced in 2. This we shall call the solution set and it exists for games played on compact manifolds and is stable under "reasonable" motions.

We look to the points of stat C as solution sets of the game, as in 1. However, even for W compact stat C may be empty. We have however the following topological result.

Theorem 3.1 (1) stat C is empty $\Rightarrow$ Euler characteristic of W is zero.

Thus if the Euler characteristic of W is non-zero we know that stat C is non empty. However the set of asymptotically stable points of stat C may still be empty. In fact asymptotic stability seems too strong a criterion. It demands stability under a large class of motions, many of which are unreasonable choices for the players. We introduce optimality into the game to reduce the class of motions for which the point of stat C must be stable.

Definition An admissible vector field, V, for the g.c.s. C is a piecewise smooth vector field $V : W \to T(W)$ such that $V(w)$ is in $C(w)$ and

$$V(w) = 0 \iff C(w) = \{0\}$$

Definition An admissible vector field, V, is called C-optimal if :

$$(\pi_i)_* V(w).\text{grad } f_i(w) \geq (\pi_i)_* U(w).\text{grad } f_i(w) \text{ for all}$$

$U(w)$ in $C(w)$, all i and all $w \in W$.

Definition The solution curve of a C-optimal vector field is called a C-optimal path.

Theorem 3.2 There is a unique C-optimal vector field on W, namely

$$((\pi_1)_* \text{ grad } f_1, (\pi_2)_* \text{ grad } f_2, \ldots, (\pi_N)_* \text{ grad } f_N)$$

Clearly the C-optimal vector field vanishes at stat C, and any sink of this C-optimal vector field will be stable under optimal motions. Whilst we do not have, necessarily, the existence of such sinks, we do have, on compact manifolds, the existence of an asymptotically stable limit set of the C-optimal vector field.

Definition The asymptotically stable limit sets of the C-optimal vector field are called the solution sets of the game.

Example 4 If we look to example 2 we see that the Nash equilibrium point is not stable even under the restricted class of optimal motions.

Example 5 If we look to example 3, we see that the Nash equilibrium point of the original game is optimally stable and thus a solution set of the game. As we perturb to Fig. 2b the same point, though not a Nash equilibrium point remains a solution set of the game. If we continue the perturbation, in like manner however, the solution set exhibits the Hopf bifurcation (4) from a stable point into an unstable point of stat C and a stable periodic orbit as solution set around the one time Nash equilibrium point.

4. The General Game

We now assume that rule 2 is given for a game which allows a more general coalition structure than totally cooperative or competitive and maybe allows the formation of more than one coalition structure. We illustrate the extension of the ideas of 1, 2 and 3 to this general game by considering a particular example.

As in 1 and 2 for any permissible coalition structure we can define a g.c.s. underlying the dynamics of that coalition. To allow for several possible coalition structures we simply define $G : W \to TW$ for the general game as the "union" of each such g.c.s. As in 3 we can introduce "optimal play" for each coalition. However, if the totally competitive coalition is not permissible it may be necessary to include an initial coalition structure in the initial data. Moreover, the general-optimal (or G-optimal) vector fields will be discontinuous at "switching loci" as the players switch from one coalition structure to another.

Theorem 4.I (3) If a game $G : W \to TW$ includes more than one permissible coalition structure, then generically stat G is empty.

Thus, the solution sets of a general game, the asymptotically stable limit sets of the G-optimal vector fields would seem, necessarily, not to be isolated point sets. However the following example, chosen to illustrate the concepts above indicates that this may not necessarily be the case.

Example 6 $N = 2$, $M_1 = S^1$, $M_2 = S$, $f_1 : S^1 \times S^1 \to R$ defined by

$$f_1(x_1, x_2) = e^{\cos 2\pi x_2} \cos 2\pi(x_1 - x_2) \text{ and}$$

$f_2 : S^1 \times S^1 \to R$ defined by

$$f_2(x_1, x_2) = e^{\cos 2\pi(x_1+\frac{c}{2})} \cos 2\pi(x_1-x_2+c)$$

$$c\varepsilon(0, 1) \text{ and } (x_1, x_2)\varepsilon [0, 1] \times [0, 1]$$

We assume that rule 2 allows the formation of both the cooperative and competitive coalitions. We define $G : W \to T(W)$ by

$$G(w) = H(w) \cup C(w)$$

From 3.2 the C-optimal vector field is

$$(-2\pi c^{\cos 2\pi x_2} \sin 2\pi(x_1-x_2), \; -2\pi e^{\cos 2\pi(x_1+\frac{c}{2})} \sin 2\pi(x_1-x_2+c))$$

This vector field we find never vanishes and thus stat C is empty. However the asymptotically stable limit set of this vector field is the closed orbit.

$$x_2 = x_1 + \frac{c}{2} \bmod 1 \qquad \text{see Fig. 3.}$$

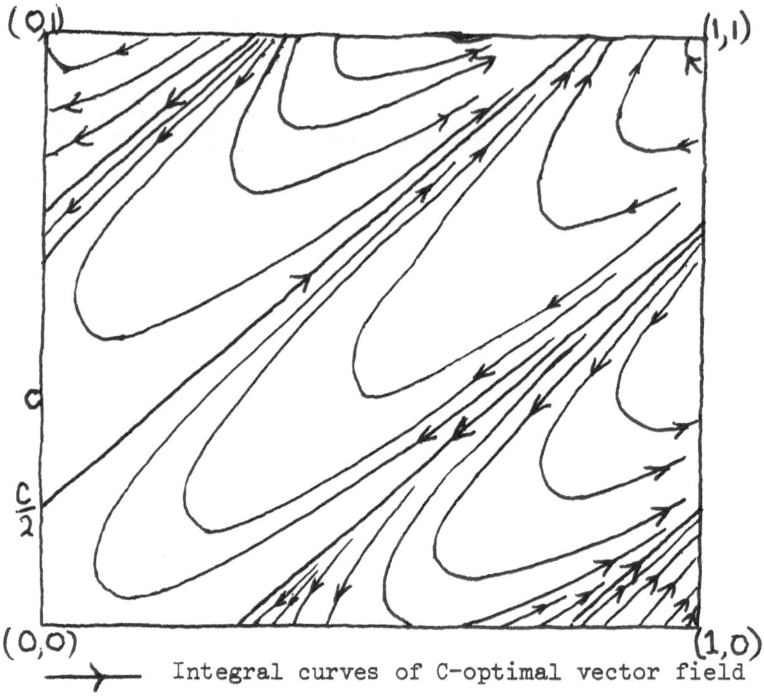

Fig. 3

Thus stat G = stat C $\cap$ stat H is already empty. stat H $\subset$ S'(F) (see example I) which in this case is given by the formula

$$\sin 2\pi\ (x_1 - x_2) \cos 2\pi(x_1 - x_2 + c) \sin 2\pi(x_1 + \tfrac{c}{2})$$

$$+ \quad \sin 2\pi\ (x_1 - x_2 + c) \cos 2\pi(x_1 - x_2) \sin 2\pi x_2$$

$$+ \quad \sin 2\pi\ (x_1 + \tfrac{c}{2}) \sin 2\pi x_2 \cos 2\pi(x_1 - x_2 + c) \cos 2\pi(x_1 - x_2) = 0$$

The structure of S'(F) depends on the value of c and is in any case difficult to calculate. We are more interested in the G-optimal vector fields and the switching locus.

A G-optimal vector field is either a C-optimal vector field or a Pareto-optimal, (2)(or H-optimal) vector field depending on which coalition is in force. We do not go into the definition of H-optimal here, but we shall look near the critical Pareto set, where H-optimality is obvious.

Let us consider where $S'(F)$ intersects $x_2 = x_1 + \frac{c}{2} \bmod 1$. We find this happens at the 2 points:

$$\left(\frac{1-c}{2}, \frac{1}{2}\right) \text{ and } \left(\frac{2-c}{2}, 1\right) = \left(\frac{2-c}{2}, 0\right)$$

Moreover on $x_2 = x_1 + \frac{c}{2} \bmod 1$ we have

$$\text{grad } f_1 = -2\pi e^{\cos 2\pi x_2}(-\sin c\pi, \sin c\pi + \cos c\pi \sin 2 x_2)$$

and

$$\text{grad } f_2 = -2\pi e^{\cos 2\pi x_2}(\sin c\pi + \cos c\pi \sin 2\pi x_2, -\sin c\pi)$$

Thus we see that $\left(\frac{1-c}{2}, \frac{1}{2}\right)$ is an unstable point of θ and $\left(\frac{2-c}{2}, 1\right) = \left(\frac{2-c}{2}, 0\right)$ is a stable point of θ, see Fig. 3a, 3b.

Moreover the H-optimal vector field is tangential to $x_2 = x_1 + \frac{c}{2}$ mod 1.

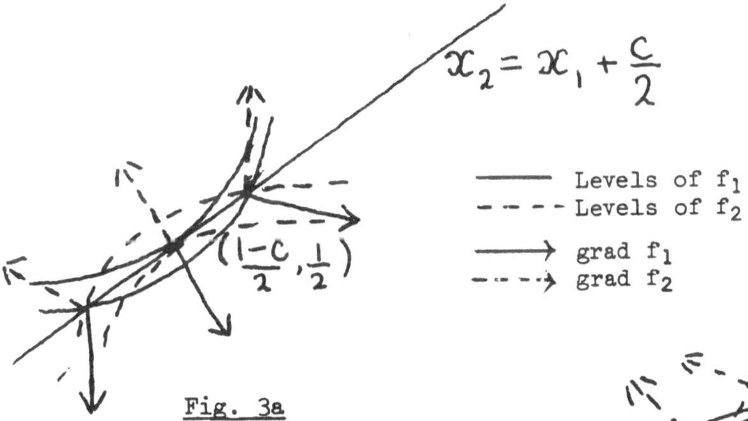

Fig. 3a

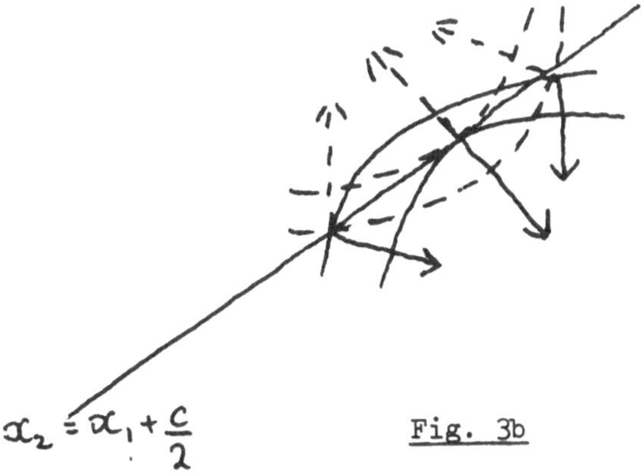

Fig. 3b

Are there then any switching loci for this game and what are the solution sets? Without explicitly calculating the H-optimal vector fields throughout S' x S' it is not possible to answer either part of the question. However we know the H-optimal vector field along $x_2 = x_1 + \frac{c}{2}$.

It is in fact

$$\frac{1}{2}\,\text{grad } f + \frac{1}{2}\,\text{grad } f$$

$$= -\pi e^{\cos 2\pi x_2} (\cos c\pi \sin 2x_2, \cos c\pi \sin 2\pi x_2)$$

Whereas the C-optimal vector field is

$$2\pi e^{\cos 2\pi x_2} (\sin c\pi\ , \sin c\pi) \text{ and thus if}$$

$$\cos c\pi \sin 2\pi\, x_2 = 2 \sin c\pi \quad \text{for some value of } x_2$$

then the switching locus intersects $x_2 = x_1 + \frac{c}{2}$.

Alternatively if $\tan^{-1}\frac{1}{2} < c\pi < \tan^{-1}(-\frac{1}{2})$, then the switching locus does not intersect $x_2 = x_1 + c/2 \bmod 1$ and the G-optimal vector field is the C-optimal vector field. Thus, by continuity, the G optimal vector field is the C-optimal vector field in a neighbourhood of $x_2 = x_1 + c/2 \bmod 1$ and thus $x_2 = x_1 + c/2 \bmod 1$ is a solution set of the general game G.

If, for example, $c\pi < \tan^{-1}(\frac{1}{2})$ we have on $x_2 = x_1 + c/2 \bmod 1$ two values of x_2 both less than $\frac{1}{2}$ at which the switching locus cuts the set. At the lower value of x_2 the G-optimal vector field along the set $x_2 = x_1 + c/2$ is directly opposed to itself at the intersection point, approaching the point from both sides. At the upper value the G-optimal vector field is similarly opposed but leaving the point. It would seem then that it is possible that the point corresponding to the lower value of x_2 is a possible point solution set of the game, despite the fact that stat G is not zero there. To establish this we would have to show the stability of the point as a point of S' x S', not just as a point of $x_2 = x_1 + c/2 \bmod 1$.

<u>Question</u> Has this candidate for a point solution set arisen from some non generic property in the example?

REFERENCES

(1) Ekeland, I., Topologie Differentielle et Theorie des Jeux, To be published, Topology, 1974.

(2) Grote, J.D., A Global Theory of Games, I, to be published, J. of Math. Econ., 1974.

(3) Grote, J.D., A Global Theory of Games, II, Control Theory Centre Report, University of Warwick, 1973.

(4) Hopf, E., Abzweigung einer Periodischen Losung eines Differential Systems, Ber. der Math. Phys. Akad. der Wiss. zu Leipzig XCIV (1942) pp. 1 - 22.

(5) Levine, H.I., Singularities of Differentiable Mappings, Proc. Liverpool Singularities Symp. 1, pp. 1 - 89, Springer-Verlag 192.

(6) Owen, G. Game Theory, Saunders, 1968.

(7) Smale, S., Global Analysis and Economics, 1., Proc. 1971, Brazil Dynamical Systems Symp.

(8) Thom, R., Stabilite Structurelle et Morphogenese, Benjamin, 1972.

(9) Milnor, J., Morse Theory, Annals of Maths. Studies 51, Princeton 1963.

ALTERNATIVES TO THE TREE MODEL FOR EXTENSIVE GAMES

H. S. Witsenhausen

Bell Laboratories
Murray Hill, New Jersey

INTRODUCTION

The interaction between the moves of the players, the information available for each move and the outcome of chance moves, is described for finite games by the well known tree model due to Kuhn [2]. Among the reasons for considering alternative models of extensive games are

(i) A desire to weaken the finiteness requirement and the need to handle the measurability problems which then arise.

(ii) The natural way in which extensive games often present themselves is an input-output description. Even in the finite case, to derive a tree model from such a description requires the general solution of the closed loop relations, which can be demanding.

(iii) For several simple but important conclusions concerning the effect of information changes in games, the tree model is already more detailed than necessary and may even hide the simplicity of the situation. On the other hand, the matrix form of the game does not contain enough detail to permit formulation of even such a concept as "open-loop strategy".

The intrinsic model using σ-fields described in [5] is motivated by (i) and (ii) above. Since it is at least as detailed as the tree model, it does not answer to motivation (iii). For that reason the simpler loop model is of interest. The main emphasis of this paper will be to derive elementary conclusions from the loop model [6] for both 2 person zero-sum and n-person games. Two of these conclusions (the valuation of competitive

J. D. Grote (ed.), The Theory and Application of Differential Games. 77-84.

information and the information monotonicity of the Shapley value) do not seem to have been explicitly stated before.

THE HEURISTICS OF THE LOOP MODEL

Each player is viewed as the strategist for an organization consisting of many agents (devices) distributed over space and time. Each agent chooses his action as a function of the information available to him. This function is prescribed by the player and such a prescription for each agent is what is meant by a pure strategy for that player.

In the intrinsic model [5] the situation of each agent is examined in detail. In the loop model, one simply observes that the effect of a pure strategy is to make the totality u of all actions taken by the members of the organization a certain function of the totality y of the observations available to one or more of the agents. This may be written $u_i = \gamma_i(y_i)$ where the subscript i refers to the player. The constraints of implementation (including the information constraints) are reflected in the set Γ_i of all possible functions γ_i.

In a deterministic n-player game, what happens is completely determined by the actions taken, that is by the n-tuple $u_1, \ldots, u_n$. The payoffs to the players and the observations can be described as fixed functions of $u_1, \ldots, u_n$. In particular $y_i = g_i(u_1, \ldots, u_n)$.

By composition $u_i = \gamma_i(g_i(u_1, \ldots, u_n)) = p_i(u_1, \ldots, u_n)$ and the set of possible γ_i is mapped to a set P_i of possible p_i. For stochastic games a random argument ω of given distribution has to be appended to $u_1, \ldots, u_n$.

If the players select pure strategies p_i $(i=1, \ldots, n)$, then one has to solve the simultaneous equations

$$u_i = p_i(u_1, \ldots, u_n) \quad i = 1, \ldots, n \tag{1}$$

for the u_i and then find the corresponding payoffs $K_i(u_1, \ldots, u_n)$. In the stochastic case, one has to solve, for each ω the system

$$u_i = p_i(u_1, \ldots, u_n, \omega) \quad i = 1, \ldots, n \tag{2}$$

for the u_i as functions of ω, which are then substituted into the payoff functions $K_i(u_1, \ldots, u_n, \omega)$ to yield functions of ω of which the expectation may be taken.

Either way, the vector of payoffs is determined from the list of pure strategies $(p_1, \ldots, p_n)$ selected in $P_1 \times P_2 \times \ldots \times P_n$. This constitutes the normal or matrix form of the loop game, assuming that the operations can be carried out.

The existence of a unique ω-measurable solution of (2) or (1) for all choices of the $p_i \in P_i$, $i = 1,\ldots,n$ depends on the detailed structure displayed in the intrinsic model. In the loop model it is simply assumed as a property of the n-tuple of sets of functions $P_1,\ldots,P_n$.

THE FORMAL LOOP MODEL

A stochastic n-player loop game is defined by n nonempty sets $U_i (i=1,\ldots,n)$, a probability space $(\Omega,\mathcal{B},P)$, n nonempty sets

$P_i (i=1,\ldots,n)$ where P_i is a subset of the set U_i^H with $H \equiv \Omega \times \prod_{i=1}^{n} U_i$,

as well as n real functions K_i on H.

The deterministic case is obtained for Ω a one-point set.

In the sequel we consider families of loop games with the same $(\Omega,\mathcal{B},P)$ and (U_i,K_i); $i = 1,\ldots,n$; but different choices of P_i subject to the two following assumptions:

A1: The set P_i^0 of all constant functions from H to U_i is contained in P_i, $i = 1,\ldots,n$.

A2: For all choices of $p_i \in P_i$ $(i=1,\ldots,n)$ the simultaneous equations $u_i = p_i(u_1,\ldots,u_n,\omega)$, $(i=1,\ldots,n)$ determine the u_i uniquely as functions of ω and when these functions are substituted into the K_i the resulting functions belong to $L_1(\Omega,\mathcal{B},P)$.

These assumptions and their possible weakening are discussed more deeply, for 2 person zero-sum games $(n=2, K_2=-K_1)$, in reference [6].

INFORMATION MONOTONICITY FOR 2 PLAYER ZERO SUM GAMES

If two versions of a game differ only in that one player receives more information in the second version, then the set of pure strategies available to that player in the first version is a subset of the one corresponding to the second version. Of course, such an inclusion could result also from the removal of other constraints and the conclusions will still follow.

For a zero-sum two person game, deterministic or stochastic, satisfying assumption A2 in both versions, the increase in pure strategies implies an increase in mixed strategies for the player receiving the extra information. As the pure and mixed strategy sets for the other player are unchanged, the definition of upper and lower values implies at once the following monotonicity:

Proposition 1. If in a two person zero sum game one of the players receives more information, then the following four quantities can move only in the direction favorable to the player receiving the extra information: the upper and lower values in pure strategies, the upper and lower values in mixed strategies.

As a corollary, the value moves in that same direction if a value exists in both versions. The general conditions for the existence of a value with properly defined mixed strategies have recently been elucidated [7].

Note that Proposition 1 does not assert anything in case the information of both players is increased. Also Proposition 1 does not require assumption A1.

A stronger but more special result is the one discussed in detail in [6]:

Proposition 2. In a deterministic two-person zero-sum game satisfying assumptions A1 and A2, let the information of both players be increased, then the upper and lower values in pure strategies can only move toward each other. In particular, a saddle-value in pure strategies, if it exists, remains unchanged.

As the detailed analysis of this phenomenon is available elsewhere [6], a heuristic explanation of Proposition 2 will suffice. If only the maximizer receives the extra information the upper and lower values can only increase by Proposition 1. However, the upper value is actually unchanged because it is obtained when the maximizer has foreknowledge of the pure strategy of his opponent. As the game is deterministic, the maximizer then faces a situation devoid of uncertainty and feedback is unnecessary for optimization. This means that the same upper value is obtained when the choice of the maximizers pure strategy is confined to the open loop strategies (which are admissible by assumption A1). Thus the upper value is the same for all strategy sets, i.e., all information patterns of the maximizer. Dually, the lower value is unchanged when the minimizers information is increased and together these results imply proposition 2.

Note that if the maximizer, say, knows the opponents pure strategy in a stochastic game, or if he knows the opponents mixed strategy in a deterministic or stochastic game, he is still facing uncertainty and can in general do better by using available information as opposed to an open loop strategy. Hence Proposition 2 cannot be strengthened.

EXTENSION TO n-PERSON GAMES

The extension of proposition 1 to n-person games yields a monotonicity property of the characteristic function. Indeed, this function is defined [3] for any possible coalition S as the value v(S) in mixed strategies, of the two person zerosum game in which coalition S acts as the maximizer, the countercoalition of all other players as the minimizer and the payoff is the sum of the payoffs to the members of S.

Assume the information available to just one player, player i, is increased. If i is a member of S then the strategy set of S is increased and proposition 1 implies that the characteristic function v(S) can only be increased. On the other hand, if S does not include i, then the strategy set of the countercoalition is increased so that by Proposition 1, v(S) can only be decreased. Thus one has the following result.

Proposition 3. Let v be the characteristic function of a cooperative n-person game and let v′ be the same function for a version in which the information of and only of player i has been increased. Then

$$v(S) \leq v'(S) \quad \text{for } i \in S$$

$$v(S) \geq v'(S) \quad \text{for } i \notin S.$$

Note that, as for proposition 1, assumption A1 is not required. If two players have increased information an inequality can only be obtained for coalitions containing neither or both players.

The principal consequence of proposition 3 is the following.

Proposition 4. If the information available to one player in a cooperative n-person game is increased then the new Shapley value of that player will be at least as large as before.

Indeed, consider a specific one among the n! orderings of the players, in which the increased player occupies position i. Let S be the coalition (possibly empty) of all players preceding i in the ordering. Then as i joins S his contribution is

$$\delta = v(S \cup \{i\}) - v(S)$$

and becomes, after the increase

$$\delta' = v'(S \cup \{i\}) - v'(S)$$

where v, v′ are the characteristic functions. By Proposition 3, $v'(S \cup \{i\}) \geq v(S \cup \{i\})$ and $v'(S) \leq v(S)$. Hence $\delta' \geq \delta$. Taking the average over all n! orderings one obtains the corresponding inequality for the Shapley value.

This easy result is in sharp contrast with the situation of the Nash equilibrium values, for which Ho et al. [1] have given examples in which information is harmful to a player. Their examples have infinite values for the characteristic function, so that the Shapley value has the form $\infty - \infty$. However, this singularity is not inherent and one must add this item to the list of shortcomings of the Nash equilibrium concept.

The extension of Proposition 2 to the n-person case, using the Nash equilibrium concept was carried out by Sandell [4] who proves the following.

Proposition 5. In a deterministic n-person game, if there exists a Nash equilibrium with all players restricted to pure open loop strategies, then the same strategies still constitute a Nash equilibrium when all players can use enlarged strategy sets satisfying assumptions A1 and A2, that is, closed loop strategies.

The reason for this result is simply that if one player departs from his equilibrium strategy in the open-loop solution, by going to a closed loop strategy, while all other players hold their open-loop equilibrium strategies, then the only inputs that can change are those applied by the departing player. Since the game is deterministic these inputs are determined by the solution of the loop equations and the same result (payoff vector) is obtained by applying these inputs as an open-loop strategy. By the assumed equilibrium condition in the realm of open loop strategies, the departing player cannot improve his situation. This shows that the equilibrium condition is satisfied in the realm of closed-loop strategies as well.

As Sandell points out, the open loop strategy n-tuple generated (in a deterministic game) by an n-tuple of closed loop Nash equilibrium strategies need not be an equilibrium.

THE VALUATION OF COMPETITIVE INFORMATION

Suppose that two firms manufacture similar television sets and that each has a patent or trade secret that would be useful to the other. Suppose that the manager of each firm is rewarded according to the relative performance of his firm as compared to the competition, so that the managers may be considered as engaged in a two person zerosum game. Suppose a long standing agreement provides that each firm must offer to sell the right to use its patent at a "fair" price. How would an arbitrator determine such a price?

In general, consider a two person zerosum game with value V_1. If the rules are changed in such a way that the maximizer's information is increased, the value becomes V_2. If it is the minimizer who receives the extra information it becomes V_3. If both players receive additional information let the value be V_4.

By Proposition 1, one has $V_3 \leq V_1 \leq V_2$, $V_3 \leq V_4 \leq V_2$. Before the game, the arbitrator announces prices a and b, where a is the amount the minimizer pays the maximizer for the extra information, and b is the amount paid by the maximizer for purchasing information from the minimizer. After this announcement, each player submits a sealed decision as to whether he wishes to buy the proffered information at the announced price. The arbitrator opens these decisions and announces them, after which the game is played with the extra information provided as purchased and the side payments for any purchases are made.

This situation is a two-stage game, the first stage of which is the decision of both players as to whether they should buy. The second stage is the actual play. The buying decision is based on the reduced matrix:

		Maximizer	
		buy	don't buy
minimizer	buy	$V_4 + a - b$	$V_3 + a$
	don't buy	$V_2 - b$	V_1

A definition of "fair" pricing is needed and the following one seems acceptable.

<u>Definition</u>: A pair of prices is "fair" if it makes it optimal for both players to mix buying and not buying with equal probability.

Then a straightforward calculation gives

<u>Proposition 6</u>. The "fair" prices are

$$a = \frac{1}{2}((V_2-V_3) + (V_1-V_4)),$$

$$b = \frac{1}{2}((V_2-V_3) - (V_1-V_4)).$$

However, the situation is much less clear if each player can buy one of many bundles of information from the other, including the empty bundle, and the arbitrator must now select two price <u>vectors</u>. Note that if there are k specific items of information for sale to a player then prices must be set for the 2^k bundles of items, because the value of information is certainly not additive in general.

The first-stage decision of a player is to decide which bundle to buy. The matrix for this decision can be set up as before for any pair of price vectors, but the definition of "fairness" is not clear. The following seems to be a minimum requirement for "fair" pricing: Each player must have an optimal mixed strategy in which each bundle, including the empty one, is chosen with positive probability. Indeed, if the empty bundle is never chosen then prices as a whole are to low while if any other bundle is avoided then it is overpriced.

However, this condition only yields a set of pairs of price vectors and it is not clear how one should choose within this set. For the situation of proposition 6, the set is a square in the (a,b) plane, of which the point defined in that proposition is the center.

Now suppose that there is no arbitrator and instead each player sets the price of the information he offers to his adversary. Then the value of the 2x2 matrix game described above must be considered as the payoff in a preliminary stage in which the players select, independently of each other, the prices a and b. The value of this preliminary game is V_1 and pure saddle-points are obtained for $a \geq V_1 - V_3$, $b \geq V_2 - V_1$. That is, without an arbitration agreement, no sale of information would take place.

REFERENCES

[1] Ho, Y. C., Blau, I. and Basar, T., A Tale of Four Information Structures, Proceedings, Colloquium on Control Theory, I.R.I.A., June 17-21, 1974. (to appear).

[2] Kuhn, H. W., Extensive Games and the Problem of Information, Contributions to the Theory of Games, Ann. Math. Stud. no. 28, Princeton University Press, 1953, pp. 193-216.

[3] Owen, G., Game Theory, W. B. Saunders, 1968.

[4] Sandell, N. R., Jr., On Open-Loop and Closed-Loop Nash Strategies, IEEE Trans. Vol. AC-19, pp. 435-436, 1974.

[5] Witsenhausen, H. S., On Information Structures, Feedback and Causality, SIAM Journal on Control, Vol. 9, pp. 149-160, 1971.

[6] ________________, On the Relations Between the Values of a Game and Its Information Structure, Information and Control, Vol. 19, pp. 204-214, 1971.

[7] Young, N. J., Admixtures of Two-Person Games, Proc. London Math. Soc. Vol. (3) 25, pp. 736-750, 1972.

COOPERATIVE AND NON-COOPERATIVE DIFFERENTIAL GAMES

G. Leitmann

Department of Mechanical Engineering
University of California, Berkeley, California, USA

ABSTRACT. Many player differential games are discussed for both a cooperative mood of play, in the sense of Pareto, and a non-cooperative one, in the sense of Nash. Necessary conditions and sufficient conditions for optimal play are considered. Two examples of collective bargaining are presented to illustrate the theory.

1. INTRODUCTION

1.1 Problem Statement

We deal with games which involve a number of players. The rules of the game assign to each player a cost function of all the players' decisions as well as the sets from which these decisions can be chosen.

Let there by N players. Let $J_i(\cdot)$ and D_i be the cost function and decision set, respectively, for player i. Then

$$J_i(\cdot) : D \to R^1 \qquad i = 1,2,\ldots,N \qquad (1.1)$$

where $D \subseteq \prod_{i=1}^{N} D_i$.

Loosely speaking, each player wishes to attain the smallest possible cost to himself. Thus, if there exists $d^* \in D$ such that for all $i \in \{1,2,\ldots,N\}$

$$J_i(d^*) \leq J_i(d) \quad \forall \quad d \in D \qquad (1.2)$$

J. D. Grote (ed.), The Theory and Application of Differential Games. 85-96.

then d^* is certainly a desirable decision. Unfortunately, such a utopian (or absolutely cooperative) decision rarely exists (see [1-3]) and the players are faced with a dilemma: What mood of play should they adopt, that is, how should an "optimal" decision be defined?

We consider two moods of play, one cooperative and the other non-cooperative, in the sense of Pareto [4] and Nash [5], respectively.

1.2 Cooperative Play

One way of stating Pareto-optimality is contained in

Definition 1.1 A decision N-tuple $d^* \in D \subseteq \prod_{i=1}^{N} D_i$ is Pareto-optimal iff for every $d \in D$ either

$$J_i(d) = J_i(d^*) \quad \forall \quad i \in \{1,2,...,N\}$$

or there is at least one $i \in \{1,2,...,N\}$ such that

$$J_i(d) > J_i(d^*)$$

Here, cooperation is embodied in a statement such as "I am willing to forego a gain if it is to be at your expense."

1.3 Non-Cooperative Play

If the players do not cooperate, but rather if each player strives to minimize his own cost regardless of the consequence to the other players, then each player is faced with a problem: In selecting his "best" decision, what should he assume about his opponents' decisions?

In the Nash equilibrium definition of optimality, every player assumes that each of the other players makes his decision only with a view toward minimizing his own cost. We have

Definition 1.2 A decision N-tuple $d^* \in \prod_{i=1}^{N} D_i$ is an equilibrium iff for all $i \in \{1,2,...,N\}$

$$J_i(d^*) \leq J_i(d^i)$$

for all $d_i \in D_i$, where $d^i = \{d_1^*, d_2^*, ..., d_{i-1}^*, d_i, d_{i+1}^*, ..., d_N^*\}$.

An important class of Nash equilibrium games is that of two-person zero-sum games. In these games, one player loses what the other gains; that is,

$$J_1(d) = -J_2(d) = J(d) \tag{1.3}$$

Hence, in terms of $J(d)$, player 1 is the minimizer and player 2 is the maximizer. For such a game we have

Definition 1.3 Decision couple $d^* \in D_1 \times D_2$ is a saddle point iff

$$J(d_1^*,d_2) \leq J(d_1^*,d_2^*) \leq J(d_1,d_2^*)$$

for all $d_1 \in D_1$, $d_2 \in D_2$.

Note that such games are played on a cross-product of decision sets, $D_1 \times D_2$. If that is the case, a number of desirable properties hold (e.g., see [6]).

2. DYNAMICAL SYSTEMS

2.1 State Equations

We are concerned with a dynamical system defined by its state, a point $x \in R^n$, which changes in a prescribed manner with the passing of time $t \in (-\infty,\infty)$; of course, any time-like variable can serve as independent variable. The evolution of the state is controlled by N players.

Given an initial state x^o at time t_o, let $\tau = t - t_o$. Furthermore, let the n-th component of x be t itself; that is, $x_n \equiv t$.

Consider functions

$$u^k(\cdot) : [0,\tau_1] \to R^{d_k}, \quad k = 1,2,\ldots,N$$

and C^1 functions

$$f(\cdot) : R^n \times R^{d_1} \times \ldots \times R^{d_N} \to R^n$$

and let dot denote differentiation with respect to τ. The evolution of the state is described by an absolutely continuous function

$$x(\cdot) : [0,\tau_1] \to R^n, \quad x(0) = x^o$$

satisfying a given state equation

$$\dot{x}(\tau) = f(x(\tau),u^1(\tau),u^2(\tau),\ldots,u^N(\tau)) \tag{2.1}$$

2.2 Controls and Strategies

The players influence the evolution of the state through their choices of the $u^k(\tau)$ for almost all $\tau \in [0,\tau_1]$. We consider two ways of making these choices: The players use either (relative) time τ or state $x(\tau)$ as the information on which to base their choices.

In the former case, applicable in cooperative games, each player selects an admissible open-loop control; that is, player k chooses a Lebesgue measurable, bounded function

$$u^k(\cdot) : [0,\tau_1] \to R^{d_k}$$

In the latter case, applicable in non-cooperative games, each player selects an admissible closed-loop or feedback strategy; that is, player k chooses a Borel measurable, bounded function

$$p^k(\cdot) : R^n \to R^{d_k}$$

so that, in (2.1),

$$u^k(\tau) = p^k(x(\tau)) \tag{2.2}$$ †

In addition, the values of the controls and strategies, respectively, may be subject to constraints. For instance, given $U^k \subseteq R^{d_k}$, it may be required that

$$u^k(\tau) \in U^k \quad \text{a.e.} \quad \tau \in [0,\tau_1]$$

and

$$p^k(x) \in U^k \quad \text{a.e.} \quad x \in R^n$$

respectively.

2.3 Playability‡

Whether playing cooperatively or not, we suppose here that all players desire to steer the state from given initial state x^o to a state in a prescribed target set $\theta \subset R^n$. We have

Definition 2.1 A control N-tuple $u(\cdot) : [0,\tau_1] \to R^{d_1} \times R^{d_2} \times \ldots \times R^{d_N}$ is playable at x^o iff it is admissible and generates a solution $x(\cdot)$ such that $x(0) = x^o$ and $x(\tau_1) \in \theta$. A strategy N-tuple $p(\cdot) : R^n \to R^{d_1} \times R^{d_2} \times \ldots \times R^{d_N}$ is playable at x^o iff it is admissible and generates at least one solution $x(\cdot)$ such that $x(0) = x^o$, $x(\tau) \notin \theta$ for $\tau \in [0,\tau_1)$, and $x(\tau_1) \in \theta$. A corresponding triple $\{x^o,p(\cdot),x(\cdot)\}$ is termed a terminating play.

---

† Since $p^k(\cdot)$ is Borel measurable and $x(\cdot)$ is absolutely continuous, $u^k(\cdot)$ defined by (2.2) is Borel and hence Lebesgue measurable.

‡ For games, such as pursuit-evasion ones, in which one player seeks termination and the other does not, see "games of kind" in [12] or "qualitative games" in [14].

2.4 Performance Index

Associated with each player there is a performance index or cost function. For a game in which players use controls, the costs depend on initial state x^o and admissible control N-tuple $u(\cdot)$, the corresponding solution $x(\cdot)$ being unique, whereas in the case of the players employing strategies, the costs depend on initial state x^o, admissible strategy N-tuple $p(\cdot)$ and a corresponding solution $x(\cdot)$. We take the cost for player i to be either

$$\left.\begin{array}{l} V_i(x^o,u(\cdot)) \\ \quad\text{or} \\ V_i(x^o,p(\cdot),x(\cdot)) \end{array}\right\} = \int_0^{\tau_1} f_o^i(x(\tau),u^1(\tau),\ldots,u^N(\tau))d\tau \tag{2.3}$$

where $f_o^i(\cdot) : R^n \times R^{d_1} \times \ldots \times R^{d_N} \to R^1$ is C^1. Of course, in the latter case $u^k(\cdot)$ is given by (2.2) on some bounded interval $[0,\tau_1]$.

3. COOPERATIVE DIFFERENTIAL GAMES

3.1 Pareto-optimality

For a dynamical system, Definition 1.1 becomes

Definition 3.1 A control N-tuple $u^*(\cdot)$, playable at x^o, is Pareto-optimal iff for every control N-tuple $u(\cdot)$, playable at x^o, either

$$V_i(x^o,u(\cdot)) = V_i(x^o,u^*(\cdot)) \quad \forall \quad i \in \{1,2,\ldots,N\}$$

or there is at least one $i \in \{1,2,\ldots,N\}$ such that

$$V_i(x^o,u(\cdot)) > V_i(x^o,u^*(\cdot))$$

3.2 Necessary Conditions

By means of the following, readily established lemma (see [7]) one can reduce the derivation of necessary conditions for Pareto-optimal control to an optimal control problem with isoperimetric constraints (e.g., see [8-9]).

Lemma 3.1 If $u^*(\cdot)$ is Pareto-optimal, then there exist a $j \in \{1,2,\ldots,N\}$ and N-1 real numbers r_i such that

$$V_j(x^o,u^*(\cdot)) \leq V_j(x^o,u(\cdot))$$

for all $u(\cdot)$ playable at x^o subject to

$$V_i(x^o,u(\cdot)) \leq r_i, \quad i \neq j, \quad i = 1,2,\dots,N$$

This lemma has the

Corollary 3.1 If $u^*(\cdot)$ is Pareto-optimal, then for all $j \in \{1,2,\dots,N\}$

$$V_j(x^o,u^*(\cdot)) \leq V_j(x^o,u(\cdot))$$

for all $u(\cdot)$ playable at x^o subject to

$$V_i(x^o,u(\cdot)) \leq V_i(x^o,u^*(\cdot)), \quad i \neq j, \quad i = 1,2,\dots,N \quad .$$

3.3 Sufficient Conditions

By means of the following, easily proven lemma (e.g., see [6]) one can reduce sufficient conditions for Pareto-optimal control to sufficient conditions for optimal control (e.g., see [10]).

Lemma 3.2 Control N-tuple $u^*(\cdot)$, playable at x^o, is Pareto-optimal if there is an $\alpha \in R^N$, $\alpha_i > 0$, $i = 1,2,\dots,N$, such that

$$\sum_{i=1}^{N} \alpha_i V_i(x^o,u^*(\cdot)) \leq \sum_{i=1}^{N} \alpha_i V_i(x^o,u(\cdot))$$

for all $u(\cdot)$ playable at x^o.

4. NON-COOPERATIVE DIFFERENTIAL GAMES

4.1 Nash Equilibrium

For a dynamical system, Definition 1.2 becomes

Definition 4.1 A strategy N-tuple $p^*(\cdot)$ is an equilibrium on $X \subseteq R^n$ iff

(i) it is playable at all $x^o \in X$,

and for all $i \in \{1,2,\dots,N\}$ and $x^o \in X$

(ii) $V_i(x^o,p^*(\cdot),x^*(\cdot)) \leq V_i(x^o,{}^ip(\cdot),x^i(\cdot))$ for all terminating plays $\{x^o,p^*(\cdot),x^*(\cdot)\}$ and $\{x^o,{}^ip(\cdot),x^i(\cdot)\}$, where ${}^ip(\cdot) = \{p^{1*}(\cdot),\dots,p^{i-1*}(\cdot),p^i(\cdot),p^{i+1*}(\cdot),\dots,p^{N*}(\cdot)\}$,

and

(iii) $V_i(x^o,p^*(\cdot),x^*(\cdot)) = V_i(x^o,p^*(\cdot),x^{**}(\cdot))$ for any two terminating plays $\{x^o,p^*(\cdot),x^*(\cdot)\}$ and $\{x^o,p^*(\cdot),x^{**}(\cdot)\}$.

For two-person zero-sum games

$$V_1(x^o,p(\cdot),x(\cdot)) = - V_2(x^o,p(\cdot),x(\cdot)) = V(x^o,p(\cdot),x(\cdot))$$

and Definition 1.3 is similarly altered; however, condition (iii) of Definition 4.1 is ensured by (ii) for such games. In view of playability requirement (i), decision N-tuples in differential games are not from a product of decision sets; hence, some of the desirable properties of classical two-person zero-sum games need not hold (e.g., see [11]).

4.2 Necessary and Sufficient Conditions

Necessary conditions and sufficient conditions for equilibrium strategies are rather lengthy and involved, and hence are beyond the scope of this brief article. For two-person zero-sum games, extensive discussions can be found in [6] and [12-16], among others. For N-person nonzero-sum games, see [6] and [17-18].

5. EXAMPLES

5.1 Cooperative Bargaining during Strike

Consider a bargaining process during a strike whose state at time t is defined by the offer by management, $x(t)$, and the demand by labor, $y(t)$, with $x(t_o) = x_o < y(t_o) = y_o$. The strike ends at time T, the first time $y(T) - x(T) = m > 0$; that is, labor accepts an offer when it is sufficiently close to its demand.

Both parties desire a quick end to the strike, but labor wants to maximize the demand at settlement time, whereas management wishes to minimize the final offer. Thus, the costs to management and labor, respectively, are

$$V_1 = k_1T + x(T) \quad \text{and} \quad V_2 = k_2T - y(T)$$

$$k_1,k_2 = \text{constant} > 0.$$

We suppose that the rates of concessions are proportional to the difference between demand and offer, and that the parties control the rates. Thus,

$$\dot{x}(t) = u(t)(y(t)-x(t)) \quad , \quad 0 \leq u(t) \leq a$$

$$\dot{y}(t) = -v(t)(y(t)-x(t)) \quad , \quad 0 \leq v(t) \leq b$$

where a and b are the maximum values of management's and labor's controls, $u(t)$ and $v(t)$, respectively.

This bargaining process is considered as a cooperative game in [19], and the full range of Pareto-optimal solutions is deduced.

Consider two parameters, γ_1 and γ_2 with $\gamma_1 + \gamma_2 = 1$, and let $\gamma = \gamma_1 - \gamma_2$ and $k = \gamma_1 k_1 + \gamma_2 k_2$. The set of Pareto-optimal controls $\{u^*(\cdot), v^*(\cdot)\}$ is found to be the following. For $\gamma_1 > \gamma_2$ (management "more important" than labor), we have two cases:

Case I. $k/\gamma b \geq m$

$$u^*(t) = \begin{cases} 0 & \text{for } y^*(t) - x^*(t) \geq k/\gamma b \,, \\ a & \text{for } y^*(t) - x^*(t) < k/\gamma b \,, \end{cases} \qquad v^*(t) \equiv b$$

Case II. $k/\gamma b < m$

$$u^*(t) \equiv 0 \quad , \quad v^*(t) \equiv b$$

The time at which management begins to make concessions (Case I) is $t_1^* = \frac{1}{b} \ln \frac{\gamma b(y_o - x_o)}{k}$. This time does not depend on management's control value a; however, the larger labor's maximum control value b, the later management begins to concede. If either b or m is sufficiently large, so that $k/\gamma b < m$ (that is, labor makes concessions rapidly or is willing to settle when offer falls appreciably below demand), then management does not concede at all.

For $\gamma_1 < \gamma_2$ (labor "more important" than management), the situation discussed above obtains with the roles of management and labor exchanged. Finally, if $\gamma_1 = \gamma_2$ (parties "equally important"), both concede at maximum rate; that is,

$$u^*(t) \equiv a \quad , \quad v^*(t) \equiv b \quad .$$

5.2 Collective Bargaining

Next consider a bargaining process which allows for a strike or its absence. Again let $x(t)$ be the offer and $y(t)$ the demand. Now, however, we make no *a priori* assumption about behavior. Rather we let

$$\dot{x}(t) = u(t) \qquad x(0) = x_o$$

$$\dot{y}(t) = -\,v(t) \qquad y(0) = y_o$$

and assume merely that the concession rates are bounded; say

$$u(t) \in [0,1] \quad , \quad v(t) \in [0,1]$$

Without loss of generality, we say that settlement is reached at the first time, T, when $x(T) = y(T)$.

We allow labor one more control on the negotiating process, namely, the possibility of calling a strike. The strike control variable, $w(t)$, is bivalued: $w(t) = 0 \Rightarrow$ no strike, and $w(t) = 1 \Rightarrow$ a strike, at time t.

We suppose that management's profit rate before payment of wages is k = constant, and we propose the following motivation: Management chooses $u(t)$, $t \in [0,T]$, so as to minimize the final offer $x(T)$ and the potential profit loss during a strike $\int_0^T w(t)[k-y(t)]dt$ and to maximize the potential wage loss during a strike $\int_0^T w(t)x(t)dt$. Conversely, labor chooses $v(t)$ and $w(t)$, $t \in [0,T]$, to accomplish the opposite goal. The terms $\int_0^T w(t)[k-y(t)]dt$ and $\int_0^T w(t)x(t)dt$ are "revenge terms" by means of which both labor and management use strike as a punishment to the other party for not accepting an offer or acceding to a demand, respectively, during a strike.

This bargaining process is analyzed as a non-cooperative game in [20]; in particular, a saddlepoint feedback strategy pair is deduced for the special case of a zero-sum game. Namely, we consider costs in which the three cost terms are equally weighted:

$$\begin{aligned} V_1 &= - V_2 = V \\ &= x(T) + \int_0^T [w(t)(k-y(t)) - w(t)x(t)]dt \end{aligned}$$

Using x and $z = y-x$ as state variables, let μ, ν, ω denote strategies from $S = \{(x,z) \in R^2: 0 < x < k, z > 0\}$ into R^2 such that

$$u(t) = \mu(x(t),z(t)), \; v(t) = \nu(x(t),z(t), \; w(t) = \omega(x(t),z(t))$$

Here, since the negotiating interval is unspecified, the strategies do not depend on time.

A particular saddlepoint strategy pair $\{\mu^*(\cdot),(\nu^*(\cdot),\omega^*(\cdot))\}$ calls for a strike when the offer rate falls below the demand rate, $x < k-y$, and for no strike when the offer rate exceeds the demand rate, $x > k-y$. Either situation is allowed when the offer and demand rates are equal.

Specifically, this saddlepoint is

$$\mu^*(x,z) = 1, \quad \nu^*(x,z) = 0, \quad \omega^*(x,z) = 1 \quad \text{for} \quad (x,z) \in S_1$$

$$\mu^*(x,z) = 1, \quad \nu^*(x,z) = 1, \quad \omega^*(x,z) = 1 \quad \text{for} \quad (x,z) \in S_2$$

$$\mu^*(x,z) = 1, \quad \nu^*(x,z) = 1, \quad \omega^*(x,z) = 1 \text{ or } 0$$

$$\text{for} \quad \bar{S}_2 \cap \bar{S}_3 \cap S$$

$$\mu^*(x,z) = 0, \quad \nu^*(x,z) = 1, \quad \omega^*(x,z) = 0 \quad \text{for} \quad (x,z) \in S_3$$

where for $k \leq 2$

$$S_1 = \{(x,z) : 2x \leq k - 2z\} \cap S$$

$$S_2 = \{(x,z) : k - 2z < 2x < k - z\} \cap S$$

$$S_3 = \{(x,z) : k - z < 2x\} \cap S$$

A somewhat more complicated decomposition of S obtains for $k > 2$. These decompositions as well as state space trajectories generated by the saddlepoint strategy pair are given in [20].

For the symmetrical situation considered above, we conclude the following. When labor is "reasonable" but management is "ungenerous," $(x,z) \in S_1$, then labor calls a strike and does not concede whereas management concedes at maximum rate. When labor is less "reasonable" but management is still "ungenerous," $(x,z) \in S_2$, then labor strikes but also concedes at maximum rate as does management. When management is sufficiently generous *vis-a-vis* labor's demand, $(x,z) \in S_3$, there is no strike and labor concedes at maximum rate whereas management is intransigent. The boundary between S_2 and S_3 is the set of states for which the "revenge terms" cancel each other; the presence or absence of a strike does not influence the overall cost.

REFERENCES

1. T. L. Vincent and G. Leitmann, Control Space Properties of Cooperative Games, J. of Optimization Theory and Appl., Vol. 6, No. 2, 1970.
2. G. Leitmann, S. Rocklin and T. L. Vincent, A Note on Control Space Properties of Cooperative Games, J. of Optimization Theory and Appl., Vol. 6, No. 2, 1972.
3. P. L. Yu and G. Leitmann, Compromise Solutions, Domination Structures and Salukvadze's Solution, J. of Optimization Theory and Appl., Vol. 13, No. 3, 1974.
4. V. Pareto, Manuel d'économique politique, Girard et Briere, Paris, 1909.

5. J. Nash, Non-cooperative Games, Annals of Mathematics, Vol. 54, No. 2, 1951.
6. G. Leitmann, Cooperative and Non-Cooperative Many Player Differential Games, Springer Verlag, Vienna, 1974.
7. G. Leitmann and Wm. Schmitendorf, A Simple Derivation of Necessary Conditions for Pareto Optimality, IEEE Transactions on Automatic Control, to appear.
8. M. Hestenes, Calculus of Variations and Optimal Control Theory, John Wiley and Sons, N. Y., 1966.
9. Wm. Schmitendorf, Pontryagin's Principle for Problems with Isoperimetric Constraints and for Problems with Inequality Terminal Constraints, J. of Optimization Theory and Appl., to appear.
10. G. Leitmann and Wm. Schmitendorf, Some Sufficiency Conditions for Pareto-Optimal Control, J. of Dynamic Systems, Measurement, and Control, Vol. 95, No. 4, 1973.
11. G. Leitmann and S. Rocklin, The Effect of Playability in Differential Games on the Relation Between Best Guaranteed Costs and Saddlepoint Values, J. of Optimization Theory and Appl., to appear.
12. R. Isaacs, Differential Games, John Wiley and Sons, N.Y., 1965.
13. A. Blaquière and G. Leitmann, Jeux Quantitatifs, Gauthier-Villars, Paris, 1969.
14. A. Blaquière, F. Gérard and G. Leitmann, Quantitative and Qualitative Games, Academic Press, N.Y., 1969.
15. A. Friedman, Differential Games, John Wiley and Sons, N.Y., 1971.
16. L. D. Berkovitz, Necessary Conditions for Optimal Strategies in a Class of Differential Games and Control Problems, SIAM J. on Control, Vol. 5, No. 1, 1967.
17. J. H. Case, Toward a Theory of Many Player Differential Games, SIAM J. on Control, Vol. 7, No. 2, 1969.
18. H. Stalford and G. Leitmann, Sufficiency Conditions for Nash Equilibria in N-Person Differential Games, in Topics in Differential Games (ed. A. Blaquière), North-Holland Publ. Co., Amsterdam, 1973.
19. G. Leitmann and P. T. Liu, A Differential Game Model of Labor-Management Negotiation During a Strike, J. of Optimization Theory and Appl., Vol. 13, No. 4, 1974.
20. G. Leitmann, Collective Bargaining: A Differential Game, J. of Optimization Theory and Appl., Vol. 11, No. 4, 1973.

ADDITIONAL BIBLIOGRAPHY

L. A. Zadeh, Optimality and Non-Scalar-Valued Performance Criteria, IEEE Trans. Autom. Contr., Vol. AC-8, No. 1, 1963.

N. O. Da Cunha and E. Polak, Constrained Minimization under Vector-Valued Criteria in Linear Topological Spaces, in Mathematical Theory of Control (eds. A. V. Balakrishnan and L. W. Neustadt),

Academic Press, N.Y., 1967.
A. Blaquière, L. Juricek and K. Wiese, Geometry of Pareto Equilibria and a Maximum Principle in N-Person Differential Games, J. of Math. Analysis and Appl., Vol. 38, No. 1, 1972.
M. Athans and H. P. Geering, Necessary and Sufficient Conditions for a Differentiable Nonscalar-Valued Function to Attain Extrema, IEEE Trans. Autom. Contr., Vol. AC-18, No. 2, 1973.
W. E. Schmitendorf, Cooperative Games and Vector-Valued Ctiteria Problems, IEEE Trans. Autom. Contr., Vol. AC-18, No. 2, 1973.
A. Haurie, On Pareto Optimal Decisions for a Coalition of a Subset of Players, IEEE Trans.Autom.Contr., Vol.AC-18, No.2, 1973.
A. Haurie and M. C. Delfour, Individual and Collective Rationality in a Dynamic Pareto Equilibrium, J. of Optimization Theory and Appl., Vol. 13, No. 3, 1974.
A. W. Starr and Y. C. Ho, Nonzero-Sum Differential Games, J. of Optimization Theory and Appl., Vol. 3, No. 3, 1969.
A. W. Starr and Y. C. Ho, Further Properties of Nonzero-Sum Differential Games, J. of Optimization Theory and Appl., Vol. 3, No. 4, 1969.
Y. C. Ho and G. Leitmann, eds., Proceedings of the First International Conference on the Theory and Application of Differential Games, Amherst, Mass., 1969.
Y. C. Ho, Differential Games, Dynamic Optimization and Generalized Control Theory, J. of Optimization Theory and Appl., Vol. 6, No. 3, 1970.
M. D. Ciletti, New Results in the Theory of Differential Games with Information Time Lag, J. of Optimization Theory and Appl., Vol. 8, No. 4, 1971.
M. D. Ciletti, Canonical Equations for the Generalized Hamilton-Jacobi Equation in DGWITL, J. of Optimization Theory and Appl., Vol. 13, No. 3, 1974.
K. Mori and E. Shimemura, Linear Differential Games with Delayed and Noisy Information, J. of Optimization Theory and Appl., Vol. 13, No. 3, 1974.
D. J. Wilson, Mixed Strategy Solutions for Quadratic Games, J. of Optimization Theory and Appl., Vol. 13, No. 3, 1974.
Y. C. Ho, On the Minimax Principle and Zero-Sum Stochastic Differential Games, J. of Optimization Theory and Appl., Vol. 13, No. 3, 1974.
M. Simaan and J. B. Cruz, On the Stackelberg Strategy in Nonzero-Sum Games, J. of Optimization Theory and Appl., Vol. 11, No. 5, 1973.
P. L. Yu and G. Leitmann, Nondominated Decisions and Cone Convexity in Dynamic Multicriteria Decision Problems, J. of Optimization Theory and Appl., Vol. 14, No. 5, 1974.
S. Clemhout, G. Leitmann and H. Y. Wan, Jr., A Differential Game Model of Duopoly, Econometrica, Vol. 39, No. 6, 1971.
S. Clemhout, G. Leitmann and H. Y. Wan, Jr., A Differential Game Model of Oligopoly, J. of Cybernetics, Vol. 3, No. 1, 1973.

N-PERSON STOCHASTIC DIFFERENTIAL GAMES*

Pravin Varaiya

Department of Electrical Engineering and Computer Sciences, and Electronics Research Laboratory, University of California, Berkeley, California 94720 U.S.A.

ABSTRACT. Necessary and sufficient conditions are given for the non-cooperative equilibrium policies of N players when they are simultaneously controlling the evolution of a stochastic system described by an Ito equation. In the case of perfect information, these conditions are generalizations of the well-known Hamilton-Jacobi equations. Conditions are also indicated for the case when the players have only partial information. Sufficient conditions are derived which guarantee that an equilibrium is also Pareto-efficient.

1. INTRODUCTION AND SUMMARY

We apply the results obtained in [1] to study the equilibrium policies of N players when they are simultaneously controlling the evolution of a system described by the stochastic functional differential equation

$$dz_t = f(t,z,u_t^1,\ldots,u_t^N)dt + dB_t \ , \ t \in [0,1] \tag{1.1}$$

Here $\{z_t\}$ is the "state" process and $\{B_t\}$ is a vector of independent Brownian movements. The "drift" f depends at any time t on the past $\{z_s, s \leq t\}$ of the state and also on the controls u_t^i of the $\underline{i}$th player, i=1,...,N. u_t^i takes values in a fixed metric space U_i and depends on the past $\{y_s^i, s \leq t\}$ of the observations made by i. The <u>cost</u> incurred by i is

*Research sponsored by the Joint Services Electronics Program, Contract F44620-71-C-0087.

J. D. Grote (ed.), The Theory and Application of Differential Games. 97-107.

$$J^i(u) = E[\int_0^1 h^i(t,z,u_t)dt] \tag{1.2}$$

where $\{u_t\} = \{u_t^1,\ldots,u_t^N\}$.

A set of policies $\{u_t^*\} = \{u_t^{1*},\ldots,u_t^{N*}\}$ is a (non-cooperative) equilibrium if for all i

$$J^i(u^*) \leq J^i({}^iu^*,u^i) \text{ for all } u^i \tag{1.3}$$ [1]

Thus u^* is an equilibrium iff u^{*i} is a policy which minimizes (1.2) when for all $j \neq i$ player j adopts the policy u^{j*}. This trivial fact allows us to use the results of [1] to obtain the equilibrium conditions. u^* is efficient if for all $v = \{v^1,\ldots, v^N\}$

$$J^i(v) \leq J^i(u^*) \text{ for all } i$$

implies $J^i(v) = J^i(u^*)$ for all i. Evidently, if there exist numbers $\mu_1 > 0,\ldots,\mu_N > 0$ such that

$$\sum_i \mu_i J^i(u^*) \leq \sum_i \mu_i J^i(v) \text{ for all } v, \tag{1.4}$$

then u^* is efficient. But (1.4) means that u^* is an optimal control for the cost $\sum_i \mu_i J^i$ and so we can once again apply the results of [1] to obtain efficiency conditions.

These conditions are straightforward extensions of the well-known Hamilton-Jacobi equations when the game is of complete information i.e., when $y_t^i \equiv z_t$ for all i. When the information is incomplete the conditions are much more complex.

The paper is organized as follows. The next section introduces some background material dealing with the interpretation of the Ito equation (1.1), after which the relevant results of [1] are displayed. Sections 3 and 4 treat respectively the case of complete and incomplete information. Section 5 discusses some difficulties connected with the notion of efficiency in the case of incomplete information.

---

[1] We adopt the notation $({}^iu,v^i) = (u^1,\ldots,u^{i-1},v^i,u^{i+1},\ldots,u^N)$

2. RESULTS FROM OPTIMAL CONTROL THEORY

2.1 Specification of the dynamics

Let C^k be the set of all continuous functions from [0,1] into R^k. Let ξ^k be the evaluation functional on C^k, and, for $t \in [0,1]$, let $\mathcal{F}_t^k$ be the σ-field of subsets of C^k generated by $\{\xi_s^k, s \leq t\}$. $\mathcal{A}^k$ is the σ-field of subsets of $[0,1] \times C^k$ such that a function g on $[0,1] \times C^k$ is $\mathcal{A}^k$ measurable iff $g(t,\cdot)$ is $\mathcal{F}_t^k$ measurable for all t and $g(\cdot,x)$ is Lebesgue measurable for each $x \in C^k$; thus $\mathcal{A}^k$ measurable functions are <u>non-anticipative</u>.

The state process $\{z_t\}$ is n-dimensional. The <u>i</u>th player's observation process $\{y_t^i\}$ is a n_i-dimensional subvector of $\{z_t\}$. The components of the drift f corresponding to y^i are denoted by the vector f^i, The sample paths of $\{z_t\}$ are continuous, hence they lie in C^n whereas those of $\{y_t^i\}$ lie in C^{n_i}. We can now define the admissible control policies.

U_i is a separable metric space and its Borel field is $\mathcal{V}_i$. A <u>policy for player i</u> is a measurable function u^i: $([0,1] \times C^{n_i}, \mathcal{A}^{n_i}) \to (U_i, \mathcal{V}_i)$. The set of such policies is denoted $\mathcal{U}_i$. Let $\mathcal{U} = \mathcal{U}_1 \times \ldots \times \mathcal{U}_N$, similarly for $U, \mathcal{V}$. The following conditions are imposed on f:

(i) $f: [0,1] \times C^n \times U \to R^n$ is measurable with respect to $\mathcal{A}^n \times \mathcal{V}$,

(ii) there exists K such that $|f(t,z,u)| \leq K(1 + \|x\|)$ for all (t,z,u).

Here $|\cdot|$ is the norm in R^n and $\|\cdot\|$ is the sup norm in C^n. The functions h^i in (1.2) are assumed to satisfy the condition corresponding to (i) above and in addition the h^i are assumed non-negative and uniformly bounded.

2.2 Solutions of (1.1)

Let P be Wiener measure on $(C^n, \mathcal{F}_1^n)$. Let z be the evaluation functional on C^n so that $\{z_t, \mathcal{F}_t^n, P\}$ is a standard, n-dimensional, Brownian movement. For $u \in \mathcal{U}$ define the corresponding drift $\{\phi_t^u, \mathcal{F}_t, P\}$ by

$$\phi_t^u(z) = f(t,z,u^1(t,y^1),\ldots,u^N(t,y^N))$$

Recall that y^i is a subvector of z. For future reference let $\mathcal{Y}_t^i$ be the sub σ-field of $\mathcal{F}_t^n$ generated by $\{y_s^i, s \leq t\}$. Also for each u define the non-negative random variable ρ^u by

$$\rho^u = \exp\left[\int_0^1 \phi_t^u \, dz_t - \frac{1}{2}\int_0^1 |\phi_t^u|^2 dt\right]$$

Theorem 2.1 [2,3,4] Under the above-stated assumptions on f $\int_{C^n} \rho^u(z)\, P(dz) = 1$. Hence P^u is a probability measure on $(C^n, \mathcal{F}_1^n)$ where

$$P^u(F) = \int_F \rho^u(z)\, P(dz)\ , \quad F \in \mathcal{F}^n$$

Furthermore, the process $\{w_t^u, \mathcal{F}_t^n, P^u\}$ defined by

$$w_t^u = z_t - \int_0^t \phi_s^u\, ds$$

is a Brownian movement.

This result justifies the following definition. The solution of (1.1) corresponding to a policy $u \in \mathcal{U}$ is the process $\{z_t, \mathcal{F}_t^n, P^u\}$. Thus the impact on the system of a policy u is summarized by the probability distribution P^u.

2.3 Optimality conditions

Suppose N = 1. We can then drop the index i. For $u \in \mathcal{U}$, define the process $\{W_t^u, \mathcal{Y}_t, P^u\}$ by

$$W_t^u = \underset{v \in \mathcal{U}}{\text{infimum}}\ E^u\left[\int_t^1 h(s,z,v_s)ds \,|\mathcal{Y}_t\right].$$

u is value decreasing if $\{W_t^u\}$ is a supermartingale i.e., if

$$E^u[W_{t+\delta}^u | \mathcal{Y}_t] \leq W_t^u \quad \text{a.s. for all } t,\ \delta > 0.$$

u^* is optimal if $J(u^*) \leq J(u)$ for all $u \in \mathcal{U}$. It is known that an optimal policy is value decreasing [1, p. 242].

Theorem 2.2 [1] u^* is optimal iff there exists a constant J^* and for each value decreasing u there exist processes $\{\Lambda V_t^u\}$, $\{\nabla V_t^u\}$, taking values in R and R^m respectively (where m is the dimension of the observation process y), adapted to $\mathcal{Y}_t$, and satisfying the following conditions:

(i) $x_1^u = 0$, where

$$x_t^u = J^* + \int_0^t \Lambda V_s^u\, ds + \int_0^t \nabla V_s^u\, dy_s$$

(ii) $\Lambda V_t^u + \nabla V_t^u \hat{f}^y(t,z,u_t) + \hat{h}(t,z,u_t) \geq 0 = \Lambda V_t^{u*} + \nabla V_t^{u*} \hat{f}^y(t,z,u_t^*) + \hat{h}(t,z,u_t^*)$ for all t,z,u_t.

Then $x_t^{u*} = W_t^{u*}$ and $J^* = J(u^*)$ is the minimum cost. (Here f^y is the subvector of f corresponding to y. $\hat{f}^y(t,z,u_t) = E^u[f^y(t,z,u_t)|\mathcal{Y}_t]$, and $\hat{h}$ is defined similarly).

Theorem 2.3 [1] Suppose $y_t \equiv z_t$. u_t^* is optimal iff there exists J^* and processes $\{\Lambda V_t\}$, $\{\nabla V_t\}$, taking values in R and R^n respectively, adapted to $\mathcal{Y}_t = \mathcal{F}_t^n$, and satisfying the following conditions:

(i) $x_1 = 0$, where

$$x_t = J^* + \int_0^t \Lambda V_s \, ds + \int_0^t \nabla V_s \, dz_s$$

(ii) $\Lambda V_t + \nabla V_t f(t,z,u) + h(t,z,u) \geq 0 = \Lambda V_t + \nabla V_t f(t,z,u_t^*) + h(t,z,u_t^*)$ for all t,z,u.

Then $x_t = W_t^{u*}$ and J^* is the minimum cost.

3. EQUILIBRIUM CONDITIONS: COMPLETE INFORMATION

3.1 Equilibrium conditions

The next result is then an immediate consequence of Theorem 2.2.

Theorem 3.1 (Equilibrium condition) $\{u_t^*\} = \{u_t^{1*},\ldots,u_t^{N*}\}$ is an equilibrium iff for each i there exist J^{i*} and process $\{\Lambda V_t^i\}$, $\{\nabla V_t^i\}$ adapted to $\mathcal{F}_t^n$ satisfying the following conditions:

(i) $x_1^i = 0$, where

$$x_t^i = J^{i*} + \int_0^t \Lambda V_s^i \, ds + \int_0^t \nabla V_s^i \, dz_s \tag{3.1}$$

(ii) $\Lambda V_t^i + \nabla V_t^i f(t,z,u_t^{1*},\ldots,u^i,\ldots,u_t^{N*}) + h^i(t,z,u_t^{1*},\ldots,u^i,\ldots,u_t^{N*}) \geq 0 = \Lambda V_t^i + \nabla V_t^i f(t,z,u_t^*) + h^i(t,z,u_t^*)$, (3.2)
for all t,z,u^i.

Then $J^{i*} = J^i(u^*)$. Furthermore

$$x_t^i = \underset{u^i \in \mathcal{U}_i}{\text{infimum}}\; E^{u*}\left[\int_t^1 h^i(s,z,u_s^{1*},\ldots,u_s^i,\ldots,u_s^{N*}) \, ds \,\middle|\, \mathcal{F}_t^n\right] \tag{3.3}$$

As a special case of this result we can deduce the conditions for a saddle point policy in a 2-player, zero-sum game. So suppose N=2 and $h^2=-h^1$. A policy $u_t^* = u_t^{1*},u_t^{2*}$ is a <u>saddle point</u> if

$$J^1(u^1,u^{2*}) \geq J^1(u^{1*},u^{2*}) \geq J^1(u^{1*},u^2) \text{ for all } u^1,\ u^2. \quad (3.4)$$

<u>Theorem 3.2</u> (Saddle point condition) $\{u_t^*\} = \{u_t^{1*},u_t^{2*}\}$ is a saddle point iff there exists J^{1*} and processes $\{\Lambda V_t^1\}$, $\{\nabla V_t^1\}$ adapted to $\mathcal{F}_t^n$ satisfying the following conditions:

(i) $x_1^1 = 0$ where

$$x_t^1 = J^{1*} + \int_0^t \Lambda V_s^1\, ds + \int_0^t \nabla V_s^1\, dz_s \quad (3.5)$$

(ii) $\Lambda V_t^1 + \nabla V_t^1 f(t,z,u^1,u_t^{2*}) + h^1(t,z,u^1,u_t^{2*}) \geq 0 = \Lambda V_t^1 + \nabla V_t^1$
$f(t,z,u_t^*) + h^1(t,z,u_t^*) = 0 \geq \Lambda V_t^1 + \nabla V_t^1\, f(t,z,u_t^{1*},u^2)$
$+ h^1(t,z,u_t^{1*},u^2)$ for all t,z,u^1,u^2 (3.6)

Then $J^{1*} = J^1(u^*)$ is the value of the game and

$$x_t^1 = \inf_{u^1} E^{u^*}\left[\int_t^1 h^1(x,z,u_s^1,u_s^{2*})ds \,|\, \mathcal{F}_t^n\right] = \sup_{u^2} E^{u^*}\left[\int_t^1 h^1(s,z,u_s^{1*}, u_s^2)ds \,|\, \mathcal{F}_t^n\right] \quad (3.7)$$

is the value function.

We give this result a form similar to that which has already appeared in the literature [5-8]. Define the Hamiltonian functional $H: [0,1] \times C^n \times U \times R^n \to R$ by $H(t,z,u^1,u^2,p) = pf(t,z,u^1,u^2) + h^1(t,z,u^1,u^2)$.

Then (3.6) is the Isaacs condition,

$$H(t,z,u_t^{1*},u_t^{2*},\nabla V_t) = \underset{u^2}{\text{Max}}\ \underset{u^1}{\text{Min}}\ H(t,z,u^1,u^2,\nabla V_t) = \underset{u^1}{\text{Min}}\ \underset{u^2}{\text{Max}}\ H(t,z,u^1,u^2,\nabla V_t)$$

Next, suppose that (1.1) is a diffusion equation i.e., the dependence at time t of f on z is through z_t:

$$dz_t = f(t,z_t,u_t^1,u_t^2)\, dt + dB_t,$$

and suppose further that u_t^* has the same property i.e., $u^{i*}(t,z) = u^{i*}(t,z_t)$. Then the solution $\{z_t, \mathcal{F}_t^n, P^{u^*}\}$ is a diffusion, and hence from (3.7) it follows that the value function at time t depends only on z_t, i.e., there is a function V on $[0,1] \times R^n$ such that $x_t^1 \equiv V(t,z_t)$. Secondly, if this function is sufficiently smooth then by Ito's differential rule we can identify the processes $\{\Lambda V_t^1\}$ and $\{\nabla V_t^1\}$ as $\Lambda V_t^1 = \frac{\partial V}{\partial t}(t,z_t) + \frac{1}{2}\sum_{i,j} \frac{\partial^2 V}{\partial z_i \partial z_j}(t,z_t)$, $\nabla V_t^1 = \frac{\partial V}{\partial z}(t,z_t)$. Combining these two observations yields the well-known Hamilton-Jacobi partial differential equation for the value function,

$$\frac{\partial V}{\partial t}(t,z) + \frac{1}{2}\sum_{i,j} \frac{\partial^2 V(t,z)}{\partial z_i \partial z_j} + \operatorname*{Max}_{u^2} \operatorname*{Min}_{u^1} H(t,z,u^1,u^2,\frac{\partial V}{\partial z}(t,z)) = 0.$$

for $(t,z) \in [0,1] \times R^n$; and (3.5) yields the boundary condition $V(1,z) \equiv 0$.

3.2 Efficiency conditions

We return to the N-player game of complete information. The next result is immediate from our earlier remarks.

Theorem 3.3 (Sufficiency conditions) $\{u_t^*\} = \{u_t^{1*},\ldots,u_t^{N*}\}$ is an efficient equilibrium if for each i there exist $\mu_i > 0$, J^{i*} and processes $\{\Lambda V_t^i\}$, $\{\nabla V_t^i\}$ adapted to $\mathcal{F}_t^n$, and satisfying conditions (3.1), (3.2) and

$$\sum_i \mu_i\{\Lambda V_t^i + \nabla V_t^i f(t,z,u_t^*) + h^i(t,z,u_t^*)\} = 0$$

$$= \operatorname*{Min}_{u \in U} \sum_i \mu_i\{\Lambda V_t^i + \nabla V_t^i f(t,z,u) + h^i(t,z,u)\} \tag{3.8}$$

Condition (3.8) appears to be a very stringent condition. It turns out, however, that if a certain convexity condition is satisfied, then this condition is also necessary for efficiency. We say that the convexity assumption holds if for all t, z the (N+n) - dimensional set

$$\{(h^1(t,z,u),\ldots,h^N(t,z,u), f(t,z,u)) \mid u \in U\}$$

is convex. Now replace the original game by the following one. The dynamics of this game are given by a (N+n)-dimensional Ito equation

$$\begin{aligned} dq_t^1 &= h^1(t,z,u_t)\,dt + d\beta_t^1 \\ &\vdots \\ dq_t^N &= h^N(t,z,u_t)\,dt + d\beta_t^N \\ dz_t &= f(t,z,u_t)\,dt + dB_t \end{aligned} \tag{3.9}$$

where (β,B) is an $(N+n)$-dimensional Brownian movement. The cost incurred by the ith player is

$$J^i(u) = E\,q_1^i \tag{3.10}$$

It is evident that the two games are equivalent. What we have achieved by this transformation is to remove from (3.9) the explicit dependence on the policy $\{u_t\}$. Next, as in Section 2.2, for $u \in \mathcal{U}$ define

$$\rho^u = \exp[\sum_i \int_0^1 h^i(t,z,u_t(z))\,dq_t^i + \int_0^1 f(t,z,u_t(z))\,dz_t - \frac{1}{2}\sum_i \int_0^1 |h_t^i|^2\,dt - \frac{1}{2}\int_0^1 |f_t|^2\,dt]$$

and let the set of all such random variables be

$$\mathcal{R} = \{\rho^u \,|\, u \in \mathcal{U}\}$$

Then

$$J^i(u) = \int_{C^{N+n}} q_1^i\,\rho^u(q,z)\,P(dq,dz) = \mathcal{J}^i\,(\rho^u) \text{ say}$$

where $P(dq,dz)$ is Wiener measure on $(C^{N+n},\mathcal{A}^{N+n})$. Note that the map $\mathcal{J} : \mathcal{R} \to R^N$ defined by $\mathcal{J}(\rho) = (\mathcal{J}^1(\rho),\ldots,\mathcal{J}^N(\rho))$ is linear. The next result is proved in [3] and [4].

Lemma 3.1 If the convexity assumption holds then $\mathcal{R}$ is a convex subset of $L^1(C^{n+N},\mathcal{F}_1^{n+N},P)$

Theorem 3.4 (Efficiency conditions) Suppose the convexity assumption holds. Then $\{u_t^*\} = \{u_t^{1*},\ldots,u_t^{N*}\}$ is an efficient equilibrium iff for each i there exist $\mu_i > 0$, J^{i*} and processes $\{\Lambda V_t^i\}$, $\{\nabla V_t^i\}$ adapted to $\mathcal{F}_t^n$, and satisfying conditions (3.1), (3.2), (3.8).

Proof The sufficiency follows from Theorem 3.3. To prove the necessity let u_t^* be an efficient equilibrium and suppose J^{i*}, $\Lambda V_t^i, \nabla V_t^i$ satisfy (3.1), (3.2), and (3.3). By lemma 3.1 the set $\Gamma = \{(J^1(u),\dots,J^N(u)),\ u \in \mathcal{U}\}$ is a convex subset of R^N. By efficiency Γ is disjoint from the convex set $\{(J^{1*}+x_1,\dots,J^{N*}+x_N),\ x_i \le 0 \text{ all } i \text{ and } \{x_i \neq 0\}$. By the separation theorem for convex sets there exist $\mu_i > 0$ such that

$$\sum_i \mu_i J^{i*} \le \sum_i \mu_i J^i(u) \quad \text{for all } u \in \mathcal{U},$$

i.e., u^* is an optimal control for the cost functional $\sum_i \mu_i J^i(u)$. Hence (3.8) must hold by Theorem 2.2. □

It turns out in fact that this result is true without the convexity assumtpion. The proof is much more involved unfortunately. The result implies that an efficient equilibrium is more stable than may appear at first sight. Recall the following definition. A policy $\{u_t^*\}$ is in the core if for every subset of players $S \subset \{1,\dots,N\}$ the following property holds: for all policies $\{v_t^S\}$, whenever

$$J^i(v_t^S, u_t^{S'*}) \le J^i(u_t^*) \quad , \quad i \in S.$$

then

$$J^i(v_t^S, u_t^{S'*}) = J^i(u_t^*) \quad , \quad i \in S.$$

Here $(v_t^S, u_t^{S'*})$ is the policy $\{u_t\}$ where $u_t^i = v_t^i$, $i \in S$ and $u_t^i = u_t^{i*}$ for $i \notin S$. Theorem 3.4 immediately yields the following remarkable corollary.

Corollary 3.1 Suppose the convexity assumption holds. Then the set of efficient equilibrium policies coincides with the core.

4. EQUILIBRIUM CONDITIONS: INCOMPLETE INFORMATION

We return to the game of incomplete information. The following result is immediate from Theorem 2.2.

Theorem 4.1 $\{u_t^*\} = \{u_t^{1*}(y^1),\dots,u_t^{N*}(y^N)\}$ is an equilibrium policy iff for each i there exists a constant J^{i*} and for every value decreasing $\{u_t^i(y^i)\}$ there exist processes $\{\Lambda V_t^{iu^i}\}$, $\{\nabla V_t^{iu^i}\}$ taking values in R, R^{n_i} respectively and adapted to $\mathcal{Y}_t^i$, such that the following conditions hold:

(i) $x_1^{iu^i} = 0$ where

$$x_t^{iu^i} = J^* + \int_0^t \Lambda V_s^{iu^i} ds + \int_0^t \nabla V_s^{iu^i} dy_s^i$$

(ii) $\Lambda V_t^{iu^i} + \nabla V_t^{iu^i} \hat{f}^i(t,z,{}^iu^*,u_t^i) + \hat{h}^i(t,z,{}^iu_t^*,u_t^i) \geq 0$

$$= \Lambda V_t^{iu^{*i}} + \nabla V_t^{iu^{*i}} \hat{f}^i(t,z,u_t^*) + \hat{h}^i(t,z,u_t^*), \text{ for all } t,z,\{u_t^i\}$$

Furthermore $J^{i*} = J^i(u^*)$ and

$$x_t^{iu^{i*}} = \underset{u^i \in \mathcal{U}^i}{\text{infimum}} \; E^{u^*}[\int_t^1 h^i(s,z,{}^iu_s^*,u_s^i)\, ds|\mathcal{Y}_t^i] \tag{4.1}$$

This result is not of great interest. However some interesting observations can be deduced. We give one instance. Suppose the game is <u>constant-sum</u> i.e., suppose

$\int_t^1 \sum_i h^i(s,z,u) \equiv K_t$, a non-random function. (In particular this includes 2-person 0-sum games). But this does <u>not</u> imply that

$$\sum_i x_t^{iu^{i*}} \equiv K_t \tag{4.2}$$

This negative conclusion raises the question whether such a game should be called a constant-sum game. As is clear from (4.1), there is one special case where (4.2) holds and that is the "equal" information case, $y_t^1 = y_t^2 = \ldots = y_t^N$

5. NOTION OF EFFICIENCY IN CASE OF INCOMPLETE INFORMATION

The definition of an equilibrium policy is an attempt to embody the concept of individual rationality, whereas efficiency is a precise criterion for group rationality. Thus an efficient equilibrium is stable against group action in the sense that the players will not derive any additional individual benefits from "cooperating" as a group. Now, in the situation of incomplete information where the information available to different players is substantially different and where "cooperation" means sharing of information as well as coordination of policies, it seems quite unlikely that a (non-cooperative) equilibrium will be efficient. This may appear puzzling since on *a priori* grounds one would expect equilibrium policies in the "real world" to be efficient. It is evident that one way this apparent paradox can be resolved is if we can expand our framework to include costs of cooperation, especially of sharing of information.

REFERENCES

1. M. H. A. Davis and P. Varaiya, Dynamic Programming Conditions for Partially Observable Stochastic Systems, *SIAM J. Control* *11*(2), (1970), 226-261.
2. I. V. Girsanov, On Transforming a Certain Class of Stochastic Processes by Absolutely Continuous Substitution of Measures, *Theor. Prob. Appl.* *5* (1962), 285-301.
3. V. E. Beneš, Existence of Optimal Stochastic Control Laws, *SIAM J. Control* *9*(1971), 446-475.
4. T. E. Duncan and P. Varaiya, On the Solutions of a Stochastic Control System, *SIAM J. Control* *9* (1971), 354-371.
5. W. H. Fleming, The Convergence Problem for Differential Games, II. *Advances in Game Theory*, Annals of Math. Studies, No. 52 (1964), 195-210.
6. W. H. Fleming, The Cauchy Problem for Degenerate Parabolic Equations, *J. Math. Mech.* *13* (6), (1964), 987-1008.
7. A. Friedman, Stochastic Differential Games, *J. Diff. Eq.* *11* (1), (1972), 79-108.
8. J. Danskin, Randomized Games, (1974), unpublished.

QUANTITATIVE GAMES : PROBLEM STATEMENT AND EXAMPLES, NEW GEOMETRIC ASPECTS

A. Blaquière

Laboratoire d'Automatique Théorique,
Université de Paris VII, Paris, France

1. PROBLEM STATEMENT AND ASSUMPTIONS

In view of discussing games involving stochastic state equations, delays, partial differential equations and other functional dependence, we shall return to the definition of a quantitative game and try to extend an earlier statement of the problem [1]. We shall consider the case of N players, J_α, $\alpha = 1,2,\ldots N$.

Let there be given a set G, whose members will be denoted by x, ordered by a reflexive relation $\succeq$, such that $x' \succeq x''$ or $x'' \succeq x'$ whenever x' and x" are distinct members of the union of the domain and the range of $\succeq$. Let there be given prescribed sets S_α, $\alpha = 1,2,\ldots N$, whose members will be denoted by s_α, $\alpha = 1, 2, \ldots N$, respectively ; and let there be given a relation $R \subset D \times P(G)$, where $D = G \times S_1 \times \ldots \times S_N$ and $P(G)$ is the collection of all non empty subsets of G. We shall suppose that $(x^i, s_1, s_2, \ldots s_N) R \gamma$, $(x^i, s_1, s_2, \ldots s_N) \in D$, $\gamma \in P(G)$, implies that

(a) $x^i \in \gamma$, and

(b) $x \succeq x^i$, $\forall x \in \gamma$.

Members x of G are *states* of the game. Members s_α of S_α, $\alpha = 1, 2, \ldots N$, are *strategies* of players J_α, $\alpha = 1,2,\ldots N$, respectively. Members s of $S_1 \times \ldots \times S_N$ are *strategy* N*-tuple* s .Members γ of *Range R* are *trajectories*. If $(x^i, s) R \gamma$, we shall say that trajectory γ is *generated from* x^i *by strategy* N*-tuple* s . x^i is its *initial point.*

J. D. Grote (ed.), The Theory and Application of Differential Games. 109-120.

Now, in view of defining optimality of a strategy N-tuple, let there be given N reflexive and symmetric relations $C_\alpha \subset S^2$, $S = S_1 \times \ldots \times S_N$, $\alpha = 1,2,\ldots N$. If $sC_\alpha s'$, $(s,s') \in S^2$, we shall say that strategy N-tuples s and s' are *comparable* for player J_α.

If $x^p \in \gamma$, $\gamma \in$ Range R, we shall let $\rho(\gamma,x^p) = \{ x : x \in \gamma, x \succeq x^p \}$, and if x^p and x^q are members of γ such that $x^q \succeq x^p$, we shall let $\rho(\gamma,x^p,x^q) = \{ x : x \in \gamma, x^q \succeq x \succeq x^p \}$

Assumption 1. *If* $\gamma \in$ *Range* R, *and* $x^p \in \gamma$, *then* $(x^i,s)\, R\gamma$ *implies that* $(x^i,s)\, R\rho(\gamma,x^i,x^p)$, *and that* $(x^p,s)\, R\rho(\gamma,x^p)$ *; and*

Assumption 2.

$$\left.\begin{array}{l}(x^i,s')R\gamma', \text{ and}\\ (x^j,s'')R\gamma'', \text{ and}\\ x^j \in \gamma'\end{array}\right\} \Rightarrow \left\{\begin{array}{l}\text{There exists } s \in S \text{ such that (a)}\\ \\ (x^i,s)R(\rho(\gamma'.x^i,x^j) \cup \gamma''\end{array}\right.$$

and (b) $s'C_\alpha s''$ *implies* $sC_\alpha s'$, *and that* $sC_\alpha s''$, $\alpha \in \{1, 2, \ldots N\}$, *and* $s' = s''$ *implies that* $s = s' = s''$.

If there exists x^j, $x^j \in \gamma$, $\gamma \in$ Range R, such that $x^j \succeq x$, for any $x \in \gamma$, then we shall say that x^j is the *end point* of γ. Let there be given a subset θ of G, the *target set*. A trajectory γ will be called a *path* and denoted by $\overline{\gamma}$ if none of its members, with the possible exception of its end point when it exists, belong to θ.

Finally, let there be given N linear spaces Ω_α, $\alpha = 1,2,\ldots N$, ordered by reflexive relations $(\geq)_\alpha$, $\alpha = 1,2,\ldots N$, respectively, and N mappings V_α, $\alpha = 1,2,\ldots N$:

$$V_\alpha \left\{\begin{array}{l} R \to \Omega_\alpha \\ (x^i,s,\gamma) \mapsto V(x^i,s,\gamma)\end{array}\right. \qquad \alpha = 1,2,\ldots N$$

V_α is the *performance index* of player J_α, and $V_\alpha(x^i,s,\gamma)$ is the *pay-off* of player J_α for a trajectory γ generated from x^i by strategy N-tuple s .

We shall say that a strategy N-tuple s is *playable* at state x^i if there exists a path $\overline{\gamma}$ such that $(x^i,s)\, R\overline{\gamma}$ and $\overline{\gamma} \cap \theta \neq \phi$. We shall denote by $J(x^i)$ the set of all strategy N - tuples playable at state x^i. A path $\overline{\gamma}$ such that $\overline{\gamma} \cap \theta \neq \phi$ is called a *terminating path*. We shall denote by $I(x^i,s,\theta)$ the set of all terminating paths generated from a state x^i by a strategy N-tuple s .

Then, we shall say that a strategy N - tuple s^* is *C-optimal at state* x^i, $C = (C_1, C_2, \ldots C_N)$, if

(i) $s^* \in J(x^i)$, and

(ii) there exists $\omega = (\omega_1, \omega_2, \ldots \omega_N) \in \Omega_1 \times \ldots \times \Omega_N$, such that $V_\alpha(x^i, s^*, \gamma^*) = \omega_\alpha$, $\forall \gamma^* \in I(x^i, s^*, \theta)$, $\alpha = 1,2,\ldots N$, and

(iii) $$\left.\begin{array}{l} \omega_\alpha (\geqslant)_\alpha V_\alpha(x^i, s, \overline{\gamma}) \\ \forall s \text{ such that } s \in J(x^i) \text{ and } s C_\alpha s^*, \\ \text{and } \forall \overline{\gamma} \in I(x^i, s, \theta) \end{array}\right\} \quad \alpha = 1,2,\ldots N .$$

At last a N-player quantitative game is defined by $(G, \mathcal{E}, S, R, C, \Omega, \theta, \geqslant, V)$ where, in addition to the above definitions of $G, \mathcal{E}$, S, R and C, we let $\Omega = \Omega_1 \times \ldots \times \Omega_N$, $\geqslant = ((\geqslant)_1, (\geqslant)_2, \ldots (\geqslant)_N)$, and $V = (V_1, V_2, \ldots V_N)$.

2. EXAMPLES OF QUANTITATIVE MULTISTAGE GAMES

2.1 Problem 1, a Deterministic Game

First let us consider the two-players deterministic multistage game discussed in Ref.[1]. The states are members of $G = \bigcup_{k \in \mathbb{Z}} G(k)$, where† $G(k) = \{ x : x \in E^n, x_n = k \}$. The state equation is :

$$\tilde{x}(k+1) - \tilde{x}(k) = f(\tilde{x}(k), \tilde{u}(k), \tilde{v}(k))$$
$$\tilde{u} : k \mapsto u = \tilde{u}(k), \ \tilde{v} : k \mapsto v = \tilde{v}(k), \ k \in \{i, i+1, \ldots j-1\} ,$$
$$u = (u_1, u_2, \ldots u_r) \in \mathcal{U} \subseteq E^r, \ v = (v_1, v_2, \ldots v_q) \in \mathcal{V} \subseteq E^q$$

$f = (f_1, f_2, \ldots f_{n-1}, 1)$, where functions f_ν, $\nu = 1,2,\ldots n-1$, are defined on $G \times \mathcal{U} \times \mathcal{V}$.

Target θ is a given subset of $G(K)$. Path $\overline{\gamma}^{ij}$ from x^i to x^j, generated by controls $\tilde{u}$ and $\tilde{v}$, is the sequence of states $\overline{\gamma}^{ij} = \{ \tilde{x}(i), \tilde{x}(i+1), \ldots \tilde{x}(j) \}$, $i < j \leqslant K$, where $\tilde{x} : k \mapsto x = \tilde{x}(k)$, defined on $\{ i, i+1, \ldots j \}$, satisfies the state equation. The players choose strategies p and e from given sets of functions of state x . p and e are defined on some subset X of G, and $p(x) \in K_u(x) \subseteq \mathcal{U}$, and $e(x) \in K_v(x) \subseteq \mathcal{V}$, for all $x \in X$, where $K_u(x)$ and $K_v(x)$ are constraint sets. Controls are given by

---

† $\mathbb{Z}$ denotes the set of (positive, negative and zero) integers.

$\tilde{u}(k) = p(\tilde{x}(k)) \in K_u(\tilde{x}(k))$, $\tilde{v}(k) = e(\tilde{x}(k)) \in K_v(\tilde{x}(k))$, and the play proceeds from a given state x^i. The cost of transfer from $x^i = \tilde{x}(i)$ to $x^j = \tilde{x}(j)$ along $\bar{\gamma}^{ij}$ is taken as a sum

$$V(x^i, x^j; p, e, \bar{\gamma}^{ij}) = \sum_{k=i}^{j-1} f_o(\tilde{x}(k), \tilde{u}(k), \tilde{v}(k))$$

where function f_o is defined on $G \times \mathcal{U} \times \mathcal{V}$.

From now on, we shall make

Assumption 3. *$\mathcal{U}$ and $\mathcal{V}$ are finite sets of vectors in E^r and E^q, respectively ; and target θ is a finite set of points in $G(K)$. $K_u(x)$ and $K_v(x)$, $x \in G$, are defined by*

$$K_u(x) = \{ p^a(x) : p^a(x) \in \mathcal{U}, \quad a \in A(x) \}$$
$$K_v(x) = \{ e^b(x) : e^b(x) \in \mathcal{V}, \quad b \in B(x) \}$$

where A and B are prescribed mappings

$$A \begin{cases} G \rightarrow \mathcal{P}(\mathcal{A}) \\ x \mapsto A(x) \end{cases} \qquad B \begin{cases} G \rightarrow \mathcal{P}(\mathcal{B}) \\ x \mapsto B(x) \end{cases}$$

and $\mathcal{A}$ and $\mathcal{B}$ are finite sets of indices, and $\mathcal{P}(\mathcal{A})$ and $\mathcal{P}(\mathcal{B})$ are the collections of their non-empty subsets.

Assumption 4. *For any x in X, there exists a strategy pair (p,e) playable at x .*

We shall let $X(k) = X \cap G(k)$, $k \in \mathbb{Z}$, and $\theta = X(K)$.

2.2 Problem 2, a Stochastic Game

For any x in X, let $\Pi(x)$ and $E(x)$ denote the sets of all mappings

$$\pi(x) \begin{cases} A(x) \rightarrow [0,1] \\ a \mapsto \pi(x,a) \end{cases} \quad \text{and} \quad \varepsilon(x) \begin{cases} B(x) \rightarrow [0,1] \\ b \mapsto \varepsilon(x,b) \end{cases}$$

respectively, such that $\sum_{a \in A(x)} \pi(x,a) = \sum_{b \in B(x)} \varepsilon(x,b) = 1$; and let Π and E denote the sets of all mappings

$$\pi \begin{cases} X \rightarrow \bigcup_{x \in X} \Pi(x) \\ x \mapsto \pi(x) \end{cases} \quad \text{and} \quad \varepsilon \begin{cases} X \rightarrow \bigcup_{x \in X} E(x) \\ x \mapsto \varepsilon(x) \end{cases}$$

Moreover, let $\mathcal{X}(k)$ be the set of all functions P^k, for any $k \leq K$

such that $X(k) \neq \phi$,

$$P^k \begin{cases} X(k) \to [0,1] \\ x \mapsto P^k(x) \end{cases}$$

such that $\sum_{x \in X(k)} P^k(x) \leq 1$ and let $\mathcal{X} = \bigcup_{k \leq K} \mathcal{X}(k)$. $\pi(x,a)$ and $\varepsilon(x,b)$ can be thought of as probabilities associated with the controls $u^a = p^a(x)$ and $v^b = e^b(x)$, respectively, $a \in A(x)$, $b \in B(x)$, knowing that the state in Problem 1 is at point x†. Then, for given π and ε, a sequence $\bar{\gamma}^{ij} = \{\tilde{x}(i), \tilde{x}(i+1), \ldots \tilde{x}(j)\}$, $\tilde{x}(i) \in X(i)$, $\tilde{x}(j) \in X(j)$, $i < j$, of Problem 1 becomes a random sequence.

In Problem 2, members of $\mathcal{X}$ are states of the game, that is, a state at a stage k for which $X(k) \neq \phi$ is a function $P^k \in \mathcal{X}(k)$. Π and E are strategy sets.

Now let us derive the equation of motion of the state. Let

$$M(k+1) = \{x^{k+1} : x^{k+1} \in X(k+1),\ \exists x \in X(k),\ \exists a \in A(x),\ \exists b \in B(x),\ x + f(x),\ p^a(x),\ e^b(x)) = x^{k+1}\}$$

From the definition of $X(k)$ it follows that $X(k) \neq \phi \Rightarrow M(k+1) \neq \phi$. Then, for $x^{k+1} \in M(k+1)$, let

$$N(x^{k+1}) = \{(x,a,b) : x \in X(k),\ a \in A(x),\ b \in B(x),\ .x + f(x,\ p^a(x),\ e^b(x)) = x^{k+1}\}$$

and,

$$P^{k+1}(x^{k+1}) = \sum_{(x,a,b) \in N(x^{k+1})} P^k(x)\pi(x,a)\varepsilon(x,b),$$ and for $x^{k+1} \in X(k+1) - M(k+1)$ let $P^{k+1}(x^{k+1}) = 0$. One can see easily that $P^{k+1}(x^{k+1}) \leq 1$ for all x^{k+1} in $X(k+1)$ and that

$\sum_{x \in X(k+1)} P^{k+1}(x) \leq 1$, and so, $P^{k+1} : X(k+1) \to [0,1]$ belongs to $\mathcal{X}(k+1)$.

P^{k+1} is the state at stage k+1 for given state P^k at stage k, and given strategies π and ε. The functional relation between P^{k+1} and (P^k, π, ε) will be written

---

† $P^k(x)$ is the product of two probabilities ; namely, the probability for the state in Problem 1 to be in $X(k)$ at stage k, and the probability of being at point x knowing that it is in $X(k)$.

$$P^{k+1} = F(P^k,\pi,\varepsilon)$$

Now, we shall define the cost of transfer along a path $\overline{\Gamma}^{ij} = \{P^i, P^{i+1},\ldots P^j\}$, $i < j \leq K$, generated from P^i by strategy pair (π,ε). For $x \in X(k)$, let†

$$R(x) = \{(a,b) : a \in A(x),\ b \in B(x),\ x + f(x, p^a(x), e^b(x)) \in X(k+1)\}$$

$$F_o(P^k,\pi,\varepsilon) = \sum_{x\in X(k)} \sum_{(a,b)\in R(x)} f_o(x,p^a(x),e^b(x))P^k(x)\pi(x,a)\varepsilon(x,b)$$

Then, the cost of $\overline{\Gamma}^{ij}$ is

$$W(P^i,P^j,\pi,\varepsilon,\overline{\Gamma}^{ij}) = \sum_{k=i}^{j-1} F_o(P^k,\pi,\varepsilon)$$

We shall define target Θ as the set of all states $P^k \in X(K)$ such that $\sum_{x\in X(K)} P^K(x) \neq 0$. If $\overline{\Gamma}^{ij}$ is a terminating path, we shall denote the cost by $W(P^i,\Theta,\pi,\varepsilon) = W(P^i,P^j,\pi,\varepsilon,\overline{\Gamma}^{ij})$.

A strategy pair $(\pi^*,\varepsilon^*)$is optimal at state P^i if it is playable at P^i, and if

$W(P^i,\Theta,\pi^*,\varepsilon) \leq W(P^i,\Theta,\pi^*,\varepsilon^*) \leq W(P^i,\Theta,\pi,\varepsilon^*)$ for all strategy pairs (π^*,ε) and (π,ε^*) playable at state P^i .

2.3 Application of Dynamic Programming to Problem 2

First, we shall define by recurrence a function $\Phi^* : X \to \mathbf{R}$. Suppose that $X(k) \neq \phi$, in which case $X(k+1) \neq \phi$, and that $\Phi^*(x)$ is defined for all $x \in X(k+1)$. Then, for any x in $X(k)$ and (a,b) in $R(x)$, let

$$m^{ab} = f_o(x,p^a(x),e^b(x)) + \Phi^*\big(x + f(x,p^a(x),e^b(x))\big)$$

and,

$$\Phi(x,\pi(x),\varepsilon(x)) = \sum_{(a,b)\in R(x)} m^{ab}\pi(x,a)\varepsilon(x,b)$$

Though the table (m^{ab}) in which m^{ab} lies at the intersection of line a and row b may not be a matrix, since some of these intersections may not be occupied by numbers, $\Phi(x,\pi(x),\varepsilon(x))$ is defined

† From the definition of $X(k)$ it follows that $X(k) \neq \phi \Rightarrow R(x) \neq \phi \quad \forall x \in X(k)$.

for all $(\pi(x),\varepsilon(x))$ in $\Pi(x)\times E(x)$, and it follows directly from a theorem of Von Neumann that there exist functions $\pi^*(x) \in \Pi(x)$ and $\varepsilon^*(x) \in E(x)$ such that

(1) $\Phi(x,\pi^*(x),\varepsilon(x)) \leqslant \Phi(x,\pi^*(x),\varepsilon^*(x)) \leqslant \Phi(x,\pi(x),\varepsilon^*(x))$

for all $\varepsilon(x) \in E(x)$ and for all $\pi(x) \in \Pi(x)$. Moreover, the value $\Phi(x,\pi^*(x),\varepsilon^*(x))$ is unique. We shall let

(2) $\Phi^*(x) = \Phi(x,\pi^*(x),\varepsilon^*(x)), \quad x \in X(k)$

Hence, the value of Φ^* at any point of $X(k)$ can be deduced from the knowledge of Φ^* at stage $k+1$. We shall complement this recurrence relation by the boundary condition

(3) $\Phi^*(x) = 0 \qquad \forall x \in X(K)$

This boundary condition and the recurrence relation give Φ^*.

Now, consider the strategies π^* and ε^* whose values at any x in X are $\pi^*(x)$ and $\varepsilon^*(x)$, respectively, and suppose that (π^*,ε^*) generates a terminating path $\bar{\Gamma}^*$ from some prescribed initial state $P^1 \in X(i)$.

Let

(4) $\Psi(P^k,\pi,\varepsilon) = \sum_{x\in X(k)} P^k(x)\Phi(x,\pi(x),\varepsilon(x))$

(5) $\Psi^*(P^k) = \sum_{x\in X(k)} P^k(x)\Phi^*(x)$

$(\pi,\varepsilon) \in \Pi\times E, \quad P^k \in X(k), \quad i \leqslant k \leqslant K$

One can prove readily that

(6) $\Psi(P^k,\pi,\varepsilon) = F_o(P^k,\pi,\varepsilon) + \Psi^*(F(P^k,\pi,\varepsilon)), \qquad i \leqslant k \leqslant K-1$

Furthermore, from (1)(2)(4)(5), and since $P^k(x)$ is non negative for all $x \in X(k)$, $i \leqslant k \leqslant K$, it follows that

(7) $\Psi(P^k,\pi^*,\varepsilon) \leqslant \Psi^*(P^k) \leqslant \Psi(P^k,\pi,\varepsilon^*)$

$\Psi^*(P^k) = \Psi(P^k,\pi^*,\varepsilon^*), \qquad i \leqslant k \leqslant K$

for all $\varepsilon \in E$ and for all $\pi \in \Pi$.

From (3) and (5) we have

(8) $\Psi^*(P^K) = 0 \qquad \forall P^K \in X(K)$

At last, from (6)-(8), by similar arguments as in deterministic multistage games, it follows that $\overline{\Gamma}^*$ is an optimal path and that $\Psi^*(P^i)$ is the value of the game at state P^i, in Problem 2.

3. SOME GEOMETRIC ASPECTS OF QUANTITATIVE GAMES

3.1 Surface of the Game

In the following we shall suppose that there exists a strategy N-tuple s^* which is C-optimal at state x^i, and we shall denote by X the set of all states x at which s^* is C-optimal. Moreover we shall consider the mappings $V^*_\alpha : X \to \Omega_\alpha$, $\alpha = 1,2,\dots N$, such that $V^*_\alpha(x) = V_\alpha(x,s^*,\gamma^*)$ for all x in X, and we shall let $V^*(x) = (V^*_1(x), V^*_2(x),\dots V^*_N(x))$. $V^*(x)$ is the *value of the game* at state $x \in X$.

We shall define $(>)_\alpha$, $\alpha \in \{1,2,\dots N\}$, by $\omega'_\alpha (>)_\alpha \omega''_\alpha \iff (\omega'_\alpha \ (\geqslant)_\alpha \omega''_\alpha$ and *not* $\omega''_\alpha \ (\geqslant)_\alpha \omega'_\alpha),(\omega'_\alpha,\omega''_\alpha) \in \Omega^2_\alpha$.

Then we shall let

$$\omega' > \omega'' \iff \omega'_\alpha (>)_\alpha \omega''_\alpha \ , \ \alpha = 1,2,\dots N$$
$$\omega' = (\omega'_1, \omega'_2,\dots \omega'_N) \in \Omega, \ \omega'' = (\omega''_1, \omega''_2,\dots \omega''_N) \in \Omega.$$

Now let us define the set $\Sigma(C)$, $C \in \Omega$, by

$$\Sigma(C) = \{y = (x_o,x) : (x_o,x) \in \Omega \times X, \ x_o + V^*(x) = C\}$$

$\Sigma(C)$ is the *surface of the game* for given C. Let us define also a set $\Sigma_\alpha(C_\alpha)$, $C_\alpha \in \Omega_\alpha$, namely

$$\Sigma_\alpha(C_\alpha) = \{y_\alpha = (x_{o\alpha},x) : (x_{o\alpha}, x) \in \Omega_\alpha \times X, \ x_{o\alpha} + V^*_\alpha(x) = C_\alpha\}$$
$$\alpha \in \{1,2,\dots N\} .$$

$\Sigma_\alpha(C_\alpha)$ will be called an *α-surface of the game.*

In connection with the definition of $\Sigma(C)$, $C = (C_1,C_2,\dots C_N) \in \Omega$, we shall define the sets $A/\Sigma(C)$ and $\overline{B/\Sigma(C)}$, namely

$$A/\Sigma(C) = \{y = (x_o,x) : (x_o,x) \in \Omega \times X, \ x_o + V^*(x) > C\}$$
$$\overline{B/\Sigma(C)} = \{y = (x_o,x) : (x_o,x) = (x_{o1}, x_{o2},\dots x_{oN},x) \in \Omega \times X,$$
$$\exists \ \alpha \in \{1,2,\dots N\}, \ C_\alpha (\geqslant)_\alpha x_{o\alpha} + V^*_\alpha(x)\}$$

A point $y \in A/\Sigma(C)$ is called an A-point, and a point $y \in \overline{B/\Sigma(C)}$ is called a B-point, relative to $\Sigma(C)$.

Likewise we shall define $A/\Sigma_\alpha(C_\alpha)$ and $\overline{B/\Sigma_\alpha(C_\alpha)}$

$$A/\Sigma_\alpha(C_\alpha) = \{y_\alpha = (x_{o\alpha},x) : (x_{o\alpha},x) \in \Omega_\alpha \times X, \quad x_{o\alpha} + V^*_\alpha(x)(>)_\alpha C_\alpha\}$$

$$\overline{B/\Sigma_\alpha(C_\alpha)} = \{y_\alpha = (x_{o\alpha},x) : (x_{o\alpha},x) \in \Omega_\alpha \times X, \quad C_\alpha(\geqslant)_\alpha x_{o\alpha} + V^*_\alpha(x)\}$$

$\alpha = 1,2,\ldots N$.

A point $y_\alpha \in A/\Sigma_\alpha(C_\alpha)$ is called an A-point, and a point $y_\alpha \in \overline{B/\Sigma_\alpha(C_\alpha)}$ is called a B-point, relative to $\Sigma_\alpha(C_\alpha)$.

From now on, we shall let $Z = \Omega \times G$, and $Z^* = \Omega \times X$, and $Z_\alpha = \Omega_\alpha \times G$, and $Z^*_\alpha = \Omega_\alpha \times X$, $\alpha = 1,2,\ldots N$.

3.2 Trajectories and Paths in $P(Z)$

For any $(x^i,s) \in D$ and for any $\gamma \in$ Range R such that $(x^i,s)R\gamma$, and for any $C \in \Omega$, we shall define a *trajectory* $\Gamma(\gamma,y^i,s)$ *in* $P(Z)$ by

$$\Gamma(\gamma,y^i,s) = \{y = (x_o,x) : x_o \in \Omega,\ x \in \gamma, \quad x_o + V(x,s,\rho(\gamma,x)) = C\}$$

$$y^i = (x^i_o,x^i), \qquad C = x^i_o + V(x^i,s,\gamma)$$

We shall say that $\Gamma(\gamma,y^i,s)$ is generated from y^i by strategy N-tuple s . A trajectory $\Gamma(\gamma,y^i,s)$ will be called a *path in* $P(Z)$ if γ is a path.

Likewise we shall define an α-*trajectory* $\Gamma_\alpha(\gamma,y^i_\alpha,s)$ *in* $P(Z_\alpha)$ by

$$\Gamma_\alpha(\gamma,y^i_\alpha,s) = \{y_\alpha = (x_{o\alpha},x) : x_{o\alpha} \in \Omega_\alpha,\ x \in \gamma \quad x_{o\alpha} + V_\alpha(x,s,\rho(\gamma,x)) = C_\alpha\}$$

$$y^i_\alpha = (x^i_{o\alpha},x^i), \qquad C_\alpha = x^i_{o\alpha} + V_\alpha(x^i,s,\gamma)$$

Again, we shall say that $\Gamma_\alpha(\gamma,y^i_\alpha,s)$ is generated from y^i_α by s, and $\Gamma_\alpha(\gamma,y^i_\alpha,s)$ will be called an α-*path in* $P(Z_\alpha)$ if γ is a path. We shall say that a path $\Gamma(\gamma,y^i,s)$, or an α-path $\Gamma_\alpha(\gamma,y^i_\alpha,s)$, is *optimal* if $s = s^*$ and $\gamma = \gamma^* \in I(x^i,s^*,\theta)$.

3.3 Lemma 1

Let us introduce :

Assumption 5. *For any* (x^i,s',γ') *and* (x^j,s'',γ'') *such that* $(x^i,s')\,R\,\gamma'$ *and* $(x^j,s'')\,R\,\gamma''$ *and* x^j *is the end point of* γ'

$$V(x^i,s',\gamma') + V(x^j,s'',\gamma'') = V(x^i,s,\gamma)$$

where $\gamma = \gamma' \cup \gamma''$ *and* s *is a strategy* N-*tuple such that* $(x^i,s)R\gamma$ *; and condition* (b) *of assumption* 2.

Assumption 6. *For any* ω_α, $\omega_\alpha \in \Omega_\alpha$, $\alpha \in \{1,2,\ldots N\}$,

$$(\omega'_\alpha + \omega_\alpha)(\geqslant)_\alpha(\omega''_\alpha + \omega_\alpha) \iff \omega'_\alpha(\geqslant)_\alpha\omega''_\alpha$$

Then one can easily prove

Lemma 1. *If* s^* *is C-optimal at state* x^i, *and* $\gamma^* \in I(x^i,s^*,\theta)$,*and* $x^j \in \gamma^*$, *then* s^* *is C-optimal at state* x^j .

3.4 A Fundamental Property of Game Surfaces

By similar arguments as in Ref. [1], one can prove

Theorem 1. *No point of an* α-*path* $\Gamma_\alpha(\overline{\gamma},y^i_\alpha,s)$, *generated from* $y^i_\alpha = (x^i_{o\alpha},x^i)$ *by strategy* N-*tuple* s, *such that* $(x^i,s)\,R\,\overline{\gamma}$, $sC_\alpha s^*$, $\alpha \in \{1,2,\ldots N\}$, *is an* A-*point relative to the* α-*surface of the game through* y^i_α *; and, furthermore, if* $y^j_\alpha = (x^j_{o\alpha},x^j) \in \Gamma_\alpha(\overline{\gamma},y^i_\alpha,s) \cap Z^*_\alpha$, *then* y^j_α *is a* B-*point relative to that* α-*surface.*

Corollary 1. *No point of a path* $\Gamma(\overline{\gamma},y^i,s)$ *generated from* $y^i = (x^i_o,x^i)$ *by strategy* N-*tuple* s, *such that* $(x^i,s)\,R\,\overline{\gamma}$, $sC_\alpha s^*$, $\alpha \in \{1,2,\ldots N\}$, *is an* A-*point relative to the game surface through* y^i, *and, furthermore, if* $y^j \in \Gamma(\overline{\gamma},y^i,s) \cap Z^*$ *then* y^j *is a* B-*point relative to that game surface.*

Corollary 1 is a direct consequence of Theorem 1, and the definitions of $A/\Sigma(C)$ and $\overline{B/\Sigma(C)}$.

Theorem 2. *An optimal* α-*path in* $P(Z_\alpha)$ *emanating from* y^i_α *has all of its points in the* α-*surface of the game through* y^i_α .

Theorem 2 is a direct consequence of Lemma 1.

Corollary 2. *An optimal path in* P(Z) *emanating from* y^i *has all of its points in the game surface through* y^i .

Corollary 2 is a direct consequence of Theorem 2 and the definition of a game surface.

REFERENCE

1. A. Blaquière, F. Gérard, and G. Leitmann, Quantitative and Qualitative Games, Academic Press, 1969 .

APPENDIX 1. Proof of Lemma 1

Clearly, s^* is playable at state x^j. Let $\bar{\gamma}'' \in I(x^j,s^*,\theta)$ and $\bar{\gamma}' = \rho(\gamma^*,x^i,x^j)$ and $\bar{\gamma} = \bar{\gamma}' \cup \bar{\gamma}''$. From Assumption 2 we have $(x^i,s^*)\,R\,\bar{\gamma}$, and indeed $\bar{\gamma} \in I(x^i,s^*,\theta)$. From Assumption 5 and condition (ii) of paragraph 1 we have

$$V(x^i,s^*,\bar{\gamma}) = V^*(x^i) = V(x^i,s^*,\bar{\gamma}') + V(x^j,s^*,\bar{\gamma}'')$$

$$V(x^i,s^*,\gamma^*) = V^*(x^i) = V(x^i,s^*,\bar{\gamma}') + V(x^j,s^*,\rho(\gamma^*,x^j))$$

and accordingly $V(x^j,s^*,\bar{\gamma}'') = V(x^j,s^*,\rho(\gamma^*,x^j))$.

Now let $\bar{\gamma}'' \in I(x^j,s'',\theta)$ and $s''C_\alpha s^*$ for $\alpha \in \{1,2,\ldots N\}$. Let $\bar{\gamma}' = \rho(\gamma^*,x^i,x^j)$ and $\bar{\gamma} = \bar{\gamma}' \cup \bar{\gamma}''$. From Assumption 2, there exists a strategy N-tuple s, $sC_\alpha s^*$, such that $(x^i,s)\,R\,\bar{\gamma}$, and indeed $\bar{\gamma} \in I(x^i,s,\theta)$. From Assumption 5 we have

$$V_\alpha(x^i,s,\bar{\gamma}) = V_\alpha(x^i,s^*,\bar{\gamma}') + V_\alpha(x^j,s'',\bar{\gamma}'') \text{ ; and}$$

$$V_\alpha(x^i,s^*,\gamma^*) = V_\alpha(x^i,s^*,\bar{\gamma}') + V_\alpha(x^j,s^*,\rho(\gamma^*,x^j))$$

From Condition (iii) of paragraph 1, and Assumption 6 , we obtain

$$V_\alpha(x^j,s^*,\rho(\gamma^*,x^j))\,(\geqslant)_\alpha V_\alpha(x^j,s'',\bar{\gamma}''),\ \alpha \in \{1,2,\ldots N\}$$

Hence Lemma 1 is proved.

APPENDIX 2. Proof of Theorem 1

Let $x^j \in \bar{\gamma}$. If $s^* \notin J^*(x^j)$ the conclusion of Theorem 1 is trivial. Let us suppose that $s^* \in J^*(x^j)$ and let $\gamma^* \in I(x^j,s^*,\theta)$. We have $V_\alpha(x^j,s^*,\gamma^*) = V^*_\alpha(x^j)$. Let $\bar{\gamma}' = \rho(\bar{\gamma},x^i,x^j)$ and $\bar{\gamma}'' = \bar{\gamma}' \cup \gamma^*$. From Assumption 2 there exists s'' such that $(x^i,s'')\,R\,\bar{\gamma}''$ and $s''\,C_\alpha\,s^*$. From Assumption 5 we have $V_\alpha(x^i,s'',\bar{\gamma}'') = V_\alpha(x^i,s,\bar{\gamma}') + V^*_\alpha(x^j)$; and from condition (iii) of paragraph 1, we have $V^*_\alpha(x^i)\ (\geqslant)_\alpha V_\alpha(x^i,s'',\bar{\gamma})$. Then from

Assumption 6 we obtain

$$V_\alpha^*(x^i) - V_\alpha^*(x^j)(\geqslant)_\alpha V_\alpha(x^i,s,\overline{\gamma}')$$

From the definition of $\Gamma_\alpha(\overline{\gamma},y_\alpha^i,s)$, and Assumption 5, one deduces easily that $V_\alpha(x^i,s,\overline{\gamma}') = x_{o\alpha}^j - x_{o\alpha}^i$, and accordingly

$$V_\alpha^*(x^i) - V_\alpha^*(x^j)(\geqslant)_\alpha x_{o\alpha}^j - x_{o\alpha}^i$$

From which follows that

$$x_{o\alpha}^i + V_\alpha^*(x^i)(\geqslant)_\alpha \; x_{o\alpha}^j + V_\alpha^*(x^j)$$

Hence $y_\alpha^j \in \overline{B/\Sigma_\alpha(C_\alpha)}$, where $\Sigma_\alpha(C_\alpha)$ is the α-surface of the game through y_α^i . It follows that $y_\alpha^j \notin A/\Sigma_\alpha(C_\alpha)$. Hence Theorem 1 is proved.

STRATEGIC "PREJUDGMENTS" AND DECISION CRITERIA IN N-PERSON GAMES

P. Lévine

Laboratoire d'Econométrie, Université de Paris VI,
Paris, France

ABSTRACT. The concept of strategic "prejudgment" is first introduced. A concept of solution is then obtained by assuming that the players use decision criteria under uncertainty to choose their strategy. An existence theorem is proved about the solutions under convexity assumptions and two examples are given. The first one shows how one can find the Nash solution to the bargaining problem by making clear the player's strategic "prejudgments". The second one is an application to a differential game model of oligopoly.

INTRODUCTION

The concepts presented in this paper are derived principally from the two following criticisms. The first one is due to Schelling [4] (p. 119) : "By abstracting from communication and enforcement systems and by treating perfect symmetry between players as the general case rather than a special one, game theory may have overshot the level at which the most fruitful work could be done and may have defined away some of the essential ingredients of typical non zero-sum games". Luce and Raiffa are the authors of the second criticism [2] (p. 164) : "We should judge this omission of sociological assumptions at the level of the normal form to be one of the two major practical faults of present-day n-person theory, the other being the previously mentionned static character".

Thus our purpose is first to introduce a new basic concept for games in normal form, which we shall call the players' strategic "prejudgments". This way we try to answer the Schelling's criticism about communication and enforcement systems and Luce's and Raiffa's criticism about the omission of sociological assumptions. On the other hand, we shall try to escape the static analysis which consists

J. D. Grote (ed.), The Theory and Application of Differential Games. 121-132.

in studying if an outcome of the game is stable or not : we shall follow the way the players choose their strategy, in order to characterize the result of their choice, by supposing that they use decision criteria under uncertainty.

In sections I and II we define the concept of strategic "prejudgment" and the solutions which correspond to this concept. In the sequel we prove an existence theorem about the solutions under convexity assumptions. Finally we give two examples. The first one shows that we can find again well-known concepts of game theory, as the Nash's solution to the bargaining problem, by making clear the players' strategic "prejudgments". The other one shows how to use strategic "prejudgments" in order to formalize economic behaviours between complete cooperation and pure conflict in a differential game model of oligopoly.

I - STRATEGIC "PREJUDGMENT" FUNCTIONS AND STRONG POSITIONS OF ORDER ZERO

Let J be the n-person game in normal form $((X_i)_{1\leqslant i\leqslant n},(f_i)_{1\leqslant i\leqslant n})$ where $\forall$ i, i = 1,...,nX_i denotes the player i's space of strategies and f_i his payoff, which is a numerical function defined on $\prod_{i=1}^{n} X_i$.

Throughout the sequel if $(a_1,\dots,a_n)$ is a n-tuple of strategies we shall denote $a_{\hat{i}}$ the n-1-tuple $(a_1,\dots,a_{i-1},a_{i+1},\dots,a_n)$ and if A is the cartesian product $A_1 \times\dots\times A_n$ $A_{\hat{i}}$ shall denote $A_1 \times\dots\times A_{i-1} \times A_{i+1} \times\dots\times A_n$.

Moreover, if R is a multivalued function mapping Y into Z and if S is a subset of Y, we shall denote by R(S) the set $\bigcup_{x\in S} R(x)$.

Finally we shall suppose that :

(a) $\forall$ i, i = 1,...,n X_i is a compact topological space

(b) $\forall$ i, i = 1,...,n f_i is u.s.c. on X_i $\forall x_{\hat{i}} \in X_{\hat{i}}$.

1 - Strategic "prejudgment" functions of order zero

Let us imagine how player i may choose his strategy. He will, for example, first ask himself what strategies the other players can play. Thus, he will try to obtain information in order to exclude a priori some possibilities $y_{\hat{i}} \in X_{\hat{i}}$. Then he will decide his strategy, taking account of his commitments with the other players.

Thus, to formalize the player i's behaviour, we have to define precisely the information and commitments exchanged at the beginning of the game, which we shall call the players' strategic "prejudgments".

Now, these strategic "prejudgments" usually have a special structure which is of the following kind : "If the other players do this, I shall do that". This is clear for the threats, promises, commitments to cooperate and, more generally, for "strategic moves"[4]. But we can likewise formalize the players' reasonings involved by sociological norms or stable psychological attitudes known by every player.

Therefore we shall call player i's <u>strategic "prejudgment" function of order zero</u> ($i = 1,\ldots,n$) a u.s.c. multivalued function R_i° mapping $X_{\hat{i}}$ into X_i such that $\forall\, x_{\hat{i}} \in X_{\hat{i}}$ $R_i^\circ(x_{\hat{i}})$ is the <u>non empty</u> set of the strategies that player i decides a priori he can play, thinking his opponents will play $x_{\hat{i}}$.

Consequently if player i thinks that the other players will play $x_{\hat{i}}$ then he will give up the idea of playing any $x_i \notin R_i^\circ(x_{\hat{i}})$. And he will definitely renounce to play x_i if $\forall\, x_{\hat{i}} \in X_{\hat{i}}$ $x_i \notin R_i^\circ(x_{\hat{i}})$; in this case we can in fact delete x_i in X_i.

We shall suppose here complete information on the players' strategic "prejudgment" functions. Thus every player will know at the beginning of the game the R_i° as well as the X_i and the $f_i (i=1,\ldots,n)$.

<u>Remark</u> Using this formalization, there is no difference between non cooperative or implicitely cooperative games and cooperative games.

2 - Decision criteria and strong positions of order zero

Let us now consider the way player $i_\circ$ prepares his decision. With the $R_i^\circ (i=1,\ldots,n)$ he knows surely that the other players will play strategies $x_j \in R_j^\circ(X_{\hat{j}})$ $\forall\, j \neq i_\circ$. But, since he knows that player i $(i=1,\ldots,n)$ will play in $R_i^\circ(X_{\hat{i}})$, he can deduce that, in fact, the strategy x_j belongs to $T_j^\circ(1)$ $\forall\, j \neq i_\circ$, where $\forall\, i$, $i=1,\ldots,n$ $T_i^\circ(1)$ is the set

$$\{y_i \mid y_i \in R_i^\circ(z_{\hat{i}}) \text{ with } z_k \in R_k^\circ(X_{\hat{k}})\ \forall\, k \neq i\}.$$

Remark that $T_i^\circ(1)$ is non empty and compact since $\forall\, i$, $i=1,\ldots,n$ R_i° is u.s.c., $X_{\hat{i}}$ is compact and $R_i^\circ(x_{\hat{i}})$ is non empty $\forall\, x_{\hat{i}} \in X_{\hat{i}}$. Then by induction on n, player $i_\circ$ can construct the non empty compact and decreasing sets $T_i^\circ(n) = R_i^\circ(T_{\hat{i}}^\circ(n-1))$ and he can deduce that player j, $\forall\, j \neq i_\circ$, will play in the non empty set S_j°, if $\forall\, i$, $i=1,\ldots,n$ $S_i^\circ = \bigcap_{n \geq 1} T_i^\circ(n)$.

But this is the only fact he can be sure of with the information at his disposal since $\forall i$, $i=1,\ldots,n$ $S_i^\circ = R_i^\circ(S_{\hat{i}}^\circ)$, R_i° being u.s.c. and X_i being compact.

Thus, let us assume that i will consider the choice of a $x_{\hat{i}}$ by his opponents in $S_{\hat{i}}^\circ$ as an event under uncertainty. Moreover suppose that he has a decision criterion under uncertainty satisfying the "complete ignorance" axiom since he knows nothing about his opponents but the information given by the R_i°. This criterion will be here the "Max Inf" but for this purpose one could obviously use others. Therefore, if i wants to assume his commitments whatever happens, he must choose a strategy $\bar{x}_i$ satisfying $\bar{x}_i \in Q_i^\circ$ where $Q_i^\circ = \bigcap_{x_{\hat{i}} \in S_{\hat{i}}^\circ} R_i^\circ(x_{\hat{i}})$ and maximizing $\operatorname{Inf}_{x_{\hat{i}} \in S_{\hat{i}}^\circ} f_i(x_i, x_{\hat{i}})$ on the set Q_i° since his criterion is the "Max Inf".

We shall call such a strategy a player i's <u>strong position of order zero</u> and we shall denote $\hat{Q}_i^\circ$ the set of these positions.

$\hat{Q}_i^\circ$ is compact since $R_i^\circ(x_{\hat{i}})$ is compact $\forall\, x_{\hat{i}} \in X_{\hat{i}}$, R_i° being u.s.c., and since the mapping ϕ_i defined by $\phi_i(x_i) = \inf_{x_{\hat{i}} \in S_{\hat{i}}^\circ} f_i(x_i, x_{\hat{i}})$ $\forall x_i \in X_i$ is u.s.c. on X_i, f_i being u.s.c. on X_i $\forall x_{\hat{i}} \in X_{\hat{i}}$. Finally, if $\hat{Q}_i^\circ$ is non empty $\forall i$, $i=1,\dots,n$ we shall say that the game is <u>rationally playable at the order zero</u>.

If the players stop reasoning at this point, they will play when the game is rationally playable at the order zero, a strong position of order zero. This is particularily satisfactory when $\hat{Q}_i^\circ$ is reduced to a single strategy $\forall i$, $i=1,\dots,n$. Otherwise, we can suppose that the players will go on elaborating their strategy at a second level.

II STRATEGIC "PREJUDGMENT"FUNCTIONS AND STRONG POSITIONS OF HIGHER ORDER

1 - Strategic "prejudgment" functions and strong positions of order one

Now, assume that the game is rationally playable at the order zero and that the players have to choose between several strategies in the $\hat{Q}_i^\circ (i=1,\dots,n)$. It may be of their interest to exchange new information in order to decide at best their strategy in these sets.

Thus, we shall suppose that each player i knows the decision criteria under uncertainty of his opponents (here we shall assume to simplify that every player has a "Max Inf" behaviour) and that they communicate to each other the following information called strategic "prejudgment" functions of order one.

We shall call player i's <u>strategic "prejudgment" function of order one</u> the data of, for each $Z = \prod_{i=1}^{n} Z_i$ (where Z_i is a non empty compact subset of X_i $\forall i, i=1,\dots,n$, a u.s.c. multivalued function $R_i^1(.) / Z$ mapping $Z_{\hat{i}}$ into Z_i such that $R_i^1(x_{\hat{i}}) / Z$ is non empty $\forall x_{\hat{i}} \in Z_{\hat{i}}$.

These functions will be given at the beginning of the game to every player and they will be interpreted as follows : if at the second level player i thinks that every player will in fact play the game in

$$Z = \prod_{i=1}^{n} Z_i$$

and if i presumes that his opponents will play a strategy $x_{\hat{i}}$ in $Z_{\hat{i}}$, then he decides to play in $R_i^1(x_{\hat{i}}) / Z$.

Now, $\forall$ i, i=1,...,n let us define the sequence $T_i^1(k)$ by :

(a) $T_i^1(0) = R_i^1(\hat{Q}_{\hat{i}}^\circ) / \prod_{i=1}^{n} \hat{Q}_i^\circ$

(b) $T_i^1(k) = R_i^1(T_{\hat{i}}^1(k-1)) / \prod_{i=1}^{n} \hat{Q}_i^\circ \qquad \forall\, k \geq 1.$

and let us denote by S_i^1 the set $\bigcap_{k\geq 0} T_i^1(k)$ which is non empty for the same reason as at the order zero.

With all the previous assumptions, player i (i=1,...,n) can draw the following conclusions : since he has to take account of the case where his opponents, at the opposite of him, do not use their decision criteria under uncertainty to choose their strategy at the order zero, he must play in $\hat{Q}_i^\circ$. But taking advantage of the fact that if his opponents reason like him, they will play in $\hat{Q}_{\hat{i}}^\circ$, he will choose among the $x_i \in \hat{Q}_i^\circ$, a strategy satisfying :

$\bar{x}_i \in Q_i^1$ with $Q_i^1 = \bigcap_{x_{\hat{i}} \in S_{\hat{i}}^1} R_i^1(x_{\hat{i}}) / \prod_{i=1}^{n} \hat{Q}_i^\circ$

and maximizing $\operatorname{Inf}_{x_{\hat{i}} \in S_{\hat{i}}^1} f_i(x_i, x_{\hat{i}})$ on the set Q_i^1.

We shall call these strategies the player i's strong positions of order one and we shall denote $\hat{Q}_i^1$ the compact set they define. If the $\hat{Q}_i^1$ are non empty $\forall$ i, i=1,...,n, we shall say that the game is rationally playable at the order one.

2 - Strategic "prejudgment" functions of higher order

The study done in the preceding section can be obviously repeated as many times as one wants if the players consider strategic "prejudgments" of order m+1 to perfect their decision of order m $\forall$ m$\in$N.

Thus, we shall assume that $\forall$ m$\in$N player i (i=1,...,n) is endowed for each $Z = \prod_{i=1}^{n} Z_i$ (where Z_i is a non empty compact subset of X_i $\forall$i, i=1,...,n) with a u.s.c. multivalued function $R_i^m(.)/Z$ mapping $Z_{\hat{i}}$ into Z_i such that $R_i^m(x_{\hat{i}})/Z$ is non empty $\forall x_{\hat{i}} \in Z_{\hat{i}}$. We shall call this family of multivalued functions (denoted by R_i^m) the player i's strategic "prejudgment" function of order m.

Under these conditions we can define by induction the concepts defined in the previous section, at the order m (m>1). Assume for this purpose that the concepts of strong position of order m-1 and of rationally playable game at the order m-1 have been defined and let us denote $\hat{Q}_i^{m-1}$ the compact set of player i's strong positions of order m-1 (i = 1,...,n). Then, when the game is rationally playable at the order m-1, let us define the sets $T_i^m(k)$ by :

$$T_i^m(0) = R_i^m(\hat{Q}_{\hat{i}}^{m-1}) \,/\, \prod_{i=1}^{n} \hat{Q}_i^{m-1}$$

and $T_i^m(k) = R_i^m(T_{\hat{i}}^m(k-1)) \,/\, \prod_{i=1}^{n} \hat{Q}_i^{m-1} \quad \forall\, k \geq 1.$

Let us denote S_i^m the non empty set $\bigcap_{k\geq 0} T_i^m(k)$. We shall say that $\bar{x}_i$ is a player i's strong position of order m(i=1,...,n) if :

$\bar{x}_i \in Q_i^m$ where $Q_i^m = \bigcap_{x_{\hat{i}} \in S_{\hat{i}}^m} R_i^m(x_{\hat{i}}) \,/\, \prod_{i=1}^{n} \hat{Q}_i^{m-1}$ and $\bar{x}_i$ maximizes

$\operatorname{Inf}_{x_{\hat{i}} \in S_{\hat{i}}^m} f_i(x_i, x_{\hat{i}})$ on the set Q_i^m.

We shall denote $\hat{Q}_i^m$ the compact set of player i's strong positions of order m and we shall say that the game is rationally playable at the order m if $\hat{Q}_i^m$ is non empty $\forall$ i, i = 1,...,n.

Finally, $\forall$ m$\in$N, $\forall$ i, i=1,...,n we shall call

$\bar{u}_i^m = \operatorname{Sup}_{x_i \in Q_i^m} \operatorname{Inf}_{x_{\hat{i}} \in S_{\hat{i}}^m} f_i(x_i, x_{\hat{i}})$ the player i's value of the game of order m.

3 - Resolutions - Solutions - Value

We shall call resolution of the game J for i the family of strategic "prejudgment" functions R_i^m, m$\in$N, that player i has at his disposal. If the game J, with such resolutions, is rationally playable at the order m, $\forall$ m$\in$N, to each player corresponds a non empty set of strong positions of order m, $\forall$ m$\in$N. Therefore the only rational way to play in this case is to play a strategy which is a strong position of order m, $\forall$ m$\in$N and we shall call a solution for player i, a strategy $\bar{x}_i$ such that : $\forall$ m$\in$N $\bar{x}_i \in \hat{Q}_i^m$. We shall denote Σ_i the set of these solutions.

Moreover, we shall define the value of the game for player i as the limit, whenever it exists, of the sequence $(\bar{u}_i^m)_{m\in N}$. To justify this definition, refer to theorems 1 and 2 of III - 2.

III - EXISTENCE OF SOLUTIONS

We shall assume in this section that the game $J = ((X_i)_{1\leq i\leq n}, (f_i)_{1\leq i\leq n})$ has also the following properties :

(a) $\forall$ i, i=1,...,n the space X_i is a convex subset of a locally convex Hausdorff space E_i and its topology is induced by the topology of E_i.

(b) $\forall$ i, i=1,...,n f_i is quasi-concave on X_i $\forall\, x_{\hat{i}} \in X_{\hat{i}}$ and l.s.c. on

$X_{\hat{i}}$ $\forall$ $x_i \in X_i$.

1 - Regular multifunctions

Let Y and Z be two compact convex subsets of locally convex Hausdorff spaces F and G and R a multivalued function mapping Y into Z. We shall denote by R^* the multivalued function mapping Z into Y defined

by $R^*(z) = \{y \mid z \notin R(y)\}$ $\forall$ $z \in Z$.

And we shall say that R is weakly regular if :
(a) R(y) is convex compact $\forall$ $y \in Y$.
(b) $R^*(z)$ is convex $\forall$ $z \in Z$.

R will be regular if there exists a sequence $(R_n)_{n \in N}$ of weakly regular multivalued functions mapping Y into Z such that :

(a) $\forall$ $n \in N$ $R_{n+1}(y) \subset R_n(y)$ $\forall$ $y \in Y$.

(b) $R(y) = \bigcap_{n \in N} R_n(y)$ $\forall$ $y \in Y$.

(c) $R^*(z) = \bigcup_{n \in N} \overline{R_n^*(z)}$ $\forall$ $z \in Z$.

It is easy to see that a regular multivalued function is weakly regular and that a weakly regular multivalued function R, such that $R^*(z)$ is closed in Y $\forall$ $z \in Z$, is regular.

In the sequel we shall only be interested in the following property of regular multivalued functions :

Theorem 1 - If R is regular and if $R(y) \neq \emptyset$ $\forall$ $y \in Y$, $\bigcap_{y \in Y} R(y)$ is non empty.

Example - Let F be a real valued function defined on Y×Z, quasi-concave u.s.c. on Y $\forall$ $z \in Z$ and quasi convex l.s.c. on Z $\forall$ $y \in Y$. Then, R, defined by :

$$R(y) = \{z \mid F(y,z) \leqslant 0\} \quad \forall\ y \in Y,$$

is regular.

2 - Existence of solutions

Assume that the players are endowed with resolutions $(R_i^m)_{m \in N}$. We shall say that the player i's strategic "prejudgment" function R_i^m is regular if :
(a) When m = 0 : R_i° is a regular multivalued function mapping $X_{\hat{i}}$ into X_i.
(b) When m>0 and $Z = \prod_{i=1}^{n} Z_i$, Z_i being a non empty convex compact

subset of X_i $\forall$ i, i=1,...,n : R_i^m (.) / Z is a regular multivalued function mapping $Z_{\hat{i}}$ into Z_i.

Theorem 1 - If $\forall$ m, m$\in$N, the players have regular strategic "prejudgment" functions of order m, each player has a convex compact non empty set of solutions and a finite value.

Theorem 2 - If v_i is the value of the game for player i (i=1,...,n) we have :

$$\forall\ x_i \in \Sigma_i \quad \underset{x_{\hat{i}} \in \Sigma_{\hat{i}}}{\mathrm{Min}} \quad f_i(x_i, x_{\hat{i}}) = v_i .$$

IV - EXAMPLES

1 - The Nash's bargaining theory

We shall recall briefly the Nash's bargaining model [3] to show that it can be considered as a special case of game with strategic "prejudgments". For this purpose we shall build strategic "prejudgment" functions in order to find again the Nash's solution.

Let S be a convex compact subset of $\mathbb{R}^2$ which represents the outcomes (u_1, u_2) that the two players 1 and 2 can obtain by acting together. Assume that if they do not come to an agreement, then they must use their threat strategies x_1° and x_2°, the corresponding outcomes u_1° and u_2° satisfying :

$\exists\,(u_1, u_2) \in S$ such that $u_1 > u_1^\circ$ and $u_2 > u_2^\circ$.

Thus, we know that $\bar{u}_1$ and $\bar{u}_2$ are the Nash solution to this bargaining problem, where :

$\bar{u}_1$ and $\bar{u}_2$ maximize $(u_1 - u_1^\circ)(u_2 - u_2^\circ)$ on the set $S \cap \{(u_1, u_2) \mid u_1 \geqslant u_1^\circ\}$.

Let us now consider this game in normal form. A strategy being defined as the outcome demanded, the set of player 1's strategies is

$$X_1 = \{u_1 \mid \exists\, u_2 \text{ such that } (u_1, u_2) \in S\}$$

and the set of player 2's strategies is :

$$X_2 = \{u_2 \mid \exists\, u_1 \text{ such that } (u_1, u_2) \in S\}.$$

Thus the payoffs are given by :

$f_1(u_1, u_2) = u_1$ if $(u_1, u_2) \in S$, u_1° otherwise, for player 1 and

$f_2(u_1, u_2) = u_2$ if $(u_1, u_2) \in S$, u_2° otherwise, for player 2.

Then, let us set $\forall\ u_1 \in X_1$ $a(u_1) = \mathrm{Max}\ u_2$, $(u_1, u_2) \in S$ and

$\forall\ u_2 \varepsilon X_2 \quad b(u_2) = \text{Max}\ u_1,\ (u_1,u_2)\varepsilon S.$

We can now define the strategic "prejudgment" functions R_1° and R°_2 for the two players as follows :

$$\text{if } u_2 \geq u_2^\circ \quad R_1^\circ(u_2) =$$

$$\{u_1 \varepsilon X_1 \mid u_1 \geq u_1^\circ, a(u_1) \geq u_2^\circ, (u_1-u_1^\circ)(a(u_1)-u_2^\circ) \geq (b(u_2)-u_1^\circ)(u_2-u_2^\circ)\}$$

otherwise $R_1^\circ(u_2) = \{u_1 \varepsilon X_1 \mid u_1 \geq u_1^\circ, a(u_1) \geq u_2^\circ\}$;

$$\text{if } u_1 \geq u_1^\circ \quad R_2^\circ(u_1) =$$

$$\{u_2 \varepsilon X_2 \mid u_2 \geq u_2^\circ, b(u_2) \geq u_1^\circ, (b(u_2)-u_1^\circ)(u_2-u_2^\circ) \geq (u_1-u_1^\circ)(a(u_1)-u_2^\circ)\}$$

otherwise $R_2^\circ(u_1) = \{u_2 \varepsilon X_2 \mid u_2 \geq u_2^\circ, b(u_2) \geq u_1^\circ\}$.

According to Harsanyi [1] we can interprete in this way R_1° (resp. R_2°) : if player 1 (resp. 2) thinks that player 2 (resp. 1) will play u_2 (resp. u_1), he takes the decision to play the strategies u_1 (resp. u_2) such that he will not concede unilaterally in a negotiation where 1 would demand $(u_1, a(u_1))$ and where 2 would demand $(b(u_2), u_2)$.

It is easy to prove, under these assumptions, that :

$$\hat{Q}_1^\circ = \{\bar{u}_1\} \quad \text{and} \quad \hat{Q}_2^\circ = \{\bar{u}_2\}.$$

Thus, whatever will be the strategic "prejudgment" functions of higher order for the two players, the player 1's (resp. 2's) solution is $\bar{u}_1$ (resp. $\bar{u}_2$) and the corresponding value is also $\bar{u}_1$ (resp. $\bar{u}_2$).

Finally, we see that this method is useful in the way that it gives explicitly the only information that players need to cooperate. On the other hand, this example shows that strategic "prejudgments" may lead the players to Pareto optimal outcomes. But we shall see in the following example that it is not the general case.

2 - A differential game model of oligopoly

Let n producers provide a market with a commodity C during the period $[o,T]$. We shall suppose that we can consider a strategy of producer i (i=1,...,n) as the data, for almost every $t \varepsilon [o,T]$, of a feasible rate of change of his activity level. Thus we shall define the set X_i of producer i's strategies by :

$$\{u_i \varepsilon L^\infty[o,T] \mid a(t) \leq u_i(t) \leq b(t) \quad \text{a.e.}\}$$

where a and b denote two elements of $L^\infty[o,T]$.

In the sequel we shall assume that $L^\infty[o,T]$ is <u>endowed with the $\sigma(L^\infty[o,T],L^1[o,T])$ - topology</u>. Therefore X_i is a convex compact subset of $L^\infty[o,T]$ $\forall$ i, i=1,...,n.

If player i's strategy is $u_i \varepsilon X_i$ his production x_i will satisfy over $[o,T]$:

$$\frac{dx_i}{dt} = m(t)\, u_i(t) \ , \ x_i(o) = x_o$$

where m is a continuous strictly positive function on $[o,T]$. Moreover we shall suppose

$$x_o + \int_o^t m(\tau)a(\tau)d\tau \geqslant o \quad \forall\, t\varepsilon [o,T]$$

in order to avoid negative production.

When the producers adopt the programs of production $(x_1,\ldots,x_n)$, i obtains the profit :

$$f_i(x_1,\ldots,x_n) = \int_o^T x_i(t)F\left[\sum_{i=1}^n x_i(t)\right] - K\left[x_i(t)\right] dt$$

where F is a non increasing numerical function which gives for each amount x of C the price F(x) at which this quantity will be sold on the market and where K is the cost function of C (identical for all the producers) which gives for each amount x of C its production cost K(x).

We shall also assume that F and K are continuous and that $\forall\, z \geqslant o$ the function P_z defined by :

$$P_z(x) = xF(x+z) - K(x)$$

is concave on $\mathbb{R}^+$.

Now, for each convex compact subset C of $L^\infty[o,T]$ let be α_C and β_C, two elements of C such that :

$$||\alpha_C-\beta_C||_{L^\infty} \leqslant k \sup_{u,v\varepsilon C} ||u - v||_{L^\infty} \text{ with } o \leqslant k < 1.$$

Furthermore, for each intervalΘ of $\mathbb{R}$, let us define an interval $[I_\Theta, J_\Theta] \subset \Theta$. Let us denote $[I(C),J(C)]$ the interval

$$[I_{\Theta_C}, J_{\Theta_C}], \text{ where } \Theta_C = \{\theta\varepsilon\,\mathbb{R} \mid \int_o^T \alpha_C(t)dt \leqslant \theta \leqslant \int_o^T \beta_C(t)dt\}$$

and Φ the mapping from $L^\infty[o,T]$ into $\mathscr{C}[o,T]$ defined by :

$$\Phi(u)(t) = \int_o^t m(\tau)u(\tau)d\tau + x_o \text{ on } [o,T] \ \forall\, u\varepsilon L^\infty[o,T].$$

We are now able to explicit the strategic "prejudgment" functions of order zero R_i^o(i=1,...,n).

$\forall\, u_{\hat{i}} \varepsilon X_{\hat{i}}$ $\quad u_i \varepsilon R_i^o(u_{\hat{i}})$ if and only if :

(a) $u_i \varepsilon X_i$

(b) $\forall\, t\, \varepsilon [o,T]$ $\Phi(u_i)(t)$ belongs to :

$\left[\operatorname{Min}(\Phi(\alpha_{X_i})(t), \frac{1}{n-1}\sum_{j\neq i}\Phi(u_j)(t)), \operatorname{Max}(\Phi(\beta_{X_i})(t), \frac{1}{n-1}\sum_{j\neq i}\Phi(u_j)(t))\right]$

(c) $\int_0^T u_i(t)dt$ belongs to :

$\left[\operatorname{Min}(I(X_i), \frac{1}{n-1}\int_0^T \sum_{j\neq i} u_j(t)dt), \operatorname{Max}(J(X_i), \int_0^T \frac{1}{n-1}\sum_{j\neq i} u_j(t)dt)\right]$

In order to simplify these functions are identical for all the players. They can be communicated explicitly during a negotiation or, the most frequently, be the result of a tacit agreement. But in both cases one can interpret $\Lambda_i(t) = [\Phi(\alpha_{X_i})(t), \Phi(\beta_{X_i})(t)]$ as the interval where, at time t, the oligopolist i (i=1,...,n) judges that everyone might hold his production to keep a sufficient profit. In fact this interval depends on external constraints like anti-trust laws, profits in other sectors...

Thus (b) means that $\forall\ t\in[o,T]$:

(1) If i's competitors keep their average production in $\Lambda_i(t)$, i decides to remain in this interval too.

(2) If i's competitors keep their average production under $\Phi(\alpha_{X_i})(t)$ then i reminds them of his initial proposal in producing less than $\Phi(\beta_{X_i})(t)$ but takes the right to produce less than $\Phi(\alpha_{X_i})(t)$ and more than $\frac{1}{n-1}\sum_{j\neq i}\Phi(u_j)(t)$, which is a threat for his competitors since in this case, i can take advantage of this to increase his sales with respect to the sales of the firms $j\neq i$.

(3) If i's competitors produce on the average more than $\Phi(\beta_{X_i})(t)$ then i threatens them to play a similar strategy but, on the other hand, i reminds them of his proposal by showing he might play in $\Lambda_i(t)$.

It is easy to see that one can interpret in a similar way the interval $\Delta_i = [I(X_i), J(X_i)]$ and (c).

Furthermore, since $S_i^\circ = X_i\ \forall i, i=1,\ldots,n$ it is obvious that $Q_i^\circ = \bigcap_{u_{\hat{i}}\in S_{\hat{i}}^\circ} R_i^\circ(u_{\hat{i}}) = \{u_i\in X_i \mid \forall t\in[o,T]\ \Phi(u_i)(t)\in\Lambda_i(t) \text{ and } \int_0^T u_i(t)dt\in\Delta_i\}$ $\forall\ i$, i=1,...,n. Thus Q_i° is a non empty compact convex subset of X_i (i=1,...,n) since Φ is continuous from $L^\infty[o,T]$ into $\mathcal{C}[o,T]$ endowed with the norm topology.

Therefore $\hat{Q}_i^\circ$ is the set (identical for every player) of the $\bar{u}_i$ which maximize $\int_0^T \Phi(u_i)(t)F[\Phi(u_i)(t)+(n-1)\Phi(b)(t)] - K[\Phi(u_i(t))]$ under the constraint $u_i\in Q_i^\circ$.

Since $\forall z \geqslant o$ P_z is concave and F and K are continuous, $\hat{Q}_i^o$ is a compact convex and non empty subset of X_i $\forall$ i, i=1,...,n.

Often, $\hat{Q}_i^o$ is not reduced to a single strategy. So it is useful to consider strategic "prejudgment" functions of higher order. Thus, $\forall m \geqslant 1$, $\forall$ i, i=1,...,n and for each convex compact and non empty subset C of X_i, let us define $R_i^m(u_{\hat{i}}) / C^n$ as follows :
$u_i \varepsilon C$ belongs to $R_i^m(u_{\hat{i}}) / C^n$ if and only if :

(a) $\forall t \, \varepsilon [o,T]$ $\Phi(u_i)(t)$ belongs to :

$$\left[\mathrm{Min}(\Phi(\alpha_C)(t), \frac{1}{n-1}\sum_{j\neq i}\Phi(u_j)(t)), \mathrm{Max}(\Phi(\beta_C)(t), \frac{1}{n-1}\sum_{j\neq i}\Phi(u_j)(t))\right]$$

(b) $\int_o^T u_i(t)dt$ belongs to :

$$\left[\mathrm{Min}(I(C), \frac{1}{n-1}\int_o^T \sum_{j\neq i} u_j(t)dt), \mathrm{Max}(J(C), \frac{1}{n-1}\int_o^T \sum_{j\neq i} u_j(t)dt)\right]$$

Then we can define by induction on m the convex compact and non empty subsets $\hat{Q}_i^m$ of $L^\infty[o,T]$ $(m \geqslant 1)$:
$u_i \varepsilon \hat{Q}_i^m$ if and only if u_i maximizes

$$\underset{v_i \varepsilon \hat{Q}_i^{m-1}}{\mathrm{Min}} \int_o^T \Phi(u_i)(t) F[\Phi(u_i)(t)+(n-1)\Phi(v_i)(t)] - K[\Phi(u_i)(t)] \, dt$$

under the constraints : $u_i \, \varepsilon \, \hat{Q}_i^{m-1}$

$$\Phi(u_i)(t) \varepsilon \left[\Phi(\alpha_{\hat{Q}_i^{m-1}})(t), \Phi(\beta_{\hat{Q}_i^{m-1}})(t)\right] \quad \forall \, t \, \varepsilon [o,T]$$

$$\int_o^T u_i(t)dt \varepsilon \left[I(\hat{Q}_i^{m-1}), \quad J(\hat{Q}_i^{m-1}) \right]$$

Since, by assumption these compact sets have diameters in $L^\infty[o,T]$ decreasing to zero, there exists a single $\bar{u} \varepsilon \bigcap_{k \geqslant o} \hat{Q}_i^m$ $\forall$i,i=1,...,n.

Hence, each oligopolist has $\bar{u}$ as a solution. Thus, when they only exchange the R_i^m as an information, the players will choose, if they play rationally, the strategy $\bar{u}$ and will obtain a level of profit and a rate of change of their activity level in the intervals that they judge reasonable a priori.

REFERENCES

[1] J.C. HARSANYI, Approaches to the bargaining problem before and after the theory of games : a critical discussion of Zeuthen's, Hick's and Nash's theories. Econometrica Vol. 24 (1956).

[2] R.D. LUCE and H. RAIFFA, Games and Decisions. Wiley (1957).

[3] J. NASH, The bargaining problem. Econometrica Vol. 18 (1950).

[4] T.C. SCHELLING, The strategy of conflict. Harvard University Press 4th edition (1970).

MANY PLAYER DIFFERENTIAL GAMES WITH COALITIONS

W. E. Schmitendorf

Department of Mechanical Engineering and Astronautical Sciences, Northwestern University, Evanston, Illinois USA

ABSTRACT. Results are discussed for differential games with k players where a subset of the players form a coalition. Two solution concepts are presented: coalitive Pareto optimal and Nash coalitive optimal. Through an example it is shown that the concept of coalitive Pareto optimality can be useful to a coalition of all the players in determining which Pareto optimal solutions are rational.

1. INTRODUCTION

In a k player nonzero sum differential game there is no unique definition of optimal and many moods of play are possible. For games in which the players do not cooperate, Nash equilibrium solutions or minimax solutions are usually sought. Yet there are situations where all the players can improve their performance if they are willing to cooperate. Most interest has been focused on coalitions involving all k players where the players use solutions with the Pareto optimal property. This is a k player coalition. However, it is also possible for a coalition to be formed with less than k players.

We begin by reviewing the concepts of a minimax solution and a Pareto optimal solution and then proceeding to solution concepts for coalitions of more than one, but less than k, players.

J. D. Grote (ed.), The Theory and Application of Differential Games. 133-144.

2. PROBLEM FORMULATION

Consider the differential system

$$\dot{x}(t) = f(x(t),u_1(t),\cdots,u_k(t)) \tag{1}$$

where the state $x(t) \in R^n$, Player i's control variable $u_i(t) \in R^{m_i}$, $i = 1,2,\cdots,k$, and $f:R^n \times R^{m_1} \times \cdots \times R^{m_k} \to R^n$ is C^1. The game starts at specified initial point (x_o,t_o) and terminates at a prescribed time t_f.

Let $\mathfrak{h}_i$, $i=1,\cdots,k$, denote the set of all piecewise C^1 functions from $[t_o,t_f]$ into R^{m_i}. A control u_i is admissible if, and only if, $u_i \in \mathfrak{h}_i$ and $u_i(t) \in U_i$ where U_i is a given subset of R^{m_i}. $\mathfrak{m}_i$ will be the set of all admissible controls for Player i.

We assume that for each admissible k-tuple $u \triangleq (u_1,\cdots,u_k)$ there exists a unique solution of (1) on $[t_o,t_f]$ passing through (x_o,t_o).

Player i's cost, which he wants to minimize, is

$$J_i(u) = g_i(x(t_f)) + \int_{t_o}^{t_f} L_i(x(t),u_1(t),\cdots,u_k(t))dt \tag{2}$$

where $L_i:R^n \times R^{m_1} \times \cdots \times R^{m_k} \to R$ is C^1.

3. ONE PLAYER COALITIONS

Here the players do not cooperate and the minimax definition of optimality is used. Player i assumes all the other players are playing against him and choosing their controls to maximize his cost. Let u_{K-I} denote the control (k-1)-tuple $(u_1,\cdots,u_{i-1}, u_{i+1},\cdots,u_k)$ and $\mathfrak{m}_{K-I}$ the set of admissible u_{K-I}.

<u>Definition 1</u>. The control $u_i^* \in \mathfrak{m}_i$ is a <u>minimax control</u> for Player i if, and only if, for all $u_i \in \mathfrak{m}_i$

$$\sup_{u_{K-I} \in \mathfrak{m}_{K-I}} J_i(u_i^*,u_{K-I}) \le \sup_{u_{K-I} \in \mathfrak{m}_{K-I}} J_i(u_i,u_{K-I})$$

With the assumption that all admissible k-tuples u are playable, if there is a pure strategy saddle point solution $(\tilde{u}_i,\tilde{u}_{K-I})$

to the two player zero sum game with cost $J_i(u_i, u_{K-I})$ then u_i is a minimax control for Player i. However, in the more general situation where not all u are playable and where there are more general terminal conditions, this is no longer true [1]. The control $\tilde{u}_i$ may not be a minimax control. Also, in some games a minimax control may exist for Player i even though there is no pure strategy saddle point solution for the game with cost $J_i(u_i, u_{k-I})$. Sufficient conditions for a minimax control without the playability assumption of Sec. 2 and with general terminal conditions are presented in [2].

4. k-PLAYER COALITIONS

In some situations all k players may agree to cooperate. Then each player may be able to achieve a lower cost than he would if all the players played their minimax controls or their Nash equilibrium controls. For this case, an attractive solution concept is Pareto optimality.

Definition 2. An admissible k-tuple u^* is Pareto optimal if, and only if, for every admissible k-tuple u

$$J_i(u) - J_i(u^*) = 0 \qquad \forall \qquad i \in \{1,2,\cdots,k\}$$

or there exists at least one $i \in \{1,2,\cdots,k\}$ such that $J_i(u) - J_i(u^*) > 0$.

A simple derivation of necessary conditions for Pareto optimality is presented in [3]. The derivation is based on the result that if u^* is Pareto optimal, then, for all $\ell \in \{1,2,\cdots,k\}$, u^* is a solution to the optimal control problem of minimizing $J_\ell(u)$ subject to (1) and the constraints

$$J_i(u) \le J_i(u^*), \; i \ne \ell, \quad i = 1,2,\cdots,k$$

Applying the results of [4] to this optimal control problem leads to

Theorem 1. If u^* is Pareto optimal then there exist an $\alpha \in R^k$, $\alpha \ge 0$, a continuous $p: [t_o, t_f] \to R^n$ with $(p(t), \alpha) \ne 0$ on $[t_o, t_f]$, such that p is a solution of

i) $$\dot{p}^T(t) = -\frac{\partial H(x^*(t), u^*(t), p(t), \alpha)}{\partial x}$$

$$p^T(t_f) = \sum_{i=1}^{k} \alpha_i \frac{\partial g_i(x^*(t_f))}{\partial x}$$

and

ii) $$H(x^*(t), u^*(t), p(t), \alpha) \le H(x^*(t), u, p(t), \alpha) \quad \text{for all } u \in U$$

where $H(x,u,p,\alpha) = \sum_{i=1}^{k} \alpha_i L_i(x,u) + p^T f(x,u)$ and $U = \prod_{i=1}^{k} U_i$.

These necessary conditions are the same as those for the optimal control problem of minimizing

$$\hat{J}(u) \triangleq \sum_{i=1}^{k} \alpha_i J_i(u)$$

subject to (1). Sufficient conditions can also be stated in terms of this scalarization [5].

Theorem 2. If there exists an $\alpha \in R^k$, $\alpha > 0$, such that $\hat{J}(u^*) \le \hat{J}(u)$ for all admissible u then u^* is Pareto optimal. If there exists an $\alpha \in R^k$, $\alpha \ge 0$, $\alpha \ne 0$, such that $\hat{J}(u^*) < \hat{J}(u)$ for all admissible $u \ne u^*$ then u^* is Pareto optimal.

As α varies, an infinite number of Pareto optimal solutions are obtained. Before a particular one can be agreed upon, further rules of negotiation must be specified. Theorems 1 and 2 are valid without the playability assumption of Sec. 2 and also apply to problems with more general terminal conditions, see [3] and [5].

5. COALITIONS WITH LESS THAN k PLAYERS

Coalitions need not be restricted to those involving all k players. A subset of the players may form a coalition. For a criterion of optimality, we shall use coalitive Pareto optimality, first introduced for differential games in [6].

Let Players $1,\cdots,s$ form a coalition. Denote the control s-tuple $(u_1,\cdots,u_s)$ by u_S and the (k-s)-tuple $(u_{s+1},\cdots,u_k)$ by u_{K-S}. Let $\mathfrak{M}_S$ be the set of admissible u_S, i.e. $u_S \in \mathfrak{M}_S$ iff $u_j \in \mathfrak{M}_j$, $j=1,2,\cdots,s$. A similar definition applies to $\mathfrak{M}_{K-S}$.

Definition 3. The control s-tuple $u_S^* = (u_1^*,\cdots,u_s^*)$ is coalitive Pareto-optimal (C.P.O.) at (x_o,t_o) if, and only if, for all admissible u_S either

$$\Delta J_i \triangleq \sup_{u_{K-S} \in \mathfrak{M}_{K-S}} J_i(u_S,u_{K-S}) - \sup_{u_{K-S} \in \mathfrak{M}_{K-S}} J_i(u_S^*,u_{K-S}) = 0 \quad (3)$$

for all $i \in \{1,\cdots,s\}$ or there is at least one $i \in \{1,\cdots,s\}$ such that $\Delta J_i > 0$.

For a coalition of all k players, the definition becomes that of Pareto optimality whereas if the coalition consists of one player, the definition becomes the definition of a minimax control for the player in the coalition. In this definition each player in the coalition assumes all the players not in the coalition are playing strictly against him and he wants to determine the cost he is guaranteed if he and the members of his coalition use u_S^*.

We now present a sufficient condition in terms of a corresponding two player zero sum game [7]. Consider the following two player zero sum differential game. The system is described by

$$\dot{x}_1 = f(x_1, u_S, u^1_{K-S}) \quad , \quad x_1(t_o) = x_o \tag{4}$$

$$\dot{x}_2 = f(x_2, u_S, u^2_{K-S}) \quad , \quad x_2(t_o) = x_o$$

$$\vdots \qquad\qquad \vdots$$

$$\dot{x}_s = f(x_s, u_S, u^s_{K-S}) \quad , \quad x_s(t_o) = x_o$$

$$u_S \in \mathcal{M}_S \ , \quad u^i_{K-S} \in \mathcal{M}_{K-S} \ , \quad i=1,\cdots,s \tag{5}$$

$$J(u_S, u^1_{K-S}, \cdots, u^s_{K-S}) = \sum_{i=1}^{s} \alpha_i J_i(u_S, u^i_{K-S}) \tag{6}$$

where $J_i(u_S, u^i_{K-S}) = g_i(x_i(t_f)) + \int_{t_o}^{t_f} L_i(x_i(t), u_S(t), u^i_{K-S}(t))dt$

In this two player zero sum game, one player controls the s-tuple u_S while the other chooses the s(k-s)-tuple $(u^1_{K-S}, \cdots, u^s_{K-S})$ which will be denoted by u_M.

<u>Theorem 3</u>. Let (u_S^*, u_M^*) be an open loop saddle point solution for the two-player zero sum game defined by (4)-(6) with $\alpha_i > 0$, $\sum_{i=1}^{s} \alpha_i = 1$. Then u_S is a C.P.O. solution at (x_o, t_o) for the coalition consisting of Players $1, \cdots, s$.

Sufficient conditions for open loop saddle point solutions can be found in [7]-[9]. These results can be used for the game defined by (4)-(6) and, hence, for coalitive Pareto optimality. The assumption that for each admissible k-tuple $(u_1, \cdots, u_k)$ there exists a unique solution through (x_o, t_o) on $[t_o, t_f]$ is a strong one. It is satisfied if the system (1) is linear. Additional conditions assuring that this assumption is met are contained in Theorem 1.2.1 of [10]. For a particular problem one may be able to verify Assumption 1 without resorting to a general theorem.

It is also possible to derive sufficient conditions for C.P.O. solutions without this assumption. This result is contained in the following corollary [7].

Corollary 1. If u_S^* is an open loop minimax control for the game defined by (4)-(6) with $\alpha_i > 0$ and $\sum_{i=1}^{s}\alpha_i = 1$, then u_S^* is a C.P.O. solution at (x_o,t_o) for the coalition consisting of Players $1,\cdots,s$.

In [2] there is sufficient condition for a minimax control. It does not require that Assumption 1 be satisfied. Instead a weaker playability condition must be satisfied. Theorem 3 and Corollary 1 are sufficient conditions for coalitive Pareto optimality. Candidates for a solution with the C.P.O. property can be obtained by applying the necessary conditions for a saddle point solution to the game given by (4)-(6). These candidates can be checked by using either Theorem 3 or Corollary 1. We emphasize, however, that it has not been established that it is necessary for a C.P.O. solution to be a solution of the game of the form (4)-(6). In fact, it is conjectured that this will be true only under very stringent assumptions. The concept of coalitive Pareto optimality can be used by a player in determining with which of the other players he should attempt to form a coalition.

6. INDIVIDUAL OPTIMALITY, COLLECTIVE OPTIMALITY AND THE CORE

For the k-player coalition where all the players agree to cooperate and use a Pareto optimal solution there is usually an infinite number of solutions. Before one of these solutions can be agreed upon, further rules of negotiation must be determined.

One approach to the negotiation problem is through the concepts of individual optimality and collective optimality [11]. A Pareto optimal solution is <u>individually optimal</u> if no one player can unilaterally deviate and assure himself a cost smaller than his cost with the Pareto optimal solution. Let $\bar{u} = (\bar{u}_1,\bar{u}_2,\cdots,\bar{u}_k)$ be a Pareto optimal solution and $\bar{J}_i = J_i(\bar{u}_1,\cdots,\bar{u}_k) = J_i(\bar{u}_i,\bar{u}_{K-I})$. Then $\bar{u}$ is individually optimal if, and only if, for all $i\epsilon\{1,2,\cdots,k\}$

$$\sup_{u_{K-I}\,\epsilon\,\mathcal{M}_{K-I}} J_i(u_i,u_{K-I}) \geq \bar{J}_i \qquad \forall \qquad u_i\,\epsilon\,\mathcal{M}_i$$

The players would not agree to use a Pareto optimal solution which is not individually optimal since at least one of them could use a different control and guarantee a smaller cost for himself no matter what the other players do. Each player must do better with the Pareto optimal solution than his minimax value if it is to be expected that each player will adhere to the Pareto optimal solution. The results described in Sec. 3 can be applied to obtain the minimax values for each player and these values used to determine if a Pareto optimal solution is individually optimal.

The concept of collectively optimal is similarly defined. A Pareto optimal solution $\bar{u}$ is optimal with respect to coalition S if for all $u_S \in \mathfrak{M}_S$

$$\sup_{u_{K-S} \in \mathfrak{M}_{K-S}} J_i(u_S, u_{K-S}) = \bar{J}_i$$

for all $i \in S$ or, for at least one $i \in S$,

$$\sup_{u_{K-S} \in \mathfrak{M}_{K-S}} J_i(u_S, u_{K-S}) > \bar{J}_i$$

If this is true for all coalitions S, then $\bar{u}$ is collectively optimal.

If the players agree to cooperate and are rational, they will not agree to a Pareto optimal solution unless it is collectively optimal. A rational criterion for the players to apply in choosing a particular Pareto optimal solution is that it must be collectively optimal. If a Pareto optimal solution is collectively optimal then it is secure against deviations by any individual or by any coalition. To determine if a Pareto optimal solution is collectively optimal, the C.P.O. solutions must be found for all coalitions.

In classical game theory, the concepts of individual optimality and collective optimality are often called individual rationality and collective rationality, respectively [12]. Also, if a Pareto optimal solution $\bar{u}$ is collectively optimal then the vector $[J_1(\bar{u}), \cdots, J_k(\bar{u})]^T$ belongs to the core.

We now illustrate these concepts with an example.

7. A SIMPLE EXAMPLE

We consider a three player game with

$$\dot{x}(t) = u_1(t) + u_2(t) + u_3(t), \quad x(0) = 0$$

$$|u_i(t)| \le 1, \quad i=1,2,3$$

$$J_1(u) = -x(1) + \frac{1}{2}\int_0^1 u_1^2(t)dt, \quad J_2(u) = x(1) + \frac{1}{2}\int_0^1 u_2^2(t)dt$$

$$J_3 = \frac{1}{2}x(1) + \frac{1}{2}\int_0^1 \left[u_3^2(t) - u_1^2(t) - u_2^2(t)\right] dt$$

The minimax values for the players are

$$J_{1m} = 1.5, \quad J_{2m} = 1.5, \quad J_{3m} = 0.125$$

For the coalition consisting of Players 1 and 2, the C.P.O. costs are shown below in Fig. 1

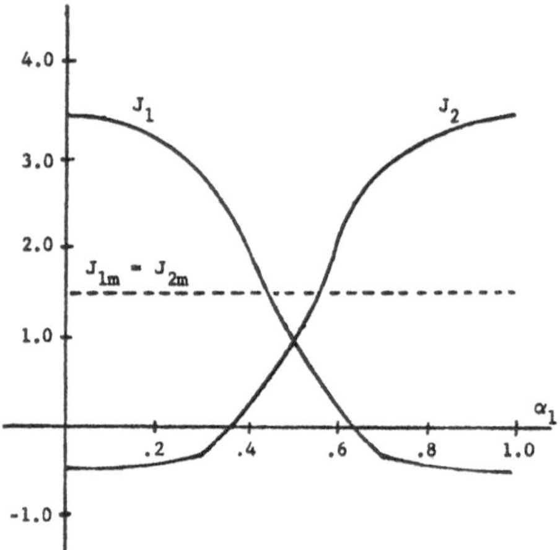

FIG. 1. C.P.O. COSTS FOR THE COALITION OF PLAYERS 1 & 2

Players 1 and 2 would agree to cooperate only if $\alpha_1 \in [.45,.55]$[†]. For $\alpha_1 < .45$, Player 1 can do better with his minimax control and he would not agree to cooperate with Player 2. For $\alpha_1 > .55$, Player 2 does better by playing minimax.

The C.P.O. costs for the coalition of Players 1 and 3 and the coalition of Players 2 and 3 are shown in Fig. 2 and 3 respectively.

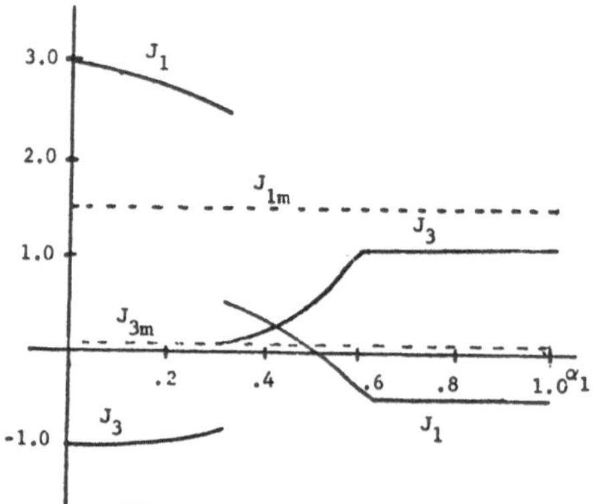

FIG. 2. C.P.O. COSTS FOR THE COALITION OF PLAYERS 1 & 3

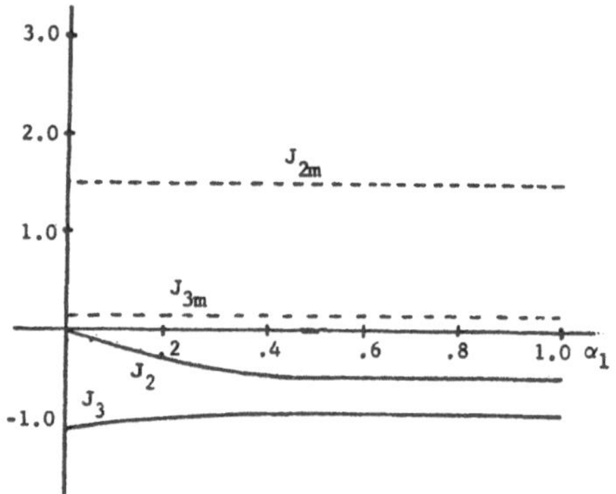

FIG. 3. C.P.O. COSTS FOR THE COALITION OF PLAYERS 2 & 3

† This interval is approximate.

Fig. 2 indicates that Players 1 and 3 will not cooperate unless $\alpha_1 = 1/3$ since for any other value of α_1, at least one of the players can do better by playing his minimax control. At this value of α_1, there are two C.P.O. solutions, but for only one of these does neither player do worse than his minimax value. Even for this solution only Player 1 receives a strictly lower cost and Player 3 may not agree to cooperate. In Fig. 2 there is a discontinuity in J_1 and J_3 at $\alpha_1 = 1/3$. However, the sum $\alpha_1 J_1 + \alpha_2 J_3$ is continuous. For the coalition of Players 2 and 3 they both do better for all values of α_1.

Pareto optimal costs are presented in Tables 1 and 2 for two values of α_3.

TABLE 1

$\alpha_3 = .25$ $\alpha_1+\alpha_2 = .75$

α_1	J_1	J_2	J_3
0	3.5	-2.5	-2.0
.1	3.5	-2.5	-2.0
.2	3.5	-2.5	-2.0
.3	3.5	-2.5	-2.0
.4	1.675	-1.268	-1.136
.5	-1.875	2.5	.5
.6	-2.5	3.42	1.03
.7	-2.5	3.5	1.0

TABLE 2

$\alpha_3 = .05$ $\alpha_1+\alpha_2 = .95$

α_1	J_1	J_2	J_3
.1	3.468	-2.5	-1.95
.2	3.321	-2.48	-1.75
.4	1.975	-1.79	- .62
.47	.864	- .864	- .194
.48	.375	- .371	- .141
.5	-1.12	1.12	1.056
.6	-2.08	2.44	1.34
.7	-2.5	3.47	2.47

None of the Pareto optimal solutions in Table 1 are individually optimal. For $\alpha_1 \leq .4$, $J_1 > J_{1m}$ and Player 1 is not satisfied while for $\alpha_1 \geq .5$, $J_2 > J_{2m}$, and $J_3 > J_{3m}$ and Players 2 and 3 are unsatisfied. In Table 2, the solutions corresponding to $\alpha_1 = .47$ and $\alpha_1 = .48$ are individually optimal, but only the former is collectively optimal. For $\alpha_1 = .48$, Players 2 and 3 can form a coalition and guarantee themselves -.5 and -.875. Thus it would be irrational for them to agree to the Pareto optimal solution corresponding to $\alpha_1 = .48$. The solutions for the other values of α_1 shown are not individually optimal.

Among all the solutions shown in the two tables, only the one corresponding to $\alpha_1 = .47$ and $\alpha_3 = .05$ would be a rational choice for the players and it is the only one belonging to the core. Of course, an investigation of Pareto optimal solutions for other values of α_3 would probably lead to additional solutions that are collectively optimal. Nevertheless, this example does illustrate that the concept of collective rationality can greatly reduce the number of Pareto optimal solutions among which the players should negotiate. Also, it shows that the concept of C.P.O. solutions is helpful to a player in determining with whom he should negotiate.

8. NASH COALITIVE OPTIMAL

The final approach to the coalition problem which we shall discuss is the concept of Nash coalitive optimality [13]. Since this concept is most attractive when k-1 players form a coalition we shall limit our discussion to problems with three players where Players 1 and 2 cooperate.

Definition 4. The control 2-tuple (u_1^n, u_2^n) is Nash coalitive optimal (N.C.O.) at (x_o, t_o) if, and only if, for every admissible 2-tuple (u_1, u_2)

$$\Delta J_i^n \triangleq J_i(u_1, u_2, u_3^n) - J_i(u_1^n, u_2^n, u_3^n) = 0 \quad \forall \quad i \in \{1,2\}$$

or for at least one $i \in \{1,2\}$, $\Delta J_i^n > 0$ where u_3^n is determined from

$$J_3(u_1^n, u_2^n, u_3^n) \leq J_3(u_1^n, u_2^n, u_3) \quad \forall \quad \text{admissible } u_3.$$

With this concept of optimality, Players 1 and 2 are assuming Player 3 plays only to satisfy his own self interest and is not vindictive and they want to determine the "best" controls to use under this assumption. The attractiveness of this scheme is that once Players 1 and 2 announce they will play (u_1^n, u_2^n), Player 3 cannot unilaterally deviate from u_3^n without penalizing himself.

Necessary conditions for N.C.O. controls can be obtained from the following theorem [13] in conjunction with the results for isoperimetric problems in [4]. The approach is similar to that of Sec. 4 for Pareto optimality.

Theorem 4. If (u_1^n, u_2^n) is N.C.O. then (u_1^n, u_2^n) is a solution to the problem of minimizing $J_1(u_1, u_2, u_3^n)$ subject to (1) and

$$J_2(u_1, u_2, u_3^n) \leq J_2(u_1^n, u_2^n, u_3^n)$$

Sufficient conditions for N.C.O. controls can be obtained through a related two player Nash equilibrium game [13].

Theorem 5. If there exists a $\beta_1 > 0$, $\beta_2 > 0$ such that for all admissible (u_1, u_2, u_3)

$$J^n(u_1^n, u_2^n, u_3^n) \leq J^n(u_1, u_2, u_3^n) \text{ and } J_3(u_1^n, u_2^n, u_3^n) \leq J_3(u_1^n, u_2^n, u_3)$$

where

$$J^n(u_1, u_2, u_3) = \beta_1 J_1(u_1, u_2, u_3) + \beta_2 J_2(u_1, u_2, u_3)$$

then (u_1^n, u_2^n) is N.C.O. at (x_o, t_o).

Sufficient conditions for Nash equilibrium solutions of two player differential games are available in [13] and [14].

9. ANOTHER EXAMPLE

$$J_1 = \frac{1}{2} x^2(1) + \int_o^1 (u_1^2 - 2u_3^2)dt$$

$$J_2 = x^2(x) + \frac{1}{2} \int_o^1 (u_2^2 - 4u_3^2)dt \ , \quad J_3 = 20x(1) + \frac{1}{2} \int_o^1 u_3^2 \ dt$$

$$\dot{x} = u_1 + u_2 + u_3 \ , \quad x(o) = 10$$

For this case an N.C.O. solution for a coalition of Players 1 and 2 is $u_1 = u_2 = 4.28$. If Players 1 and 2 announce this choice of controls to Player 3 and if Player 3 is solely interested in minimizing his own cost, he will play $u_3 = -20$ which results in $J_1 \simeq -390$ and $J_2 \simeq -790$. However, if Player 3 is vindictive, he can use a control against the above u_1 and u_2 which results in $J_1 \simeq +350$ and $J_2 \simeq +698$. Of course, Player 3 suffers a large degradation in his own performance by this retaliatory move. If Players 1 and 2 believe Player 3 is vindictive, they should use a C.P.O. solution. For example, they can guarantee a cost of no more than 11.23 and 11.84, respectively, with a particular C.P.O. solution. Thus, Players 1 and 2 must decide on the nature of their adversary before choosing their controls.

REFERENCES

1. G. Leitmann and S. Rocklin, The Effect of Playability in Differential Games on the Relation Between Best Guaranteed Costs and Saddlepoint Values, J. of Optimization Theory and Appl., to appear.
2. W. Schmitendorf, Differential Games Without Pure Strategy Saddle Point Solutions, J. of Optimization Theory and Appl., to appear.
3. W. Schmitendorf and G. Leitmann, A Simple Derivation of Necessary Conditions for Pareto Optimality, IEEE Trans. on Automatic Control, to appear.
4. W. Schmitendorf, Pontryagin's Principle for Problems with Isoperimetric Constraints and for Problems with Inequality Terminal Constraints, J. of Optimization Theory and Appl., to appear.
5. G. Leitmann and W. Schmitendorf, Some Sufficiency Conditions for Pareto-Optimal Control, J. of Dynamic Systems, Measurement, and Control, Vol. 95, No. 4, 1973.

6. A. Haurie, On Pareto Optimal Decisions for a Coalition of a Subset of Players, IEEE Trans. Autom. Contr., Vol. AC-18, No. 2, 1973.
7. W. Schmitendorf and G. Moriarty, Coalitions in a Differential Game, J. of Optimization Theory and Appl., to appear.
8. Z. V. Rekasius, On Closed-Loop and Open-Loop Solutions of Differential Games, Paper presented at the Purdue Centennial Symposium on Information Processing, Lafayette, Indiana, 1969.
9. W. Schmitendorf, Existence of Optimal Open-Loop Strategies for a Class of Differential Games, J. of Optimization Theory and Appl., Vol. 5, No. 5, 1970.
10. A. Friedman, Differential Games, John Wiley and Sons, N.Y., 1971.
11. A. Haurie and M. C. Delfour, Individual and Collective Rationality in a Dynamic Pareto Equilibrium, J. of Optimization Theory and Appl., Vol. 13, No. 3, 1974.
12. R. Aumann, A Survey of Cooperative Games Without Side Payments, in Essays in Mathematical Economics, M. Shubik, ed., Princeton, 1969.
13. G. Moriarty, Ph.D. Thesis, Illinois Institute of Technology, Chicago, Illinois, 1974.
14. M. Foley and W. Schmitendorf, On a Class of Nonzero Sum Linear Quadratic Games, J. of Optimization Theory and Appl., Vol. 7, No. 5, 1969.

EXTENSION OF 2 PERSON ZERO SUM GAME

H. MOULIN

Université Paris 9. Mathématiques de la Décision. PARIS

Let X and Y be the pure strategy sets of 2 players Xavier and Yves. Then every payoff function $g : X \times Y \longrightarrow \mathbb{R}$ defin es a 2 person zero sum game where Xavier maximize and Yves minimize. An extension of the games with pure strategy set X and Y is a "way of playing" the game, which associates to every payoff function g a "value", in the duality interval

$$\left[\sup_{x \in X} \inf_{y \in Y} g(x,y) , \inf_{y \in Y} \sup_{x \in X} g(x,y)\right]$$

The classical example of extension is the mixed one. It will be shown it is the prototype of the "extensions with out exchange of information". (part 2). We then study the general case (part 3) and typical example of Iterated games.

1. Extension. Definition and examples.

The pure strategy sets X and Y are fixed in the whole paper. Let $\mathcal{A}(Z)$ be the Banach space of bounded functions defined on Z, with supremum norm. We denote by $\theta_Z \in \mathcal{A}(Z)$ the constant function with

J. D. Grote (ed.), The Theory and Application of Differential Games. 145-156.

value 1.

Definition (1-1)

An extension of the games with pure strategy set X and Y is a 5-tuple $(\mathcal{X}, \mathcal{Y}, i, j, \pi)$ where :

i) $i : X \longrightarrow \mathcal{X}$ and $j : Y \longrightarrow \mathcal{Y}$ are 2 one to one mappings.

ii) π is a linear operator, which respects θ

$$(1\text{-}1) \quad \pi : \mathcal{A}(X \times Y) \longrightarrow \mathcal{A}(\mathcal{X} \times \mathcal{Y}) \ ; \ \pi\theta_{X \times Y} = \theta_{\mathcal{X} \times \mathcal{Y}}$$

such that

$$(1\text{-}2) \quad \forall g \in \mathcal{A}(X \times Y) \ [g \geqslant 0 \Longrightarrow \pi g \geqslant 0]$$

and

$$(1\text{-}3) \quad \forall (x,y) \in X \times Y \ \forall g \in \mathcal{A}(X \times Y) \ \pi g(i(x), j(y)) = g(x,y)$$

and finally :

$$(1\text{-}4) \quad \forall g \in \mathcal{A}(X \times Y) \ \sup_X \inf_Y g \leqslant \sup_{\mathcal{X}} \inf_{\mathcal{Y}} \pi g \leqslant \inf_{\mathcal{Y}} \sup_{\mathcal{X}} \pi g \leqslant \inf_Y \sup_X g$$

We call $\mathcal{X}$ and $\mathcal{Y}$ the extended strateg y sets and the one-to-one mappings i,j with assumption (1-3) identify the pure strategy sets X and Y with subsets of $\mathcal{X}$ and $\mathcal{Y}$. Assumptions (1-1) and (1-2) are quite natural in view of assumption (1-4) which ensures the duality interval is "non increasing" form "initial game" g on $X \times Y$ to extended game πg on $\mathcal{X} \times \mathcal{Y}$.

Example (1-1)

The main example is the mixed extension of the games with pure strategy sets X and Y. We denote by $\mathcal{A}'_1(Z)$ the simplex of the dual of the space $\mathcal{A}(Z)$. Then there is a natural one-to-one mapping I_Z :

(1-5) $$I_Z : Z \longrightarrow \mathcal{A}'_1(Z) \qquad I_Z(z) = \delta_z .$$

where the "Dirac measure" δ_z is defined by :

(1-6) $$\forall h \in \mathcal{A}(Z) \qquad \langle \delta_z, h \rangle = h(z)$$

The <u>mixed extension</u> is $(\mathcal{A}'_1(X), \mathcal{A}'_1(Y), I_X, I_Y, \pi)$ where π is defined by :

(1-7) $$\forall g \in \mathcal{A}(X \times Y) \quad \forall (\mu, \nu) \in \mathcal{A}'_1(X) \times \mathcal{A}'_1(Y) :$$

$$\pi g(\mu, \nu) = \langle \mu \otimes \nu, g \rangle .$$

It is well known that for every g belonging to $\mathcal{A}(X \times Y)$, πg has a value, the "mixed value" of g, which we note : $v_m(g)$.

<u>Example</u> (1-2)

Let $\mathfrak{X} = X^Y$ be the set of mappings $P : Y \longrightarrow X$ and $i : X \longrightarrow \mathfrak{X}$ the following one to one mapping :

(1-8) $$\forall x \in X \quad \forall y \in Y \qquad i(x)(y) = x$$

symetrically, let $\mathfrak{Y} = Y^X$ with elements $Q : X \longrightarrow Y$ and natural one ot one mapping j. We call $\mathfrak{X}$ and $\mathfrak{Y}$ the sets of decision rules of the 2 players. Fix now any y_o in Y. The <u>ergodic iterated extension</u> with start y_o is $(\mathfrak{X}, \mathfrak{Y}, i, j, \pi)$ where :

(1-9)
$$\forall g \in \mathcal{A}(X \times Y) \forall (P,Q) \in \mathfrak{X} \times \mathfrak{Y} : \pi g\,(P,Q) = \lim_{N \to +\infty} \frac{1}{2N} \sum_{i=1}^{N} g(x_i, y_{i-1}) + g(x_i, y_i)$$

with

$$(1\text{-}10) \qquad x_i = \underbrace{PQP \dots QP}_{2i-1} y_o \quad ; \quad y_i = \underbrace{QPQ \dots QP}_{2i} y_o$$

(in relation (1-9) the existence of the limit is given by Hahn Banach theorem).

In the ergodic iterated way of playing the game Xavier plays firts and then the 2 players play one after the other indefinitely. The payoff function is the mean payment over time. It is known (see [1] or [2]) that for every g belonging to $\mathcal{A}(X \times Y)$ (with some topological assumption if X or Y is infinite) πg has a value, which we call the ergodic value of g and note $v_e(g)$. Furthermore, this value do not depends on y_o neither on the firts moving player.

We give now on other form (obtained by transposition) to the extension operator π:

<u>Proposition</u> (1-1)

<u>Let</u> $(X,Y,\mathfrak{X},\mathfrak{Y},i,j)$ <u>fixed verifying condition i) of the definition</u> (1-1). <u>Then to every operator</u> π <u>satisfying</u> (1-1) (1-2) (1-3) <u>and</u> (1-4) <u>we can associate by relation</u> (1-11) :

$$(1\text{-}11) \qquad \forall g \in \mathcal{A}(X \times Y) \quad \forall (\xi,\eta) \in \mathfrak{X} \times \mathfrak{Y} :$$
$$\pi g(\xi,\eta) = \langle \pi^*(\xi,\eta), g \rangle$$

<u>an unique mapping</u> $\pi^* : \mathfrak{X} \times \mathfrak{Y} \longrightarrow \mathcal{A}_1^{!}(X \times Y)$ <u>satisfying</u> :

$$(1\text{-}12) \qquad \forall (x,y) \in X \times Y ; \quad \forall (\xi,\eta) \in \mathfrak{X} \times \mathfrak{Y} \; \pi^*(i(x),\eta) \in \delta_x \otimes \mathcal{A}_1^{!}(Y)$$
$$\pi^*(\xi, i(y)) \in \mathcal{A}_1^{!}(X) \otimes \delta_y$$

<u>and this correspondence is bijective</u>.

2 Extensions without exchange of information.

We denote by $\mathcal{A}_X$ (resp. $\mathcal{A}_Y$) the subspace of $\mathcal{A}(X \times Y)$ of this functions which do not depend on y (resp. x).

<u>Definition</u> (2-1)

<u>The extension</u> $(\mathfrak{X}, \mathfrak{Y}, i, j, \pi)$ <u>is said to be without exchange of information (briefly w.e.i) if</u> :

(2-1) $\pi(\mathcal{A}_X) \subset \mathcal{A}_{\mathfrak{X}}$ and $\pi(\mathcal{A}_Y) \subset \mathcal{A}_{\mathfrak{Y}}$

If the initial payoff function g belong to $\mathcal{A}_X$ it is clear that Yves is a dummy in this game. If the extension is w.e.i, the same is then true for the extended payoff function πg.

We denote by P_X (resp. P_Y) the projection of $\mathcal{A}_1'(X \times Y)$ onto $\mathcal{A}_1'(X)$ (resp. $\mathcal{A}_1'(Y)$). Thus P_X (resp. P_Y) is the adjoint operator to the canonical one to one mapping from $\mathcal{A}(X)$ into $\mathcal{A}(X \times Y)$ (resp. from $\mathcal{A}(Y)$ into $\mathcal{A}(X \times Y)$).

<u>Proposition</u> (2-1)

<u>An extension</u> $(\mathfrak{X}, \mathfrak{Y}, i, j, \pi)$ <u>is w.e.i. if and only if the mapping</u> π^* <u>verify</u> :

(2-2) $P_X[\pi^*(\xi, \eta)]$ do not depend on η

(2-3) $P_Y[\pi^*(\xi, \eta)]$ do not depend on ξ .

Then if the extension is w.e.i we can defin e the two mappings π^*_X and π^*_Y :

(2-4) $\pi^*_X(\xi) = P_X[\pi^*(\xi,\eta)] \in \mathcal{A}'_1(X)$

(2-5) $\pi^*_Y(\eta) = P_Y[\pi^*(\xi,\eta)] \in \mathcal{A}'_1(Y)$.

The typical example of w.e.i extension is the mixed extension (indeed $\pi^*(\mu,\nu) = \mu \otimes \nu$). The following theorem shows that it is, in a certain sense, the only one :

Theorem (2-1)

It $(\mathfrak{X},\mathfrak{Y},i,j,\pi)$ is a w.e.i extension, then for every initial payoff function g :

(2-6) $\sup_{\mathfrak{X}} \inf_{\mathfrak{Y}} \pi g(\xi,\eta) \leq v_m(g) \leq \inf_{\mathfrak{Y}} \sup_{\mathfrak{X}} \pi g(\xi,\eta)$

Thus, if the extended payoff function has a value, this value is the mixed one, and if this value exists for every g, then this extension is "equivalent" to the mixed extension. Generally this fails, but we have a bound to the duality gap :

Theorem (2-2)

If $(\mathfrak{X},\mathfrak{Y},i,j,\pi)$ is a w.e.i. extension, such that π^*_X is onto $\mathcal{A}'_1(X)$ and π^*_Y is onto $\mathcal{A}'_1(Y)$, then for every g belonging to $\mathcal{A}(X \times Y)$ we have :

(2-7) $$[\inf_{\mathfrak{Y}} \sup_{\mathfrak{X}} \pi g(\xi,\eta) - \sup_{\mathfrak{X}} \inf_{\mathfrak{Y}} \pi g(\xi,\eta)] \leqslant 2\|g\| \sup_{\xi,\eta} \|\pi^*(\xi,\eta) - \pi^*_X(\xi) \otimes \pi^*_Y(\eta)\|_*$$

where $\| \ \|$ is the supremum norm on $\mathcal{A}(Z)$ and $\| \ \|_*$ the dual norm.

Example (2.1)

Suppose X and Y are finite with cardinality n and p . Define then the following sequential extension where $\mathcal{X} = X^{\mathbb{N}}$ is the space of sequence of elements of X , with canonical injection i :

$$(2.8) \quad i : X \longrightarrow \mathcal{X} \qquad i(x) = (x,x,\dots,x,\dots)$$

symetrically $\mathcal{Y} = Y^{\mathbb{N}}$ and elements of $\mathcal{X}$ and $\mathcal{Y}$ are denoted by $\tilde{x} = (x_i)_{i \in \mathbb{N}}$ and $\tilde{y} = (y_j)_{j \in \mathbb{N}}$. Define now :

$$(2.9) \quad h_{ij} = \frac{1}{np} \left(\frac{n-1}{n}\right)^{i-1} \left(\frac{p-1}{p}\right)^{j-1} \; ; \; h_{ij} \geqslant 0 \; ; \; \sum_{i,j=1}^{+\infty} h_{ij} = 1$$

the operator π of the sequential extension is defined by :

$$(2.10) \quad g \in \mathcal{B}(X \times Y), (\tilde{x},\tilde{y}) \in \mathcal{X} \times \mathcal{Y} \; : \; \pi g(\tilde{x},\tilde{y}) = \sum_{i,j=1}^{+\infty} h_{ij} \, g(x_i, y_j)$$

Clearly this defines a w.e.i. extension, and the mappings π^*_X and π^*_Y are :

$$(2.11) \quad \pi^*_X(\tilde{x}) = \frac{1}{n} \sum_{i=1}^{+\infty} \left(\frac{n-1}{n}\right)^{i-1} \delta_{x_i} \; ; \; \pi^*_Y(\tilde{y}) = \frac{1}{p} \sum_{j=1}^{+\infty} \left(\frac{p-1}{p}\right)^{j-1} \delta_{y_j}$$

It can be shown this mappings are "onto" and then by theorems (2.1) and (2.2) that this extension is equivalent to the mixed one in the sense that for every g , πg has a value and this value is $v_m(g)$.

This sequential extension is more available in a deterministic point of view : a winning strategy for Xavier is a sequence of pure strategies, and the know ledge of this sequence is helpless for Yves. Furthermore, if we replace πg by $\pi_N g$:

$$(2.12) \quad \pi_N \, g(\tilde{x},\tilde{y}) = \sum_{i,j=1}^{N} h_{ij} \, g(x_i, y_j)$$

We obtain a finite and deterministic way of playing the game which approach the mixed way with error bound $\varepsilon(N)$:

(2.13) $$\varepsilon(N) = \|g\| \cdot \left[\frac{1}{n}\left(\frac{n-1}{n}\right)^N + \frac{1}{p}\left(\frac{p-1}{p}\right)^N + \left(\frac{n-1}{n}\right)^N \left(\frac{p-1}{p}\right)^N\right]$$

③ Decomposition of an extension

We will give a large class of extensions which can be decomposed as the "product" of an extension w.e.i. and an extension "pure exchange of information" (Briefly p.e.i).

3.1. Extension pure exchange of information

Let $\mathcal{X}$ be a subset of X^Y, the set of mappings from Y into X and $\mathcal{Y}$ a subset of Y^X. We shall say $(\mathcal{X}, \mathcal{Y})$ is a feasible pair if $\mathcal{X}$ and $\mathcal{Y}$ contain the constant mappings and if we have :

(3.1) $$\forall (\xi, \eta) \in \mathcal{X} \times \mathcal{Y} \quad \exists\, !(x,y) \in X \times Y : (x = \xi(y) \text{ and } y = \eta(X))$$

If $(\mathcal{X}, \mathcal{Y})$ is a feasible pair, they define the following extension :

(3.2) $$i : X \longrightarrow \mathcal{X} \qquad \forall x \in X \quad \forall y \in Y \quad i(x)(y) = x$$

(3.3) $$j : Y \longrightarrow \mathcal{Y} \qquad \forall y \in Y \quad \forall x \in X \; j(y)\,(x) = y$$

and, if $<\xi, \eta>$ denotes the unique fixed point whose existence is assumed in (3.1) :

(3.4) $$\forall g \in \mathcal{A}(X \times Y) \; \forall (\xi, \eta) \in \mathcal{X} \times \mathcal{Y} \quad : \pi g(\xi, \eta) = g(<\xi, \eta>)$$

We call this extension the p.e.i. extension associated to the feasible pair $(\mathcal{X}, \mathcal{Y})$.

Example (3.1)

Put : $\mathfrak{X} = X^Y$, $\mathcal{Y} = j(Y)$ = subset of constant mappings from X into Y . Clearly $(\mathfrak{X},\mathcal{Y})$ is a feasible pair and the p.e.i. extension associated is such that, for every $g \in \mathfrak{B}(X \times Y)$, πg has the value $\inf_Y \sup_X g(x,y)$.

In this extension we say that Xavier is fully informed about Yves's pure strategy.

Example (3.2)

The pair $(\mathfrak{X},\mathcal{Y}) = (X^Y, Y^X)$ is not feasible. Clearly full information for both players about other's player pure strategy is not feasible !

Proposition (3.1)

Let $E = (\mathfrak{X},\mathcal{Y},i,j,\pi)$ be an extension. Then, there exists a feasible pair $(\mathfrak{X}_o,\mathcal{Y}_o)$ and a bijective mapping which identify E with the p.e.i. extension associated to $(\mathfrak{X}_o,\mathcal{Y}_o)$, if and only if :

(3.5) $\pi^*(\mathfrak{X} \times \mathcal{Y}) = I_{X \times Y}(X \times Y) = \{\delta_x \otimes \delta_y / (x,y) \in X \times Y\}$

and

(3.6) $\forall(\xi,\eta) \in \mathfrak{X} \times \mathcal{Y} \exists!\ (x,y) \in X \times Y : \pi^*(\xi,\eta) = \pi^*(\xi,j(y)) =$

$\ldots = \pi^*(i(x),\eta)$

3.2. Blind strategies and d-entensions

Let $(\mathfrak{X},\mathcal{Y},i,j,\pi)$ be an extension.
A blind strategy ξ for Xavier is an extended strategy ξ $(\xi \in \mathfrak{X})$ such that $P_X(\pi^*(\xi,\eta))$ do not depends on η $(\eta \in \mathcal{Y})$.
We have seen ((2.2)) that if the extension is w.e.i., then every

strategy in $\mathfrak{X}$ is a blind strategy. (proposition (2.1)). We call this strategy "blind" because, if g belongs to $\mathfrak{R}_X$, then $\pi g(\xi, \eta)$ do not depends on η : playing ξ, Xavier don't use any information about Yves's strategy.
We denote by $B(\mathfrak{X})$ the set of blind strategies of Xavier. Symetrically $B(\mathcal{Y})$ is the set of blind strategies of Yves, that is the set of this η such that $P_Y(\pi^*(\xi, \eta))$ do not depends on ξ.
We define now the d-extension.

Definition(3.1)

An extension $(\mathfrak{X}, \mathcal{Y}, i, j, \pi)$ is a d-extension if for every pair $(\xi, \eta) \in \mathfrak{X} \times \mathcal{Y}$:

a) There exists an unique blind strategy $\bar{\xi} \in B(\mathfrak{X})$ such that

(3.7) $\forall g \in \mathfrak{R}(X \times Y) \quad \pi g(\xi, \eta) = \pi g(\bar{\xi}, \eta)$

b) There exists an unique blind strategy $\bar{\eta} \in B(\mathcal{Y})$ such that :

(3.8) $\forall g \in \mathfrak{R}(X \times Y) \quad \pi g(\xi, \eta) = \pi g(\xi, \bar{\eta})$

c) Furthermore we have :

(3.9) $\forall g \in \mathfrak{R}(X \times Y) \quad \pi g(\xi, \eta) = \pi g(\bar{\xi}, \bar{\eta})$

Then, for a d-extension, to every fixed pair (ξ, η) of strategy, Xavier can associate a (unique) blind strategy equivalent to ξ in the sense of (3.7), and similarly for Yves.

3.3. The decomposition theorem.

Definition (3.2)

Let $E_1 = (\mathfrak{X}^1, \mathcal{Y}^1, i^1, j^1, \pi^1)$ an extension. Then π^1 maps $\mathcal{A}(X \times Y)$ into $\mathcal{A}(\mathfrak{X}^1 \times \mathcal{Y}^1)$. Now, let $E_2 = (\mathfrak{X}^2, \mathcal{Y}^2, i^2, j^2, \pi^2)$ an extension with $\mathfrak{X}^1, \mathcal{Y}^1$ as pure strategy sets.
Then π^2 maps $\mathcal{A}(\mathfrak{X}^1 \times \mathcal{Y}^1)$ into $\mathcal{A}(\mathfrak{X}^2 \times \mathcal{Y}^2)$.

The product of this 2 extensions is the extension $(\mathfrak{X}^2, \mathcal{Y}^2, i^2 \circ i^1, j^2 \circ j^1, \pi^2 \circ \pi')$ with pure strategy sets X,Y.

Theorem (3.1)

Every d-extension is the product of a w.e.i. extension (as E_1) an a p.e.i extension (as E_2). Conversely every extension which is (modulo bijection) the product of an w.e.i. extension and a p.e.i. extension, is a d-extension.

Consider for example, the following "way of playing the game" given in the extensive form : Xavier plays first and choose $x_1 \in X$. Then Yves, choose $y_1 \in Y$. And so on, each player choosing alternatively an element in his pure strategy set, the number of steps being infinite. The information structure is supposed to be with perfect recall and described by two disjoints subsets of $\mathbb{N} \times \mathbb{N}$, C_X and C_Y

a) $(i,j) \in C_X$ means that before he plays x_i, Xavier nows y_j $(j \leq i-1)$

b) $(i,j) \in C_Y$ means that before he plays y_j Yves nows x_i $(i \leq j)$.

Finally the payoff is :

(3.10) $\sum_{i,j=1}^{+\infty} a_{ij}\, g(x_i, y_j)$ (where $a_{ij} \geq 0$ and $\sum_{i,j=1}^{+\infty} a_{ij} = 1$)

it is easy to verify, that this "way of playing" the initial game g defines an extension, even a d-extension, and then the product of a w.e.i extension and a p.e.i. extension . Indeed the w.e.i extension corresponds to the case where $C_X = C_Y = \emptyset$. And the p.e.i. extension corresponds to the feasible pair $(\mathfrak{X}, \mathcal{Y})$ with elements ξ, η :

$$(3.11) \quad \xi \in \mathfrak{X} : \xi : Y^{\mathbb{N}} \longrightarrow X^{\mathbb{N}} \quad \xi(\tilde{y}) = \tilde{x}$$

such that : x_i depends on y_j if and only if $(i,j) \in C_X$.

Similarly :

$$(3.12) \quad \eta \in \mathcal{Y} : \eta : X^{\mathbb{N}} \longrightarrow Y^{\mathbb{N}} \quad \eta(\tilde{x}) = \tilde{y}$$

such that : y_j depends on x_i if and only if $(i,j) \in C_Y$

REFERENCES

[1] MOULIN H. Jeux itérés - Cahiers de Mathematiques de la Décision 7305. Mai 1973.

[2] MOULIN H. Iterated Games - International Journal of Game Theory (à paraitre).

[3] MOULIN H. Prolongements de jeux et jeux itérés colloque d'analyse convexe Saint-Pierre-de-Chartresse - Springer - à paraître.

DIFFERENTIAL GAMES WITH PARTIAL DIFFERENTIAL EQUATIONS

Emilio O. Roxin

Department of Mathematics, University of Rhode Island, Kingston, R.I.02881, U.S.A., temporarily at the University of Würzburg and the T.H. Darmstadt, Germany

ABSTRACT. Examples of differential games are given, where the state equation is a partial differential equation. They can be solved explicitly and show clearly how the values of the control functions enter in the solution. This enables us to set up a method of solving these games, which should also be applied to more complicated differential games, complementing known results about existence of solutions from the general theory.

1. THE WAVE EQUATION IN ONE DIMENSION

We shall consider differential games with the wave equation in one dimension and boundary controls. We take this equation in the form

$$u_{tt}(t,x) = u_{xx}(t,x), \tag{1}$$

where the unknown function u(t,x) is assumed to be defined in $0 \leqq x \leqq 1$, $0 \leqq t$, where it is piecewise C^2.

We take the initial conditions

$$u(0,x) = \phi(x), \quad u_t(0,x) = \psi(x) \tag{2}$$

and the boundary conditions

$$u(t,0) = f(t), \quad u(t,1) = g(t) \text{ for } t \geqq 0. \tag{3}$$

J. D. Grote (ed.), The Theory and Application of Differential Games. 157-168.

Here f(t) and g(t) will be the control variables, chosen by the two players which for simplicity we call player "f" and player "g". On these control variables we impose the restrictive conditions

$$|f(t)| \leqq a, \quad |g(t)| \leqq b. \tag{4}$$

Admissible controls will be all piecewise C^2 functions satisfying (4). Therefore we have to take the concept of solution of (1) in a generalized sense, admitting possibly discontinuous solutions (which we may think as limits of smooth C^2 solutions). Here we do not want to go into these technical difficulties, which are well known from the general theory of partial differential equations (see, for example, [2] and [5]).

We split the solution of the initial-boundary value problem (1), (2), (3) as

$$u(t,x) = u_1(t,x) + u_2(t,x), \tag{5}$$

where

$$u_1(t,x) = \frac{1}{2}\left[\phi(x-t) + \phi(x+t) + \int_{x-t}^{x+t} \psi(s)\, ds\right] \tag{6}$$

is the solution corresponding to boundary values equal to zero, and

$$\begin{aligned} u_2(t,x) = {} & f(t-x) + g(t-1+x) - g(t-1-x) - \\ & - f(t-2+x) + f(t-2-x) + g(t-3+x) - \ldots \end{aligned} \tag{7}$$

is the solution corresponding to initial values equal to zero. To apply (6), the initial functions $\phi(x)$, $\psi(x)$ should be extended outside the interval (0,1) as odd functions, periodic of period 2.

From (6) and (7) we can obtain immediately the expressions for $u_t(t,x)$; we obtain for example the formula for u_t corresponding to (7) substituting there the values of f(.) and g(.) by their derivatives f'(.) and g'(.). In problems where $u_t(t,x)$ is evaluated, as in the case the energy of the vibrating string is wanted, we must assume the controls f(.) and g(.) to be continuous or understand the derivative in the sense of distributions of L.Schwartz (in which case the energy of the vibrating string becomes infinite if u(t,x) is discontinuous).

2. DIFFERENTIAL GAMES. UPPER AND LOWER VALUE.

The following examples can be understood with a minimum of theory of differential games. In each case there will be given a "payoff functional" $P[u(T,x),u_t(T,x)]$, where $T > 0$ has the meaning of the time the game ends. Player "f" should minimize and player "g" should maximize this payoff. The game is deterministic in the sense that each player knows, at every time t, the complete evolution of the game at times τ prior to t. The question of instantaneous information can be settled in two opposite ways: in the "upper game", the minimizing player "f" must choose first his value f(t) for every t (and announce his choice); "g" can then choose his value g(t) knowing f(t) (for the same argument t). This way the maximizer has some information advantage and the resulting optimal payoff is the "upper Value" V^+. In the "lower game", the maximizing player "g" must choose first and the situation is reversed: the "lower Value" V^- is more favorable to the minimizing player.

To develop the whole theory rigorously, one must consider, according to A. Friedman, approximating δ-games which can then be played as "upper δ-games" or as "lower δ-games"(see [3], [4]).

It is easily understood that always $V^+ \geq V^-$. In case these two Values happen to be equal, we call them the "Value" of the game and denote this by V. Conceptually in this case it is irrelevant which player chooses first: the above mentioned information advantage is worthless.

3. EXAMPLES OF DIFFERENTIAL GAMES

3.1. Linear game.

For the sake of completeness we start considering the linear game (1), (2), (3), (4) with the payoff

$$P = L[u(T,.),u_t(T,.),f(.),g(.)], \qquad (8)$$

where L[...] is a linear functional as for example

$$P = \int_0^1 \alpha(x)\ u(T,x)\ dx + \int_0^T [\beta(t)\ f(t) + \gamma(t)\ g(t)]dt.$$

According to the principle of superposition the payoff can be written in form

$$P[\phi,\psi,f,g] = P_0[\phi,\psi] + P_1[f] + P_2[g]. \tag{9}$$

Hence the differential game decomposes into two independent problems of optimal control: $P_1[f] = \min!$ and $P_2[g] = \max!$.

3.2. Example Nr.2.

Let the payoff be

$$P = \int_0^1 |u(T,x)|\, dx, \tag{10}$$

with T given in advance.

It follows from (5) and (7), that the region of influence of the value $f(t_1)$ on the solution $u(T,x)$, reduces to the point $x = T-t_1-2k$ or $x = 1-(T-t_1-(2+1)k)$, where k is the integer for which $0 \leqq x \leqq 1$. For the region of influence of $g(t_1)$ a similar (symmetric) statement is also true. From this it follows that in order to optimize (10), each player should optimize $|u(T,x)|$ for every $x \in [0,1]$ independently.

For simplicity we assume homogeneous initial conditions

$$u(0,x) = u_t(0,x) = 0. \tag{11}$$

According to (7) we can write the solution $u(T,x)$ at the point $R = (T,x)$, with the notation of figure 1, as

$$u(R) = f(A_1) + g(B_1) - f(A_2) - g(B_2) + f(A_3) + \cdots \tag{12}$$

From here we obtain the following optimal play, where we have to distinguish different cases for different values of the duration T of the game.

a) For $T < \frac{1}{2}$ we have (see figure 2) as optimal strategies for the players:

$$\left.\begin{array}{l} f(t) = 0 \\ g(t) = b \text{ (or } -b) \end{array}\right\} \quad \text{for } 0 \leqq t \leqq T.$$

Hence the optimal payoff = Value is $V = T\, b$.

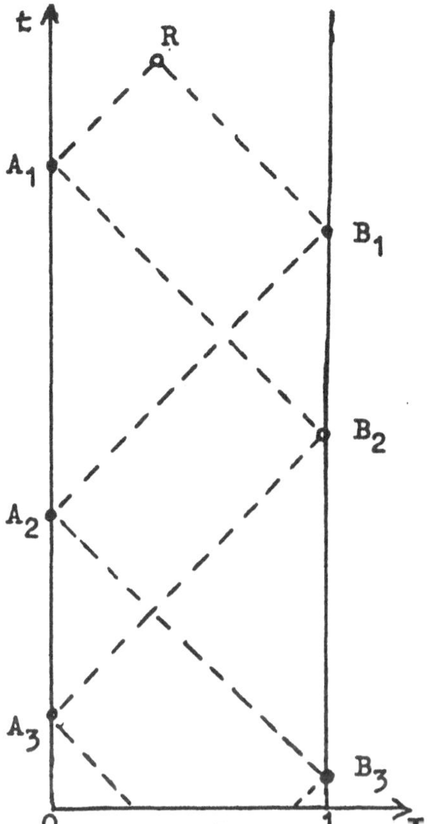

Figure 1

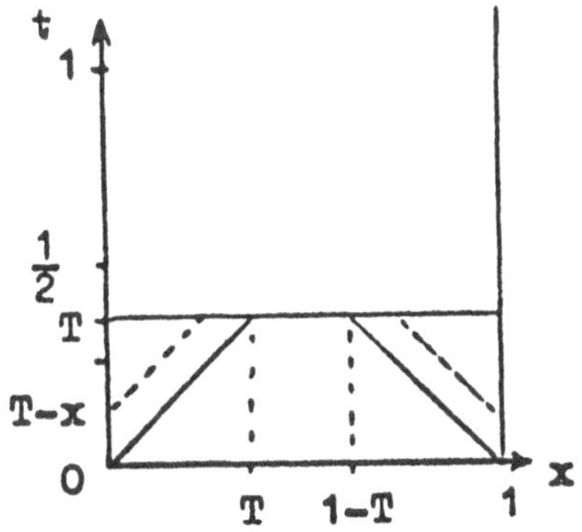

Figure 2

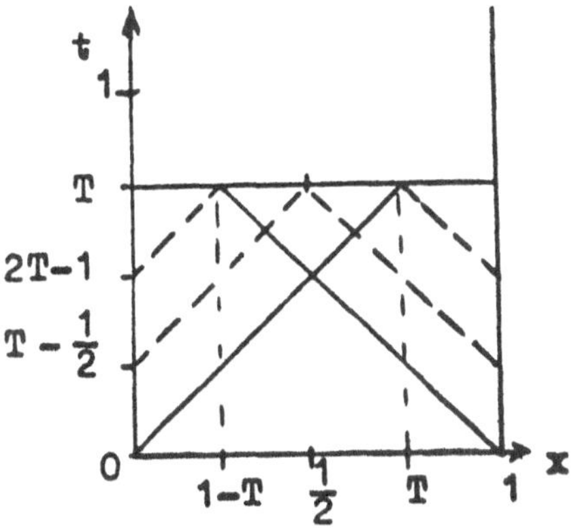

Figure 3

b) For $\frac{1}{2} < T < 1$ (see figure 3) we have the following situation. In $0 < x < 1-T$ we have $|u(T,x)| = |f(T-x)|$. As "f" minimizes, we take optimally

$$f(t) = 0 \text{ in } 2T-1 < t < T.$$

In $1-T < x < \frac{1}{2}$ we have $|u(T,x)| = |f(T-x) + g(T-1+x)|$. As "g" has to choose first his value $g(T-1+x)$, he will choose $g(T-1+x) = b$ or $-b$ indistictly. Later, "f" will try to subtract from $|b|$ as much as possible. In the case $a \geq b$ in (4), he will choose $f(T-x) = - g(T-1+x)$, obtaining $u(T,x) = 0$. Hence in this case any effort of player "g" was worthless and he may very well have chosen $g(T-1+x) = 0$. On the other hand, if $a < b$, player "f" will chosse $f(T-x) = -a$ or $+a$, obtaining as optimal result $|u(T,x)| = b-a$.

In the interval $\frac{1}{2} < x < T$ we also have $|u(T,x)| = |f(T-x) + g(T-1+x)|$, but now "f" chooses first. As he is minimizing, he must choose $f(T-x) = 0$; later player "g" will choose $g(T-1+x) = b$ or $-b$ indistinctly.

Finally, in the interval $T < x < 1$, we have $|u(T,x)| = |g(T-1+x)|$ and $g(T-1+x) = \pm b$ is optimal.

Hence we obtain the following optimal controls, where we have chosen $g(t) = 0$ when all possible values of $g(t)$ were equivalent, and $g(t) = b$ where $+b$ and $-b$ were equivalent.

In the case $a \geq b$:

$$\text{in } t \in (0,T-\tfrac{1}{2}): \quad f(t) = 0, \quad g(t) = 0;$$

$$\text{in } t \in (T-\tfrac{1}{2},2T-1): \quad f(t) = 0, \quad g(t) = b;$$

$$\text{in } t \in (2T-1,T): \quad f(t) = 0, \quad g(t) = b.$$

The Value of the game is $V = \frac{1}{2} b$.

For the case $a \leq b$:

$$\text{in } t \in (0,T-\tfrac{1}{2}): \quad f(t) = 0, \quad g(t) = b;$$

$$\text{in } t \in (T-\tfrac{1}{2},2T-1): \quad f(t) = -a, \quad g(t) = b;$$

$$\text{in } t \in (2T-1,T): \quad f(t) = 0, \quad g(t) = b.$$

The Value of the game is now $V = T\,b - (T-\frac{1}{2})\,a$.

c) For $T > 1$ the discussion of the optimal play can be made along the same lines. Figure 4 shows the optimal controls for the case $1 < T < \frac{3}{2}$, $a \leq b$. The Value of the game results in this case $V = (2b-a)(T-1) + (b-a)(\frac{3}{2}-T) + \frac{1}{2} b$.

Figure 5 shows the optimal controls for the case $1 < T < \frac{3}{2}$, $b \leqq a \leqq 2b$. The Value of the game results $V = (2b-a)(T-1) + \frac{1}{2} b$.

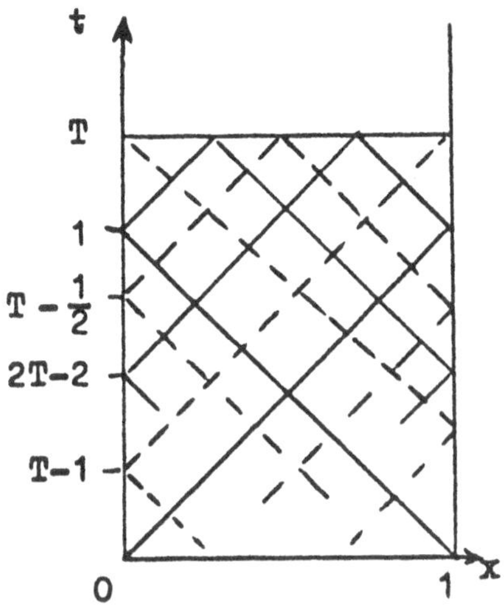

Case $a < b$

Optimal controls:

$$f(t) = \begin{cases} 0 & \text{for } t < T-\frac{1}{2} \\ -a & \text{for } t > T-\frac{1}{2} \end{cases}$$

$$g(t) = b$$

Figure 4

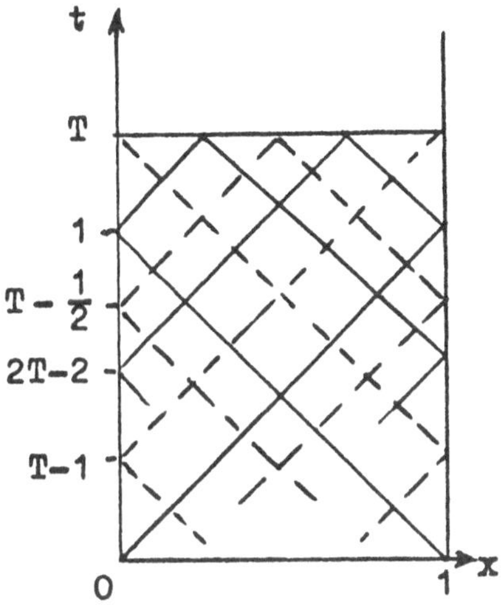

Case $b < a < 2b$

Optimal controls:

$$f(t) = \begin{cases} 0 & \text{for } t < 1 \\ -a & \text{for } t > 1 \end{cases}$$

$$g(t) = \begin{cases} 0 & \text{for } t \in (2T-2, T-\frac{1}{2}) \\ b & \text{otherwise} \end{cases}$$

Figure 5

The apparent contradiction, that where $u = f(.)+g(.)$, player "f" chooses $f(t) = 0$ and then player "g" chooses $g(.) = b$, so that $u = b$, while "f" could have chosen some negative value in order to get a smaller u, is easily explained. Indeed, if "f" would have chosen a negative value, then "g" would have chosen $g(.) = -b$ and the absolute value of u would have increased.

The case of non-zero initial values (when (11) is not valid) can be treated in a similar way.

3.3. Example Nr. 3.

Let the payoff be the energy of the vibrating string at time T > 0:

$$\mathcal{E}(T) = \frac{1}{2}\int_0^1 (u_t(T,x)^2 + u_x(T,x)^2)\ dx. \qquad (13)$$

Assuming homogeneous initial conditions (11), we obtain from (7)

$$\begin{aligned} u_t(T,x) &= f'(T-x) + g'(T-1+x) - g'(T-1-x) - \\ &\quad - f'(T-2+x) + f'(T-2-x) + \cdots \\ u_x(T,x) &= - f'(T-x) + g'(T-1+x) + g'(T-1-x) - \\ &\quad - f'(T-2+x) - f'(T-2-x) + \cdots . \end{aligned} \qquad (14)$$

This way we obtain

$$\begin{aligned} u_t(T,x)^2 &= [f'(T-x) - g'(T-1-x) + f'(T-2-x) - \cdots]^2 + \\ &\quad + [g'(T-1+x) - f'(T-2+x) + \cdots]^2 + \\ &\quad + 2\ [f'(T-x) + \cdots][g'(T-1+x) - \cdots], \\ u_x(T,x)^2 &= [f'(T-x) - g'(T-1-x) + f'(T-2-x) - \cdots]^2 + \\ &\quad + [g'(T-1+x) - f'(T-2+x) + \cdots]^2 - \\ &\quad - 2\ [f'(T-x) + \cdots][g'(T-1+x) - \cdots], \end{aligned}$$

hence

$$\begin{aligned} \tfrac{1}{2}[u_t(T,x)^2 + u_x(T,x)^2] &= \\ &= [f'(T-x) - g'(T-1-x) + f'(T-2-x) - \cdots]^2 + \\ &\quad + [g'(T-1+x) - f'(T-2+x) + \cdots]^2. \end{aligned} \qquad (15)$$

In order for the optimization problem to make sense, we must introduce restrictions of the type

$$|f'(t)| \leqq a_1, \quad |g'(t)| \leqq b_1. \qquad (16)$$

It is then possible to treat the two right-hand terms of (15) in the same way we optimized example Nr. 2. Esentially we have to take the derivatives $f'(t)$ and $g'(t)$ as control variables. Any additive constant in $u(T,x)$ is indeed irrelevant to the energy (13). Of course, once we have f' and g' chosen, we obtain f and g by integration. Here we may find some conflict with the original conditions (4), but this has little physical meaning. Indeed, a change of sign of both controls f' and g' is irrelevant in the optimization of (15) at each point x. Therefore we may change, from time to time, the sign of the controls in order to remain within the conditions (4).

3.4. Example Nr. 4.

Let the payoff be the convolution

$$P = \int_0^1 u(T,x)\, u(T,1-x)\, dx. \tag{17}$$

We again assume homogeneous initial conditions (11), obtaining (7) as expression for the solution u(T,x). In this example we will find that in some cases, the choice of f(t) and g(t) at the same instant t have a decisive influence on each other. In those cases we shall obtain an upper Value V^+ different from the lower Value V^-, as explained in section 2.

In order to discover the optimal strategies we start with small values of T.

a) Case $0 < T < \frac{1}{2}$. Here we obtain (see figure 2)

$$P = \int_0^T f(T-x)\, g(T-x)\, dx + \int_{1-T}^1 g(T-1+x)\, f(T-1+x)\, dx =$$

$$= 2 \int_0^T f(t)\, g(t)\, dt.$$

In the "upper game", player "f" chooses first. Here f(t) = 0 will be optimal, as player "g" could take advantage of any other choice of "f" (by choosing g(t) of the same sign). Therefore we obtain $V^+ = 0$.

In the "lower game", player "g" chooses first, and for esentially the same reason as before, g(t) = 0 is optimal. Hence $V^- = V^+ = V = 0$.

b) Case $\frac{1}{2} < T < 1$ (see figure 3). Here

$$P = 2 \int_0^{1-T} f(T-x)\, g(T-x)\, dx +$$

$$+ 2 \int_{1-T}^{\frac{1}{2}} [f(T-x) + g(T-1+x)][f(T-1+x) + g(T-x)]\, dx.$$

For the same reasons as above, optimal play will make the first integral equal to zero. Hence f(t) = = g(t) = 0 is optimal for 2T-1 < t < T.

The second integral brings new problems. We can write this integral as

$$\int_0^{T-\frac{1}{2}} [f(t) + g(2T-1-t)][g(t) + f(2T-1-t)]\, dt.$$

Here f(t) and g(t) have to be chosen first, and f(2T-1-t) and g(2T-1-t) later.

In the "upper game", "f" chooses first f(t), then "g" chooses g(t), after this comes the choice of f(2T-1-t) and last g(2T-1-t).

If $a > b$ in conditions (4), then $f(t) = a$, $f(2T-1-t) = -a$ is optimal (or both with reversed sign). Then "g" can optimally choose $g(t) = -b$ and $g(2T-1-t) = b$ (always in the interval $0 < t < T-\frac{1}{2}$).

This way we obtain $V^+ = -2\,(a-b)^2\,(T-\frac{1}{2})$.

If $a < b$ in conditions (4), then "g" could use any non-zero value of f(t) to his advantage. The optimal sequence of choices comes out therefore to be: $f(t) = 0$, $g(t) = b$, $f(2T-1-t) = -a$ and $g(2T-1-t) = b$ (Or all with opposite signs).

This way we obtain $V^+ = 2\,b\,(b-a)(T-\frac{1}{2})$.

For the lower game, where "g" must choose first, we obtain the following sequence of optimal choices (or all of them with reversed sign).

If $a < b$: $g(t) = b$, $f(t) = -a$, $g(2T-1-t) = b$, $f(2T-1-t) = -a$; $(0 < t < T-\frac{1}{2})$.

This way we obtain $V^- = 2\,(b-a)^2\,(T-\frac{1}{2})$.

If $a > b$: $g(t) = 0$, $f(t) = a$, $g(2T-1-t) = b$, $f(2T-1-t) = -a$; $(0 < t < T-\frac{1}{2})$.

This way we obtain $V^- = -2\,a\,(a-b)(T-\frac{1}{2})$.

c) Case $1 < T < \frac{3}{2}$ (figure 4). Here

$$\frac{1}{2} P = \int_0^{T-1} [f(t-x) + g(T-1+x) - g(T-1-x)]\cdot$$
$$\cdot[g(T-x) + f(T-1+x) - f(T-1-x)]\, dx +$$
$$+ \int_{T-1}^{\frac{1}{2}} [f(T-x) + g(T-1+x)][g(T-x) + f(T-1+x)]\, dx.$$

Decisive is, as before, the sequence of acts of choice. In the "upper game" this sequence is: f(T-1-x), g(T-1-x), f(T-1+x), g(T-1+x), f(T-x), g(T-x). In the "lower game", the roles of "f" and "g" are changed.

As before, it is possible to find the optimal choices for f(.) and g(.) for each x. This somewhat lengthy discussion will be omitted here, as the method is already clear. In the same way it is possible to analyze the game for any duration T, but this analysis becomes more and more lengthy.

In the case of nonhomogeneous initial values, the procedure is basically the same, but the results will obviously change.

4. MORE GENERAL DIFFERENTIAL GAMES

The examples given above could be solved explicitly because the solution (5), (6), (7) of the wave equation has a particularly simple structure. Already with the heat equation, or with the wave equation in more variables, the solution should be expressed by means of a Green's function in the form

$$u(T,x) = \int [G_1(T,t,x)\ f(t) + G_2(T,t,x)\ g(t)]\ dt. \quad (18)$$

If we now take a payoff of the type

$$P[u(T,x)] = \int F(u(T,x))\ dx, \quad (19)$$

where F(.) is a nonlinear function, this payoff in terms of f(t) and g(t) becomes very complicated.

The corresponding optimal control problems with partial differential equations are well known (see for example [7], [8] for the theoretical treatment and [1] for the applications). But here, in differential games the new and decisive question of sequence of decisions comes into play. This is intrinsic in differential games, but even here, the principle of optimality, which Issacs calls the tenet of transition, must be applied with great care. On the other hand, the general theory of differential games in abstract spaces, as developed for example by Friedman 4 , is applicable and gives existence theorems. What is extremely difficult is to obtain constructive methods to actually find the solution. We may hope that by solving characteristic examples some light may be thrown upon the general methods to attack this problem.

REFERENCES

1. Butkovskiy, A. G., Distributed Control Systems, Elsevier, New York-London-Amsterdam, 1969.

2. Courant - Hilbert, Methods of Mathematical Physics, English transl., Interscience (Wiley), New York, 1962

3. Friedman, A., Differential Games, Wiley-Interscience, New York, 1971.

4. Friedman, A., Differential Games, Regional Conference Series in Appl. Math., SIAM, Philadelphia, 1974.

5. Hellwig, G., Partial Differential Equations, Blaisdell, New York-Toronto-London, 1964.

6. Isaacs, R., Differential Games, Wiley, New York-London-Sydney, 1965.

7. Lions, J. L., Contrôle Optimal de Systèmes Gouvernés par des Équations aux Dérivées Partielles, Dunod, Gauthier Villars, Paris, 1968.

8. Lions, J. L., Some Aspects of the Optimal Control of Distributed Parameter Systems, Regional Conference Series in Applied Math. SIAM, Philadelphia, 1972.

THE TACTICAL AIR GAME: A MULTIMOVE GAME WITH MIXED STRATEGY SOLUTION

Leonard D. Berkovitz

Department of Mathematics, Purdue University,
West Lafayette, Indiana U.S.A.

1. INTRODUCTION

The interest in differential games is motivated to a large extent by the potential applications of the theory. We shall discuss several discrete time versions of differential games involving resource allocation. The games were developed over fifteen years ago to study the optimal employment of tactical air forces, and they are fairly realistic and complex. Complete mathematical solutions were then obtained and were of interest in the area of application. A significant feature of the solution in the most important model was that optimal play required the use of mixed strategies.

We review these results here for several reasons. First, they do represent a successful application of differential games. Second, they reemphasize the need to consider mixed strategies in differential games. Here we have realistic games whose solutions require mixed strategies. Moreover, the use of mixed strategies is plausible on "physical grounds". Finally, these games can be used as meaningful test problems. The usefulness of general characterizations of solutions of differential games can be tested to see whether they will yield the solutions of these examples. Also, proposed computational techniques for the solution of differential games can be tested against these examples.

2. FORMULATION OF GAME

The tactical air war game consists of a series of strikes, or moves, each of which consists of simultaneous counter-air, air-defense, and close-support operations by each side undertaken to accomplish

J. D. Grote (ed.), The Theory and Application of Differential Games. 169-177.

a given theater mission or payoff. We assume that at the start of the air operations the BLUE side has p planes and the opposing side, RED, has q planes. Let us look at a strike in the campaign, say the initial strike. Suppose that on this strike BLUE dispatches x planes on counter-air operations and u planes on air-defense operations, and the remaining amount, $m = p - x - u$ planes, on ground-support operations. Similarly, suppose that for his first strike RED allocates y planes to counter-air, w planes to air-defense, and the remaining number, $n = q - y - w$ planes, to support his ground forces. For this initial strike and for any future strikes, the above decisions are made by each side in ignorance of the allocation of the opposing side. It is assumed, however, that each side knows the number of planes that he and his opponent have.

Since RED allocates w planes to air-defense we can expect a reduction in the number of BLUE's planes that get through to counter-air targets. The number of interceptions by RED will be proportional to w, say cw, unless BLUE's attacking planes are saturated. The number of BLUE attacking planes that penetrate RED's defenses is $x - cw$ as long as cw is not larger than x. If cw is larger than x, no BLUE aircraft will penetrate. Hence the number of BLUE attacking planes that penetrate RED's defenses is the larger of the two numbers $x - cw$ and 0, or $\max(0,x-cw)$.

The objective of BLUE's counter-air operations is to reduce the enemy's air force by bombing certain targets, and the number of aircraft destroyed will vary with the number of attacking planes that penetrate RED's defenses. If we assume that each of BLUE's penetrating planes can destroy b planes of the enemy, then BLUE's initial counter-air strike can destroy at most $b \max(0,x-cw)$ RED planes. The number of RED planes actually destroyed by BLUE's counter-air strike will depend on the number of RED aircraft at risk at the time of the strike.

We assume that RED's air force is also reduced during the strike by such factors as accidents and antiaircraft fire. Let us assume that these losses are proportional to the number of planes used by RED during the strike. Then aq represents these losses, where a is RED's accident rate.

Finally, let us assume that RED's air force is increased during the strike by s planes. These replacements are subject to BLUE's counter-air attack, but can not be used by RED during this strike.

Thus, the number of RED planes at risk at the time of the strike is $q - aq + s$. Therefore BLUE's initial counter-air strike will destroy

$$\min[q - aq + s, b \max(0,x - cw)]$$

RED planes.

The planes used in air-defense are assumed to survive, and the RED aircraft that fail to penetrate the BLUE air-defense are assumed to return to base. That is to say, we assume that losses suffered in the air battle are negligible compared to the other losses, and that air-defense aircraft prevent attacking planes from successfully delivering their bombs without necessarily shooting them down.

If we sum the losses and add the replacements, we see that after the initial strike RED's force is reduced to

$$\begin{aligned} q_1 &= q + s - aq - \min[q - aq + s, b \max(0,x - cw)] \\ &= \max[0,q + s - aq - b \max(0,x - cw)]. \end{aligned} \tag{1}$$

In exactly the same manner we can analyze the effect of the initial strike on BLUE's inventory. We obtain that at the end of the initial strike, BLUE's inventory of planes is

$$p_1 = \max[0,p + r - dp - e \max(0,y - ku)], \tag{2}$$

where the coefficients d, e, and k have the same interpretation as a, b, and c, respectively, and r is the number of BLUE replacements

BLUE now has p_1 planes to allocate among the three tasks for the second strike, and RED has q_1 planes to allocate for the second strike. This strike will result in new inventories, p_2 and q_2, for the third strike. The process is repeated for the duration of the campaign.

We assume that BLUE's objective is to assist the ground forces in the battle area, and the results will vary with the number of planes he allocates to ground support operations. We assume that it is possible to construct a payoff function S, giving the payoff for each strike of the campaign in the form of the distance advanced by BLUE's ground forces as a function of the number m of planes allocated to ground support. We assume that it is possible to construct a similar function T for RED. Thus if BLUE allocates m aircraft in support of ground operations and RED allocates n aircraft the net advance of BLUE ground troops will be

$$S(m) - T(n).$$

The foregoing expression represents the payoff to BLUE for this one period or one strike. The payoff for the entire campaign of N strikes is the sum of these net yields for each of the N strikes, or

$$M = \sum^N [S(m) - T(n)].$$

The problem faced by each side is now apparent. At a given move, BLUE would like to allocate a large number of planes to ground-support missions and thereby increase the value of $S(m)$, yet he would like to destroy the RED air force by means of counter-air operations in order to ensure that $T(n)$ is small, or zero, for subsequent moves. Further, if he does not provide for air-defense he may suffer severe losses to his own air force if RED elects to mount a large counter-air strike. Each player has to take the future and the possibilities open to his opponent into account.

In our model of the tactical air war we shall make the further simplification that the yield functions S and T are linear, say, $S(m) = m$, $T(n) = n$. The payoff in the campaign then is

$$M(x,u;y,w) = \sum^{N}[(p - x - u) - (q - y - w)]. \tag{3}$$

BLUE wishes to make this payoff as large as possible by properly choosing the x's and u's during each of the N strikes, and RED wishes to make the payoff as small as possible by properly choosing the y's and w's.

3. OPTIMAL STRATEGIES FOR TWO TASKS

We shall begin our discussion of the tactical air war model by considering the case in which the air-defense potentials c and k are zero. In effect, we are assuming that there are only two tasks to which aircraft can be allocated, counter-air and ground-support. Equations (1) and (2) now read

$$q_1 = \max[0,q + s - aq - bx] \tag{4}$$

$$p_1 = \max[0,p + r - dp - ey] \tag{5}$$

and the payoff (3) reduces to

$$M(x,y) = \sum^{N}[(p - x) - (q - y)]. \tag{6}$$

This version of the tactical air war game was formulated by Fulkerson and Johnson [1] and was solved by them under the assumption of symmetry in the parameters, i.e., $a = d$ and $b = e$. The complete solution of the asymmetric version was later obtained by Dresher [2]. A more accessible summary of Dresher's results can be found in [3].

An outstanding characteristic of the solution of the tactical air game with two tasks is that, independent of the attrition parameters, initial conditions, and relative strengths at a given move, both sides have optimal pure strategies. Although every strike by a player is made simultaneously with his opponent, nevertheless, a player never needs to randomize, or bluff. An optimal strategy for a player can then be specified by giving, for each strike of the campaign, the number of planes he allocates to counter-air operations and the number of planes he allocates to ground-support. These optimal allocations depend on the attrition parameters, a, b, d, e, and on the number of strikes remaining in the campaign.

We shall not describe the optimal strategies for this game in detail, except to say that in most, but not all, situations the optimal strategies are bang-bang. That is, a player allocates all of his resources either to counter-air or to close support. There are some parameter combinations for which the optimal strategy is to allocate some resources to close support and some to counter air. For details we refer the reader to [3].

4. OPTIMAL STRATEGY FOR THREE TASKS

We now return to the more general model with all three tasks - counter-air, air-defense, and close-support - present. We shall see that increasing the number of air tasks to three leads to substantial changes in the character of the optimal tactics.

In order to simplify the analysis we assume that BLUE and RED have the same air-defense potential: each plane allocated to defense can prevent one attacking plane from reaching target - that is, we assume that $c=k=1$. We also assume that each attacking plane that penetrates the defense can destroy one plane in an airfield strike, or $b=e=1$, and that losses due to aborts, accidents, and antiaircraft fire are negligible. Finally, we assume that replacements are absent, i.e, $r=s=0$. Then the inventory of planes at the end of a strike will be, for BLUE and RED respectively,

$$p_1 = \max[0,p - \max(0,y - u)], \quad q_1 = \max[0,q - \max(0,x - w)].$$

We emphasize the fact that these simplifying assumptions have no effect on the general form of the optimal strategies.

The optimal strategies in the three-task model are different from the two-task model in the following two important ways: First, the optimal tactics depend upon the relative strengths of the two sides. Second, optimal play requires one player to use a mixed strategy. We shall give a complete description of the optimal employment of tactical air forces in terms of the number of strikes remaining and the relative strengths of the two sides. In our descriptions of the optimal allocations we shall always assume that at the move in question BLUE is the stronger side and RED is the weaker, that is to say $p \geq q$. It should be emphasized that this is merely a convention to facilitate the description of optimal tactics, and is not meant to imply that a side which is the stronger side at a given stage of the game will always remain the stronger for all subsequent moves. Of course, if a player who is initially stronger plays optimally, then he will remain the stronger throughout the campaign.

Our discussion will begin with a qualitative description of the optimal strategies. The optimal tactics have the following properties:

Campaign ends with ground support. The campaign always ends with a series of strikes on ground-support - i.e., during the closing period of the campaign both RED and BLUE concentrate all their forces on ground-support missions. In this terminal period both sides have the same optimal tactics, regardless of their initial forces.

BLUE (stronger) splits his forces. At all times other than the closing phase of the campaign, RED and BLUE have very different optimal tactics. During any of these early strikes, the stronger side, BLUE, has a pure strategy. That is, there exists a best allocation of BLUE's air force among the three air tasks. There is a critical value (about 2.7) of the ratio of the BLUE force size to the RED force size that governs BLUE's allocation during the early period in the following manner: If the force ratio is less than this critical value, then the optimal allocation in the early period consists of splitting the stronger air force between two air tasks, counter-air and air-defense, and neglecting the ground-support task. The size of split depends on the relative strengths of the two air forces and the number of strikes left in the campaign. However, if BLUE's strength relative to RED's is greater than the critical value, then BLUE should divide his force in a fixed way, regardless of his strength, among the three tasks, counter-air, air-defense, and ground-support. The number of aircraft allocated to each mission, however, is still dependent on the number of strikes remaining.

RED (weaker) mixes his tactics and concentrates his forces. The weaker combatant cannot use a single strategy, but must bluff during all the strikes other than those of the terminal phase. Unlike his opponent, the weaker combatant does not have a single allocation that is best. He must use a mixed strategy and gamble for high payoffs. If he is not too weak - i.e., if the force ratio is less than the critical value - then he concentrates his entire force either on counter-air or on air-defense; but which of these tasks receives the full effort is decided by some chance device. However, if RED is very weak (force ratio larger than critical value), then he allocates his entire air force to any one of the three air tasks with the particular task again chosen at random. In other words, if a player is very weak relative to the opponent, then he takes a chance on an early payoff. Of course, he must bluff correctly - i.e., the random device should select the tasks with the proper relative frequencies.

Mix and split the same tasks. It is of interest to note that on each strike RED, the weaker side, bluffs with the same tasks that BLUE uses in his allocation. Thus if RED is very weak he bluffs with each of the three tasks, and BLUE splits his forces among each of the three tasks. However, if RED is moderately weak, he bluffs with two tasks - counter-air or air-defense - and BLUE splits his forces between the same two tasks, counter-air and air-defense.

BLUE's defense decreases during campaign. As was noted above, prior to the closing phase of the campaign, BLUE splits his forces among his air tasks. The actual split is a function of the force sizes of BLUE and RED and the number of strikes left in the campaign. However, as the campaign proceeds, the fraction of BLUE's force allocated to air-defense will decrease. At the same time, the fraction allocated by BLUE to counter-air will increase. During this time, the chance that RED will attack BLUE also decreases, but the chance that RED will defend himself increases.

BLUE's defense in a long campaign. In the early stages of a relatively long compaign, the stronger side defends itself against a concentrated attack by the weak side. During this period, BLUE dispatches on air-defense a force of planes approximately the size of RED's entire force. Recall that we assumed a particular value for the air-defense effectiveness.

5. THE THEOREM

The description of the optimal tactics in Section 4 is based upon the following theorem, whose proof is to be found in [4]. This theorem enables us to compute the optimal allocations inductively. That is, for any N move game, given the i-th move, $1 \le i \le n$ and the force sizes p_i for BLUE and q_i for RED, the optimal allocations for BLUE and RED can be found explicitly. In [3] they are tabulated for all N move games with $N \le 8$.

Let N denote the length of the campaign, and m the number of a particular move, counted from the end of the campaign.

THEOREM: If $N = 1$ or 2, the value of the game is given by $V_N(p_N, q_N) = N(p_N - q_N)$. BLUE has an optimal pure strategy: $x_m = u_m = 0$ for $m \leqq N$. RED has an optimal pure strategy: $y_m = w_m = 0$ for $m \leqq N$. If $N \geqq 3$ the value of the game is given by the $(N - 2)$-piecewise linear function:

$$V_N(p_N, q_N) = A_N^i\, p_N - B_N^i\, q_N, \qquad (i = 1,2,\ldots,N-2)$$

where the constants A_N^i and B_N^i are positive and monotone decreasing in i for fixed N; the value of the superscript i is determined by the ratio p_N/q_N. The optimal strategies for the two players are as follows:

(i) At move $m = 1,2$ (counting from the end) the players choose

$$x_m = u_m = y_m = w_m = 0.$$

(ii) At move $m = 3$, if $p_3 = q_3$, then BLUE chooses x_3, u_3 such that

$$q_3 = x_3 = \min\left(\frac{p_3 + q_3}{2}, \frac{3q_3}{2}\right)$$

$$u_3 = x_3 - q_3.$$

Red chooses either $y_3 = q_3$ or $w_3 = q_3$, each with probability $\frac{1}{2}$.

(iii) At the $(m + 1)$st move, where $3 \leqq m \leqq N - 1$, if $p_{m+1} \geqq q_{m+1}$, then the ratio p_{m+1}/q_{m+1} determines an integer i, $1 \leqq i \leqq m - 1$, and BLUE chooses

$$x_{m+1} = \frac{(2m - A_m^i)p_{m+1} - (m - 2B_m^i)q_{m+1}}{m + B_m^i}, \quad u_{m+1} = p_{m+1} - x_{m+1},$$

for $i = 1,2,\ldots,m - 2$, and

$$x_{m+1} = (2 - 1/B_m^{m-2})q_{m+1}, \quad u_{m+1} = (1 - 1/m)q_{m+1},$$

for $i = m - 1$, where the constants A_m^i and B_m^i are those associated with a game of length m and initial condition p_m, q_m. RED chooses either y_{m+1} or $w_{m+1} = q_{m+1}$ with probabilities $\alpha_m^i = B_m^i/(m + B_m^i)$ and $\beta_m^i = m/(m + B_m^i)$, respectively for $i = 1,2,\ldots,m - 2$; however if $i = m - 1$, RED chooses $y_{m+1} = q_{m+1}$ with probability $\alpha_m^i = 1/m$, or $w_{m+1} = q_{m+1}$ with probability $\beta_m^i = 1/B_m^{m-2}$, or $y_{m+1} = w_{m+1} = 0$ with probability $1 - 1/m - 1/(B_m^{m-2})$.

The constants A_N^i and B_N^i of the theorem are defined inductively as follows:

$$A_3^1 = 3, \qquad A_{n+1}^{n-1} = (A_n^{n-2} + 1), \qquad (n \geqq 3)$$

$$B_3^1 = 3, \qquad B_{n+1}^{n-1} = (4 - 1/A_n^{n-2} - 1/B_n^{n-2}), \qquad (n \geqq 3)$$

$$B_n^0 = A_N^0 = 0, \qquad (n \geqq 3)$$

$$A_{n+1}^i = \frac{A_n^{n-2}(2B_n^i + A_n^i)}{B_N^i + A_n^{n-2}}, \quad B_{n+1}^i = \frac{3A_n^{n-2}B_n^i}{B_n^i + A_n^{n-2}} \quad (i \geqq 1;\ n = i+2, i+3, \ldots$$

The superscript i is associated with the ration p_N/q_N by means of a step function $\phi(p_N/q_N)$. For each N, the jump points of ϕ are determined by a sequence λ_N^i $(i = 1,\ldots,N - 2)$ to be defined presently. We have $\phi(p_N/q_N) = i$ whenever $\lambda_N^i \leqq p_N/q_N \leqq \lambda_N^{i+1}$. The sequences λ_N^i are defined as follows:

$$\lambda_3^1 = 1,$$

$$\lambda_n^{n-1} = +\infty \qquad (n = 3,4,5,\ldots)$$

$$\lambda_n^{n-2} = B_n^{n-2} - 1 \qquad (n = 4,5,6,\ldots)$$

$$\lambda_n^i = \frac{B_n^{i-1} - B_n^i}{A_n^{i-1} - A_n^i} \qquad (n = 4,5,6,\ldots;\ i = 1,2,3,\ldots,m-3)$$

6. A TWO AIRCRAFT MODEL

In [5] the tactical air war model is modified as follows. Each side has two types of aircraft, fighters and bombers. The bombers can be used either on close support or on counter air operations. The fighters can be used either on close support or on air defense operations. The interactions are as before, with the obvious modifications. The payoff is again measured in terms of close support sorties. The complexity of the model is such that a complete solution only exists for games with length $N \leq 3$. The solutions are given in [5].

REFERENCES

1. D. R. Fulkerson and S. M. Johnson, A tactical air game, Opns. Res. 5 (1957), 704-712.
2. M. Dresher, Optimal Tactics in a Multistrike Air Campaign, The RAND Corporation, Research Memorandum RM-1335 (1954).
3. L. D. Berkovitz and M. Dresher, A game theory analysis of tactical air war, Opns. Res. 7 (1959), 599-620.
4. L. D. Berkovitz and M. Dresher, A multimove infinite game with linear payoff, Pacific J. Math. 10 (1960), 743-765.
5. L. D. Berkovitz and M. Dresher, Allocation of two types of aircraft in tactical air war: a game theoretic analysis. Opns. Res. 8 (1960), 694-706.

IMPULSIVELY CONTROLLED D.G.

J. Case

Department of Mathematical Sciences,
The Johns Hopkins University, Baltimore, Maryland

1. An Example: I shall speak today about a new class of differential games. Let me begin with a discussion of a simple example involving only one player. Consider the owner of a roadside inn who has but a few regular customers. The bulk of his business must come from strangers, of whom many pass by on the road each day. The ability to attract new customers into the inn depends heavily on the appearance thereof, which we assume to be indexed by a number x between zero and 1.

The index x of the inn's appearance is clearly a function of time t, since paint does get old and needs to be renewed. Thus we shall assume that x(t) decays according to the law

$$\dot{x} = -kx \tag{1.1}$$

during time intervals $a < t < b$ between paint-jobs. Here k is a positive constant. Also we shall assume that the inkeeper's total profit in the planning period $0 \leq t \leq T$ is

$$J(T) = A \int_0^T x(t)dt - CN(T) , \tag{1.2}$$

N(T) being the number of times he has the inn repainted during the period, C the cost (in $) of each paint job, and A another positive constant.

The state of such a game is a vector (τ,x), where x is as above, and $\tau = T - t$ is the time remaining. The options in any such state are two: to repaint, in which case the state jumps immediately to $(1,\tau)$, or not to, in which case x continues to decay according to (1.1) which may now be rewritten

J. D. Grote (ed.), The Theory and Application of Differential Games. 179-187. *All Rights Reserved.*

$$\frac{dx}{d\tau} = kx \qquad (1.1')$$

The set of all possible states is then the semi-infinite strip $0 \le x \le 1$, $\tau \ge 0$, the desideratum an optimal repainting policy.

Clearly no infinite number of repaintings could ever be optimal in a finite period $0 \le t \le T$, for such would render (1.2) negative and infinite. Therefore there must be a last paint job. We seek first the states at which that might occur.

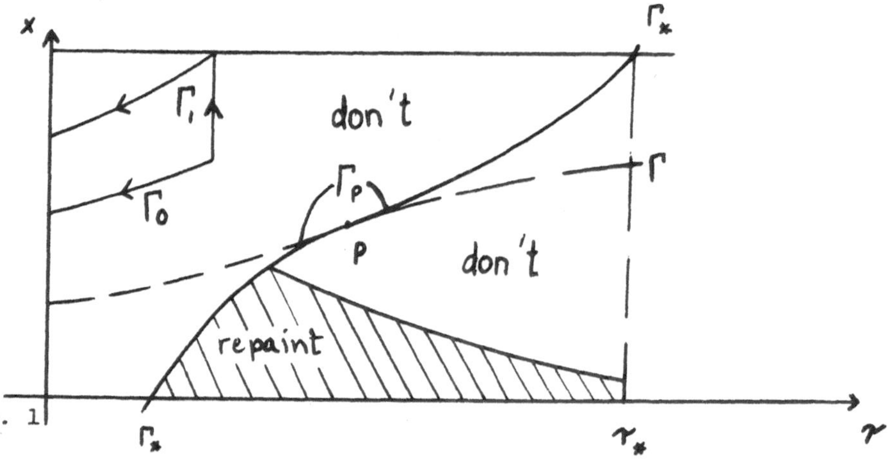

Fig. 1

Suppose that (τ,x) is such a state. Then the payoff to be got by following the upper path Γ_1 in figure 1 is no less than that from Γ_o. That is

$$Ax(1-e^{-k\tau})/k \le A(1-e^{-k\tau})/k - C \,, \qquad (1.3)$$

or equivalently

$$x \le \frac{Ck}{A(1-e^{-k\tau})} \qquad (1.4)$$

Above the curve Γ whereon (1.4) holds with equality, repainting immediately is not worth the expense. But below Γ it is.

Next, consider the set of states (τ,x) such that the solution of (1.1') thru (τ,x) does not meet Γ. That set is bounded below by a single solution curve Γ_p of (1.1') which meets Γ at just one point P. Both P and Γ_p are indicated in figure 1. Finally let Γ_* be the curve consisting of the part of Γ below P and the part of Γ_p above it. It is optimal never to repaint if the game begins in a state above Γ_*, for no state below Γ_* can subsequently be reached therefrom.

If, however, the game begins in a state (τ, x) which lies vertically under Γ_*, it is optimal to repaint once before the end. For the payoff achieved by repainting at an intermediate instant $0<\sigma<\tau$ is

$$A(1 - e^{-k\sigma})/k - C + A \int_{\sigma}^{\tau} x(s)ds \quad , \tag{1.5}$$

which is maximised by $\sigma = \tau$ if $x \leq e^{-k\tau}$; and by some $\sigma < \tau$ if not. Thus if (τ_*, x_*) are the coordinates of the intersection of Γ_* with the line $x = 1$, the optimal repainting policy for the rectangle $0 \leq x \leq 1$, $0 \leq \tau \leq \tau_*$ is as indicated in figure 1.

If larger values of τ than τ_* are to be considered, the segment $0 \leq x \leq 1$, $\tau = \tau_*$ may be considered a new "terminal surface", and the process repeated. Indeed it may be repeated several times if necessary, though the computations quickly become tedious in practice.

2. Impulsive Optimal Control Problems: The above game could also be called a problem of "impulsive optimal control". This is a new subject (to me at least), and I know of only three papers ([1], [2], and [3]) on it. But as those are concerned with somewhat different questions than I, and because I agree with M. Lions that the subject is important, I shall digress at some length about it before proceeding to its many-player generalization. I hope the audience will find the digression informative.

Roughly speaking, we must specify ten things to describe an impulsive optimal control problem. We must specify a set E of possible states x, and a subset $T \subset E$ of terminal states. The (one player) game is over whenever $x \in T$. Also we need an equation

$$\dot{x} = f(x) \tag{2.1}$$

of evolution, and a (possibly state-dependent) set A(x) of alternative states, such that the controller (player) is permitted to halt the evolutionary process at any time, and move the state instantaneously from x to any $y \in A(x)$ before allowing the evolution to resume. For instance in the inn-keeper's problem, A(x) was the set

$$A(\tau,x) = \{(\tau,x), (\tau,1)\}. \tag{2.2}$$

We shall always assume that

$$x \in A(x) , \tag{2.3}$$

so that one is never forced to interrupt the system. These describe the system's dynamics completely.

Next we must specify the payoff functional. It is assumed to consist of three parts; a terminal part

$$K(x) \tag{2.4}$$

defined for each $x \in T$, an impulsive part

$$C(x,y) \tag{2.5}$$

defined for each pair (x,y) such that $x \in E$ and $y \in A(x)$ and representing the cost of the impulse $x \to y$, and an integral part

$$\int_a^b L(x(t))\,dt \tag{2.6}$$

describing the accumulation of payoff during periods $a < t < b$ wherein the system is not interrupted. It is assumed too that

$$C(x,x) = 0 \tag{2.7}$$

for every $x \in E$, so that there is no cost for not interrupting the system. Thus if the game begins at time $t_o = 0$ in state x, and a strategy σ is adopted which causes the game to be interrupted at instants $0 < t_1 < \ldots < t_r$ to receive impulses $x_1 \to y_1, \ldots, x_r \to y_r$ respectively, and to terminate at $t_{r+1} > t_r$ in state $x_{r+1} \in T$, the payoff $\mathcal{P}(x;\sigma)$ accumulated is

$$\begin{aligned} \mathcal{P}(x;\sigma) = K(x_{r+1}) + \sum_{j=1}^{r+1} \int_{t_{j-1}+0}^{t_j-0} L(\varphi(t-t_j;y_j))\,dt \\ - \sum_{j=1}^{r} C(x_j,y_j)\, I_\sigma(x_j), \end{aligned} \tag{2.8}$$

$\varphi(t;\xi)$ being the complete solution of the initial value problem

$$\dot{x} = f(x) \quad , \quad x(0) = \xi , \tag{2.9}$$

and (for the moment) $I_\sigma(x) \equiv 1$. These describe the reward structure of the game.

Finally, it is necessary to specify the class of strategies in which the optimum is to be sought. A strategy is, of course, a function σ associating with each possible state $x \in E$, an admissible alternative $\sigma(x) \in A(x)$. We require in addition that

$$\sigma(\sigma(x)) = \sigma(x) , \tag{2.10}$$

so no two jumps need ever follow one another in a single instant. The set

$$E_D(\sigma) = \{x: \sigma(x) = x\} \tag{2.11}$$

is of particular importance, as no impulses are given the system while the state vector remains therein. Those all occur as x enters the complementary set

$$E_J(\sigma) = \{x: x \neq \sigma(x) \in A(x)\} . \tag{2.12}$$

Henceforth $I_\sigma(x)$ will denote (in equation (2.8)) the characteristic function of $E_J(\sigma)$.

We impose one final condition on the functions σ in the class Σ of admissible strategies, namely that for each $x \in E_D(\sigma)$ there exist a least positive t such that

$$\varphi(t;x) \in E_J(\sigma) \cup T. \tag{2.13}$$

This does not prevent strategies $\sigma \in \Sigma$ from giving the state vector $x(t)$ an infinite number of impulses, but at least it makes the impulses countable.

For certain applications it is desirable to allow nature as well as the controller to impart impulses to the system, according to some given strategy $\nu \in \Sigma$. As long as ν is known to the controller, he is still confronted with an optimization problem, rather than a game against a genuine opponent. We shall generalize (2.8) to

$$\begin{aligned}\mathcal{P}(x,\sigma) = K(x_f) + \sum_{j=1}^{r} \int_{t_{j-1}+0}^{t_j-0} L(\varphi(t-t_j;y_j))\, dt \\ - \sum_{j=1}^{r} C(x_j,y_j)\, I_\sigma(x_j) - \sum_{j=1}^{r} G(x_j,y_j)\, I_\nu(x_j)\end{aligned} \tag{2.8'}$$

in cases where nature too gives impulses. In fact G will vanish in most cases of interest, but we include it here for technical reasons. An obvious generalization of this formula pertains to the case wherein several impulses are given "successively" in a single instant.

We may finally conclude that an impulsive optimal control problem is determined when all the elements of the set

$$\{E, T, t(\cdot), A(\cdot), K(\cdot), C(\cdot,\cdot), G(\cdot,\cdot), L(\cdot), \nu(\cdot), \Sigma\} \tag{2.14}$$

are specified. The identification of those quantities in the inn-keeper's problem is left as an exercise.

Shortly we shall state (without proof) the fundamental sufficiency theorem for impulsive optimal control problems. But first it will be instructive to illustrate, by means of a few examples, certain difficulties which arise from the notion of strategy we propose. Let us take $E = \mathbb{R}^2$, $f(x_1,x_2) \equiv (1,0)$, $x(0) = \xi = (-1,0)$, and

$$\begin{aligned} T &= \{(x_1,x_2): x_1 \geq 2.5\} \\ A &= \{(x_1,x_2), (x_1,x_2+1)\} . \end{aligned} \tag{2.15}$$

It is assumed here that both $\nu(x_1,x_2)$ and $\sigma(x_1,x_2)$ must be elements of $A(x_1,x_2)$, although in general we shall exempt nature from that restriction. Then ν and σ are determined when the sets $E_J(\nu)$ and $E_J(\sigma)$ are specified. First let (example 1)

$$E_J(\nu) = \emptyset \text{ and } E_J(\sigma) = \{(x_1,x_2): x_2 \leq \frac{x_1}{1-x_1}\} . \tag{2.16}$$

Then the trajectory $x_{\sigma,\nu}(t;\xi)$ of the system arising from the strategy pair (σ,ν) has the form indicated in figure 2a
It is defined only for $0 \leq t \leq 2$, and never reaches T.

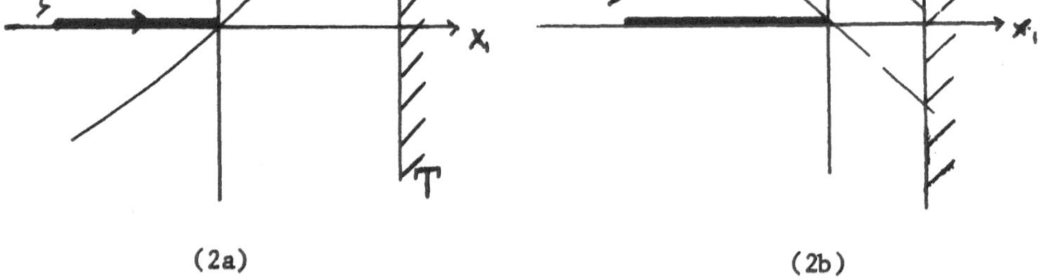

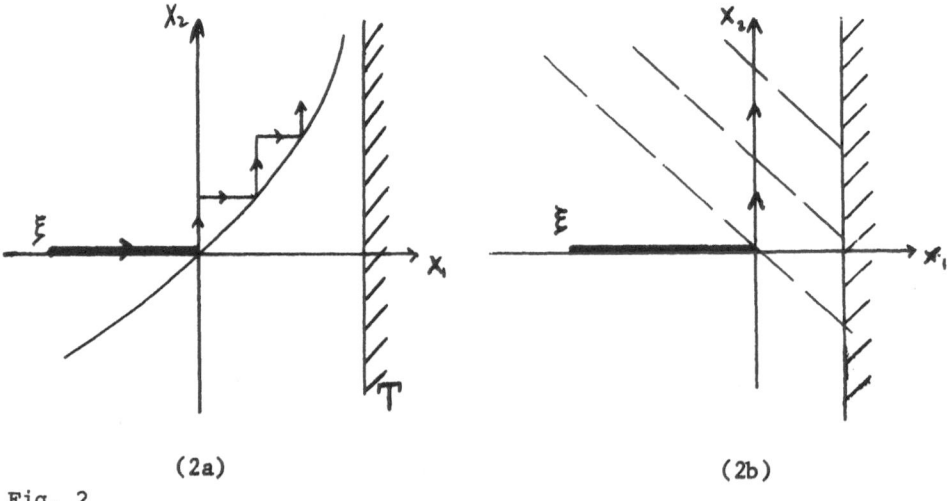

Fig. 2

Next let (example 2)

$$E_J(\sigma) = \{(x_1,x_2): [x_1+x_2] \text{ is even and } \geq 0\}$$
$$E_J(\nu) = \{(x_1,x_2): [x_1+x_2] \text{ is odd and } \geq 0\} \quad (2.17)$$

where $[\cdot]$ denotes the greatest integer function. Then $x_{\sigma,\nu}(t;\xi)$ is as shown in figure (2b). It is defined only for $0 \leq t < 1$, and also fails to reach T. Note that the difficulty in example 1 is due to the nature of σ alone, whereas that in example 2 is the result of an incompatability between σ and ν.

A third sort of difficulty arises if $\nu(x_1,x_2)$ must be chosen from $A'(x_1,x_2)$ instead of from $A(x_1,x_2)$, where

$$A'(x_1,x_2) = \{(x_1,x_2), (x_1,x_2-1)\}$$
$$E_J(\sigma) = \{(x_1,x_2): x_2 \leq x_1\} \quad (2.18)$$
$$E_J(\nu) = \{(x_1,x_2): x_1+ x_2 \geq 0\} .$$

Then either of the two trajectories indicated in figure 3 has apparently an equal claim to the title $x_{\sigma,\nu}(t;\xi)$.

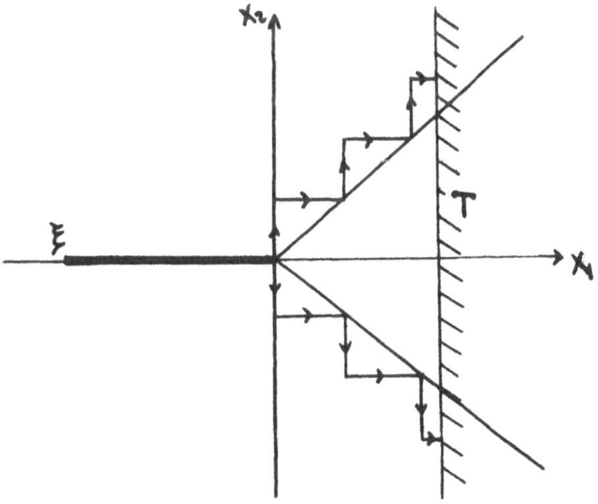

Fig. 3

From our (admittedly modest) experience with problems of the present sort, it seems that difficulties of the first two kinds illustrated do not occur if σ is some reasonable candidate for optimality, but that the non-uniqueness exhibited in example 3 is inextricably a part of the subject. Indeed it appears in even the simplest games.

3. Sufficient Conditions: Let ν be an admissible strategy for nature, let S be an open subset of E, and let $\sigma^* \in \Sigma$ transfer each $x \in \bar{S}$ (the closure of S) to T without driving x out of $\bar{S}$ after at most a finite number of interruptions whenever nature plays ν. Finally, write

$$V(x) = \mathcal{P}(x;\sigma^*) \tag{3.1}$$

for each $x \in S$. Then the following theorem provides a sufficient condition that σ^* be an optimal strategy for the problem (2.14).

Theorem Suppose that the directional derivative $D_{\mathfrak{f}(x)} V$ exists at each $x \in \bar{S}$, and suppose that the conditions

$$V(x) \geq \max_{y \in A(x)} \left[V(y) - C(x,y) \right] \tag{3.2}$$

and

$$D_{f(x)} V(x) + L(x) \leq 0 \tag{3.3}$$

hold at each point $x \in \bar{S}$ such that $\nu(x) = x$, and that

$$V(x) = V(\nu(x)) - G(x, \nu(x)) \tag{3.4}$$

holds in $E_J(\nu) \cap \bar{S}$. Moreover suppose that (3.2) holds with equality in $E_J(\sigma^*) \cap \bar{S}$ and (3.3) is an equality in $E_D(\sigma^*) \cap \bar{S}$. Finally suppose

$$V(x) = K(x) \tag{3.5}$$

for each $x \in T$. Then σ^* is optimal among all strategies σ which transfer points $x \in \bar{S}$ to T without leaving $\bar{S}$.

We shall not pause here for a proof of the theorem, but content ourselves with the remark that one can be given which closely parallels Carathéodory's "royal road" to the calculus of variations.

4. Games: A game between players 1,2,...,N may be represented as a set

$$\{E, T, f(\cdot), A_1(\cdot), \dots, A_N(\cdot), C_1(\cdot,\cdot), \dots, C_N(\cdot,\cdot), K_1(\cdot), \dots, K_N(\cdot), G_1(\cdot,\cdot), \dots, G_N(\cdot,\cdot), L_1(\cdot), \dots, L_N(\cdot), \nu, \Sigma, \dots, \Sigma_N\} \tag{4.1}$$

of $6N + 4$ elements, each having quite the same meaning it would have in a one player game. The idea is as before; the system (1.1) evolves continuously until interrupted by one or more of the players, and given impulses by them. Each player i strives, by his choice of an impulsional strategy $\sigma_i \in \Sigma_i$, to maximize the value of a functional of the form $(2.8)^i$.

A Nash equilibrium strategy N-tuple $(\sigma_1, \dots, \sigma_N)$ may be recognized by use of the theorem of section three. One need only form an "aggregate strategy" $\mu_i = (\sigma_1, \dots, \sigma_{i-1}, \sigma_{i+1}, \dots, \sigma_N, \nu)$ and verify that each σ_i is optimal against the corresponding μ_i. This of course is not always easy, but it can often be done in specific cases by constructing value functions $V_1(x), \dots, V_N(x)$ which satisfy the conditions (3.2)-(3.5) as required in the theorem.

Indeed for games involving several players, it is easier to do so with impulsive controls than in the usual setting, because there is no difficulty in locating the characteristic curves of the system of partial differential equations of which $V_1, \dots, V_N$ are solutions. They are simply the integral curves of system (2.1). Thus impulsively controlled games do not contain the obstacle which has done most to retard the development of a useful theory of conventional many player differential games. It is for this reason that I commend impulsive DG to the attention of the audience as a topic for future investigation.

REFERENCES

1. A. Bensoussan et J. L. Lions - Nouvelle Formulation de Controle Impulsionnel et Applications CRAS 1973
2. A. Bensoussan, M. Goursat, et J. L. Lions - Controle Impulsionnel et Inequations Quasi-Variationnelles Stationnaires CRAS 1973
3. A. Bensoussan et J. L. Lions - Control Impulsionnel et Inequations Quasi-Variationnelles d' Evolution CRAS 1973

THE THEORY OF HAMILTONIAN DYNAMICAL SYSTEMS, AND AN APPLICATION TO ECONOMICS

Karl Shell

Professor of Economics
University of Pennsylvania
Philadelphia, Pennsylvania 19174

INTRODUCTION

A Hamiltonian dynamical system (HDS) naturally arises in the standard control problem involving optimization over time. On this ground alone, a systematic study of the basic structure of the general HDS should be extremely useful for mathematical control theory. Applications of HDS theory extend beyond models involving optimization. The classic studies of such systems were motivated by problems in celestial mechanics. While much of the analysis of my lecture will be motivated by the normative (optimizing) model of macroeconomic growth, I will show in passing how HDS theory may be useful in analyzing many positive ("nonoptimizing") models of macroeconomic growth.

The basic approach of this lecture is to relate stability properties of the HDS to the geometry of the underlying Hamiltonian function (HF) generating that HDS. For simplicity, the present analysis is restricted to continuous-time models, although the Hamiltonian approach is equally powerful in analyzing discrete-time models.

The emphasis of this lecture on the geometry of the HF as an important determinant of (global) stability of the HDS is drawn directly from a joint paper with my colleague David Cass [4], which will appear in the Journal of Economic Theory. All of the basic propositions -- i.e., those relating global stability to Hamiltonian steepness -- are from the Cass-Shell article.

J. D. Grote (ed.), The Theory and Application of Differential Games. 189-199. *All Rights Reserved.*

LOCAL ANALYSIS FOR THE TIME-AUTONOMOUS HAMILTONIAN FUNCTION: THE CLASSICAL THEOREM OF POINCARE

Let $\xi(t)$ and $\eta(t)$ be m-dimensional vectors dependent upon time t. The HF $H(\xi(t), \eta(t))$ is said to be time-autonomous, since H depends on t solely through ξ and η. $H(\xi,\eta)$ is said to generate the HDS

$$\dot{\xi} = -\frac{\partial H}{\partial \eta}, \quad \dot{\eta} = \frac{\partial H}{\partial \xi}, \tag{1}$$

where here $\partial H/\partial \xi$ and $\partial H/\partial \eta$ are vectors of partial derivatives (later, vectors of generalized gradients) and $\dot{\eta}$ and $\dot{\xi}$ are vectors of time derivatives. I suppress the dependence on t whenever confusion will not arise. Assume that the above HDS possesses a rest point, which can be chosen as the origin without loss of generality. Then

$$\frac{\partial H(0,0)}{\partial \eta} = 0 = \frac{\partial H(0,0)}{\partial \xi}. \tag{2}$$

I follow the approach of classical mechanics; see, e.g. [10], especially pp. 76-81. Make a linear approximation to the HDS about the rest point. If λ is a characteristic root of the associated linear system, then it must satisfy the characteristic equation

$$\begin{vmatrix} H_{\eta\xi} + \lambda I & H_{\eta\eta} \\ H_{\xi\xi} & H_{\xi\eta} - \lambda I \end{vmatrix} = 0, \tag{3}$$

where $H_{\xi\xi}$, $H_{\eta\eta}$, $H_{\xi\eta}$ and $H_{\eta\xi}$ are m × m matrices of cross partials evaluated at (0,0) and I is the m × m identity matrix. I assume that $H(\cdot)$ is twice continuously differentiable. Notice the abundant symmetry in the characteristic equation (3) for the linearized HDS. $H_{\eta\xi} = H'_{\xi\eta}$, $H_{\eta\eta} = H'_{\eta\eta}$, and $H_{\xi\xi} = H'_{\xi\xi}$, where primes denote matrix transposition. Substituting in (3), transposing the determinant on the left, and interchanging rows and columns yields

$$\begin{vmatrix} H_{\eta\xi} - \lambda I & H_{\eta\eta} \\ H_{\xi\xi} & H_{\xi\eta} + \lambda I \end{vmatrix} = 0. \tag{4}$$

Comparing (3) and (4) gives Poincaré's result: If λ is a root to the characteristic equation for the linearized HDS, then $-\lambda$ is also a root.

This Poincaré theorem is at once extremely simple and also suggestive of deep fundamental results. If we could rule out roots with zero real parts, we would then have established a saddlepoint result for the linearized HDS in 2m-dimensional (ξ, η) phase space: In the neighborhood of the rest point (0,0), the manifold of (forward) solutions tending to the rest point as $t \to +\infty$ is of dimension m, and the manifold of (backward) solutions tending to the rest point as $t \to -\infty$ is also of dimension m.

Since zero real parts would seem to be something of a mathematical accident, it would seem apparent that <u>the local saddlepoint property is generic</u> in the sense of global analysis. That is, HDS's possessing this property are dense in the class of all HDS's possessing a rest point.

I know of no formal proof that the saddlepoint property is generic. There are even some reasons that might make one skeptical of the validity of saddlepoint genericity: (1) Since the λ's are solutions to a polynomial equation, "outcomes" are further constrained; e.g., if λ is a complex root, then its complex conjugate is also a root. (2) Somehow, real-life planetary motion seems to be able to sustain itself.

Another approach suggested by examples from economics (see [17,18,19]) is to seek sufficient conditions in terms of the geometry of the HF that ensure the saddlepoint property. The Poincaré result requires little in the way of restrictions on the HF, viz., the existence of a rest point and twice continuous differentiability. (It should be noted that somewhat stronger conditions may be needed for the saddlepoint property to carry over from the linearized version to the HDS itself.)

Rockafellar [12], in an important mathematical paper stimulated by problems in economics (cf., [17]), establishes -- among other things -- a global version of the saddlepoint property for the case in which the rest point (0,0) is a saddlepoint of the HF $H(\xi,\eta)$ with $H(\cdot)$ <u>strictly</u> convex in ξ and <u>strictly</u> concave in η. Therefore, if the HF is strictly convex-concave and possesses a saddlepoint (0,0), then the HDS possesses a saddlepoint (0,0).

In optimal economic growth, the HF is interpreted as the maximized value of socially imputed net national product with ξ as output prices and η as input stocks. It is thus natural to assume at least convexity in prices ξ and concavity in stocks η.

Moreover -- although various schools of economics seem to "rediscover" this in particular examples -- it is quite natural for a competitive dynamical system to possess this saddlepoint property in phase space. Such intertemporal development is shown in the phase-plane of Figure 1, where $m = 1$.

From Figure 1, we see that for each initial endowment $\eta(0)$ there exists at most one (locally, exactly one) price $\xi(0)$ for which $(\xi(t),\eta(t)) \to (\xi^*,\eta^*)$, the rest point, as $t \to \infty$. Errant paths -- those not tending to (ξ^*,η^*) -- seem to violate additional conditions for optimality, such as nonnegativity of prices or asset-market clearing, or some transversality condition, or bounded value of capital. Such arguments have, however, only been firmed-up for special cases (see, e.g., [5,17,19]) and seem to rely for proof on topological properties appropriate to the phase-plane but inappropriate for higher dimensional phase spaces.

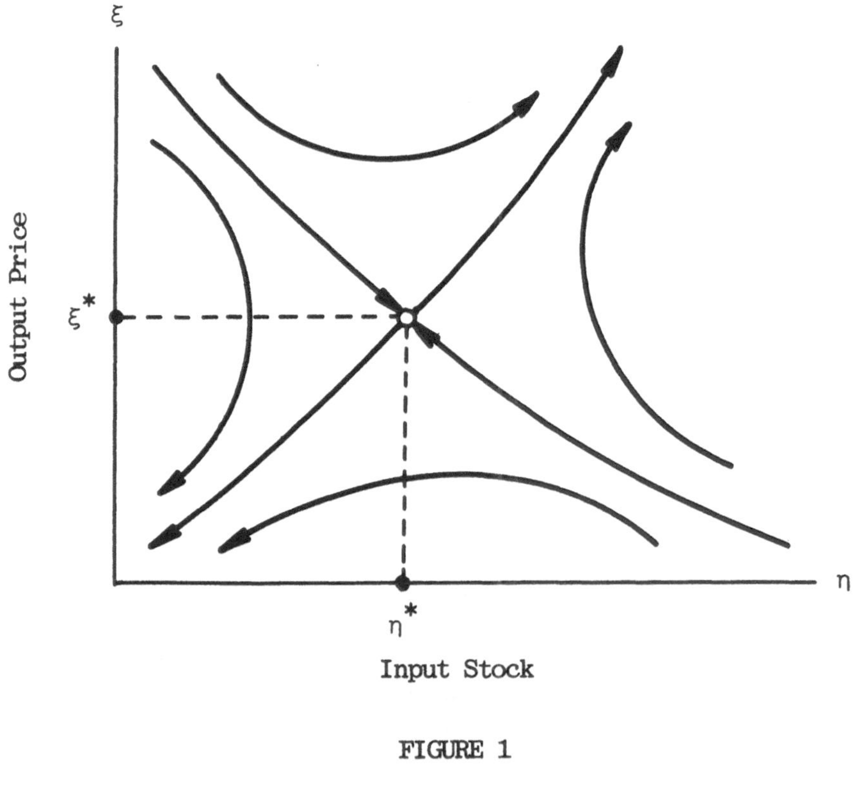

FIGURE 1

In what follows, I shall report on the Cass-Shell approach to Hamiltonian dynamics. We start with the geometry of the Hamiltonian function, but rely on global steepness properties rather than measures of convexity and concavity. We also jump over the step of studying the dimensionality of the stable manifold. Instead, we add nonnegativity and transversality conditions to our definition of the HDS and investigate the global stability of the rest point (ξ^*,η^*), or more naturally the global stability of the "real" (nonprice) component, η^*. Furthermore, we are able to accommodate an important class of non-time-autonomous HDS. Before turning to the stability analysis, I must generalize our notion of the HF and the HDS it generates.

THE HAMILTONIAN FUNCTION: AN ECONOMIC REPRESENTATION

Not only does a Hamiltonian function generate a Hamiltonian dynamical system which is convenient for dynamic economic analysis, but the HF is also an interesting construct for static economic analysis. Changing notation from the previous sections, let $k(t)$ be an m-vector of capital stocks at time t, let $z(t)$ be an m-vector of net investment goods output at time t so that $\dot{k}(t) = z(t)$, let $q(t)$ be an m-vector of present prices of investment goods at time t, let $c(t)$ be (scalar) consumption or "instantaneous social utility" at time t, and let $p(t)$ be the (scalar) present price of consumption at time t. Given $(p,q) \geq 0$, the present value of net national output is derived by maximizing $pc + qz$ subject to technology and endowments of capital k, and endowment of the sole (for convenience) fixed factor, ℓ, (interpreted as labor). If all capital stock inputs are feasible, this static optimization yields an HF (interpreted as the present value net national output)

$$H(p,q,k,\ell)$$

defined over the nonnegative orthant $\{(p,q,k,\ell) : (p,q,k,\ell) \geq 0\}$. For convex technologies with free disposal, this convenient Hamiltonian representation of technology is fully equivalent (and often easier to work with) than the usual input-output set-theoretic representation or the related production-frontier representation. The HF has the following properties:

(a) H is nondecreasing in p and ℓ;

(b) H is linear homogeneous in (p,q). Without loss in generality, since convex technologies can always be described as constant-returns technologies after introducing a fictitious commodity, we also assume that H is linear homogeneous in (k,ℓ);

(c) H is convex in (p,q) and concave in (k,ℓ);

(d) The generalized gradients satisfy

(i) $\frac{\partial H}{\partial(p,q)} = (c,z)$ and

(ii) at least for $(p,q) \geq 0$,

$\frac{\partial H}{\partial(k,\ell)} = (r,w)$, where by duality, r is the vector of competitive rental rates and w is the wage rate on labor.

For further discussion of the static aspects of the above HF, $H(p,q,k,\ell)$, see Cass [3], Lau [7], and Cass-Shell [4].

GENERALIZED HAMILTONIAN LAWS OF MOTION

In terms of the HF, $H(p(t),q(t),k(t),\ell(t))$, our generalized HDS becomes

$$\left.\begin{aligned} \dot{k}(t) &= \frac{\partial H(p(t),q(t),k(t),\ell(t))}{\partial q(t)}, \\ \dot{q}(t) &= -\frac{\partial H(p(t),q(t),k(t),\ell(t))}{\partial k(t)}. \end{aligned}\right\} \tag{5}$$

For steady state analysis to apply, $\ell(t)$ must grow at an asymptotically constant rate. Without further loss of generality, I can set $\ell(t) \equiv 1$. Of course, initial capital stock endowments are exogenously specified so that $k(0) = k_0$. If $(p(t),q(t)) \geq 0$, then (5) is consistent with the usual perfect-foresight competitive asset-market clearing equations from economics.

OPTIMAL ECONOMIC GROWTH WITH POSITIVE DISCOUNTING

As an economic example, I propose to investigate the problem of maximizing social welfare as given by

$$\int_0^\infty c(t)e^{-\rho t}\,dt, \tag{6}$$

where $\rho > 0$ is a constant scalar. (I could include the limiting case $\rho = 0$, but although analytically simpler, it would open up various time-consuming caveats about boundedness of the criterion functional and transversality, etc. Instead see [4]. Maximization is subject to technological constraints, initial endowments, k_0, and labor availability, $\ell(t) \equiv 1$. To my knowledge, for all cases in which the Maximum Principle has been worked out, HDS(5) is necessary for optimality. Similarly, I take as a necessary condition that

$$\dot{p}(t)/p(t) = -\rho$$

or without loss in generality that

$$p(t) = e^{-\rho t} \tag{7}$$

I must also add a transversality condition:

$$\lim_{t \to \infty} q(t)k(t) = 0. \tag{8}$$

In discrete-time Weitzman [20] has shown that (8) is necessary for optimality. I suspect that techniques similar to those developed in [4] should be useful in establishing the necessity of (8) in continuous time, but Cass and I have not verified this conjecture.

It will be convenient to switch from present prices to current prices. Define $Q \equiv q/p$ and $H(Q,k) \equiv H(1,Q,k,1) \equiv H(1,q/p,k,1)$. The system (5),(7), and (8) can be written as

$$\left.\begin{aligned} &\dot{k} = \frac{\partial H(Q,k)}{\partial Q}, \quad k(0) = k_0, \\ &\dot{Q} = -\frac{\partial H(Q,k)}{\partial k} + \rho Q, \quad Q(0) \geqq 0, \\ &\lim_{t \to \infty} Qe^{-\rho t}k(t) = 0. \end{aligned}\right\} \tag{9}$$

GLOBAL STABILITY OF OPTIMAL ECONOMIC GROWTH

I assume here that there exists a rest point, (Q^*,k^*), to HDS(9). For a proof of existence, see [4].

I have promised to provide geometrical conditions on the HF, $H(Q,k)$, which will ensure global stability of the steady state capital stock, i.e., that ensure $\lim_{t\to\infty} k(t) = k^*$. The stability condition, which is due to Cass-Shell [4], is

$$\left.\begin{array}{l}\underline{\text{Stability Assumption}}\text{: For every } \varepsilon > 0\text{, there is a } \delta > 0 \\ \text{such that } \|k-k^*\| > \varepsilon \Longrightarrow \\ (Q-Q^*)\dfrac{\partial H(Q,k)}{\partial Q} - \dfrac{\partial[H(Q,k)-\rho Q^*k]}{\partial k}(k-k^*) > \delta - \rho(Q-Q^*)(k-k^*).\end{array}\right\} \quad (S)$$

(S) is the basic stability assumption, although for economic problems attention is further restricted to HF's which are in addition convex in Q and concave in k, so that

$$(Q-Q^*)\frac{\partial H(Q,k)}{\partial Q} - \frac{\partial[H(Q,k)-\rho Q^*k]}{\partial k}(k-k^*) \geqq 0. \quad (10)$$

Even if the ε-δ bounds in S are removed, the weakened condition still ensures uniqueness of k*. If (S) is strengthened in the obvious way, it will provide for the uniqueness and global stability of the full rest point, (Q*,k*).

(S) is a "steepness" requirement on the HF, looking from any point, (Q,k) with $k \neq k^*$ to (Q*,k*), steepness must be bounded above that given by the quadratic form $-\rho(Q-Q^*)(k-k^*)$.

Cass-Shell [4] provides more motivation of condition (S) -- both from the economic and geometric points of view. Here, I merely remark that (S) can be thought of as a generalization of the bounded-value-loss condition of Radner [11] and reduces to Radner's condition when $\rho = 0$.

Condition (S) immediately suggests the choice of Lyapunov valuation function for stability analysis:

$$V = (Q-Q^*)(k-k^*). \quad (11)$$

Time differentiation in (11) and application of (10) yields:

$$\left.\begin{aligned} d(Ve^{-\rho t})/dt &\sim \dot{V} - \rho V \\ &= (Q-Q^*)\frac{\partial H(Q,k)}{\partial Q} - \frac{\partial[H(Q,k)-\rho Qk]}{\partial k}(k-k^*) \\ &\qquad -\rho(Q-Q^*)(k-k^*) \\ &= (Q-Q^*)\frac{\partial H(Q,k)}{\partial Q} - \frac{\partial[H(Q,k)-\rho Q^*k]}{\partial k}(k-k^*) \\ &\geqq 0. \end{aligned}\right\} \quad (12)$$

The transversality condition (9) requires that

$$\lim_{t\to\infty} Ve^{-\rho t} \leqq \lim_{t\to\infty} [Qe^{-\rho t}k + Q^*e^{-\rho t}k^*] = 0 \tag{13}$$

and hence, $V \leqq 0$ and $\lim_{t\to\infty} V = V^* \leqq 0$.

Proof of global stability. Suppose that $\lim_{t\to\infty} k(t) = k^*$ were not true. Then for some $\varepsilon > 0$ there would be a sequence of points $\{t_j\}$ such that $\|k(t_j)-k^*\| > 2\varepsilon$. Then if we assume uniform continuity[1] of $k(t)$ on the halfline $[0,\infty)$ -- there would also have to be a sequence of intervals $\{[\underline{t}_j,\bar{t}_j]\}$ such that $\bar{t}_j-\underline{t}_j > \Delta t > 0$ and $\|k-k^*\| > \varepsilon$ for $t \in [\underline{t}_j,\bar{t}_j]$. From (12), it follows that

$$\dot{V} = (Q-Q^*)\frac{\partial H(Q,k)}{\partial Q} - \frac{\partial[H(Q,k)-\rho Q^*k]}{\partial k}(k-k^*) + \rho V$$

so that (S) implies there is a $\delta > 0$ such that

$$\dot{V}(t) > \delta \text{ for } t \in [\underline{t}_j,\bar{t}_j].$$

Hence, for t' sufficiently large, it follows that both

$$V^* - \delta \leqq V(t) \leqq V^*$$

and

$$V(t) \geqq V(t') + \sum_{t' \leqq \underline{t}_j \leqq \bar{t}_j \leqq t} \delta(\bar{t}_j-\underline{t}_j) \geqq V(t')$$

$$+ [\max \{j : t' \leqq \underline{t}_j \leqq \bar{t}_j \leqq t\} - \min \{j : t' \leqq \underline{t}_j \leqq \bar{t}_j \leqq t\}]\delta\Delta t$$

for $t \geqq t'$, which are inconsistent, establishing that a solution to (9), if it exists, must exhibit stability in the sense that $\lim_{t\to\infty} k(t) = k^*$.

NOTE

[1]Uniform continuity is established in [4].

ACKNOWLEDGMENT

Research support of the National Science Foundation is gratefully acknowledged.

REFERENCES

1. W. A. Brock & J. Scheinkman, "Global Asymptotic Stability of Bounded Trajectories Generated by Modified Hamiltonian Dynamical Systems: A Study in the Theory of Optimal Growth," Journal of Economic Theory Symposium, forthcoming.
2. D. Cass, "Duality: A Symmetric Approach from the Economist's Vantage Point," Journal of Economic Theory, 7 (1974), 272-295.
3. D. Cass, paper on Hamiltonian duality, in preparation for a Buffalo conference volume.
4. D. Cass & K. Shell, "The Structure and Stability of Competitive Dynamical Systems," Journal of Economic Theory Symposium, forthcoming.
5. C. Caton & K. Shell, "An Exercise in the Theory of Heterogeneous Capital Accumulation," Review of Economic Studies, 38 (1971), 13-22.
6. F. H. Hahn, "On Some Equilibrium Paths" in (J. A. Mirrlees & N. Stern, editors) Models of Economic Growth, London: The Macmillan Press Limited, 1973, 193-211.
7. L. Lau, "A Characterization of the Normalized Restricted Profit Function," Journal of Economic Theory Symposium, forthcoming.
8. E. Malinvaud, "Capital Accumulation and Efficient Allocation of Resources," Econometrica, 21 (1953), 233-268.
9. L. W. McKenzie, "Accumulation Programs of Maximum Utility and the von Neumann Facet," in Value, Capital, and Growth, (J. Wolfe, ed.), Edinburgh University Press, 1968, 353-383.
10. H. Pollard, Mathematical Introduction to Celestial Mechanics, Englewood Cliffs, New Jersey: Prentice-Hall, Inc., 1966.
11. R. Radner, "Paths of Economic Growth that Are Optimal with Regard Only to Final States: A Turnpike Theorem," Review of Economic Studies, 28 (1961), 98-104.
12. R. T. Rockafellar, "Saddle Points of Hamiltonian Systems in Convex Problems of Lagrange," Journal of Mathematical Analysis and Applications, forthcoming.
13. R. T. Rockafellar, paper in preparation for Journal of Economic Theory Symposium, forthcoming.
14. P. A. Samuelson, "A Catenary Turnpike Theorem Involving Consumption and the Golden Rule," American Economic Review, 55 (1965), 486-496.
15. K. Shell (editor), Essays on the Theory of Optimal Economic Growth, Cambridge, Mass.: MIT Press, 1967.

16. K. Shell, "Applications of Pontryagin's Maximum Principle to Economics" in H. W. Kuhn & G. P. Szegö (eds.), Mathematical Systems Theory and Economics I, Berlin-Heidelberg-New York: Springer-Verlag, 1969, 241-292.
17. K. Shell, "On Competitive Dynamical Systems" in H. W. Kuhn & G. P. Szegö (eds.), Differential Games and Related Topics, Amsterdam: North-Holland Publishing Co., 1971, 449-476.
18. K. Shell, M. Sidrauski, & J. E. Stiglitz, "Capital Gains, Income, and Saving," Review of Economic Studies, 36 (1969), 15-26.
19. K. Shell & J. E. Stiglitz, "The Allocation of Investment in a Dynamic Economy," Quarterly Journal of Economics, 81 (1967), 592-609.
20. M. L. Weitzman, "Duality Theory for Infinite Horizon Convex Models," Management Science, 19 (1973), 783-789.

AVERAGED HAMILTONIANS IN DIFFERENTIAL GAMES

Robert J. Elliott

Department of Pure Mathematics, University of Hull,
HULL, Yorkshire. England.

1. INTRODUCTION.

Consider a differential game G whose dynamics are given by the equations $(x \in R^m)$

$$\frac{dx}{dt} = f(t,x,y,z),$$

$$x(0) = 0 \in R^m$$

Here $t \in [0,1]$ and f is continuous; to simplify the argument we suppose f satisfies a uniform Lipschitz condition in t and x. Further, we suppose the payoff has the form

$$P = g(x(1))$$

where g is real valued, twice differentiable and $\frac{\partial g}{\partial t}$, $\frac{\partial g}{\partial x_i}$, $\frac{\partial^2 g}{\partial x_i \partial x_j}$ satisfy uniform Lipschitz conditions in t and x.

Write K for the Lipschitz constant in all cases. The situation where f satisfies weaker Lipschitz and continuity conditions and the payoff has a more general form, can be treated by approximation arguments as in [3].

The game is zero sum so we suppose there is a player J_1 choosing y from a compact metric space Y with the object of maximising P, whilst a player J_2 chooses z from a compact

J. D. Grote (ed.), The Theory and Application of Differential Games. 201-207.

metric space Z with the object of minimising P.

In our first two lectures we have seen that a differential game has an upper value U and a lower value V. In fact we saw there is an upper value W^+ in the sense of Fleming, an upper value V^+ in the sense of Friedman and an upper value U defined in terms of strategies as in section 5 of our first lecture, but because all these values are limits of the unique solutions of a family of parabolic equations the upper values are all equal.

The Fleming and Friedman upper values, W^+ and V^+ respectively, are limits of the upper values of the approximating upper n-stage games, W_n^+ and V_n^+. As described in the first lecture, in these approximating games the time interval $[0,1]$ is divided into n equal subintervals and the player J_1 has an advantage because it is supposed that J_2 chooses, and makes known his choice of control first, at each step.

The idea of studying approximating differential games, in which J_1 and J_2 make their moves alternately by choosing piecewise constant controls on overlapping time intervals, is due to John Danskin [1]. The value of Danskin's approximating game lies between W_n^+ and W_n^+, and Danskin proves the existence of its limit as $n \to \infty$ by introducing "random moves".

In this talk we describe work done with A. Friedman and N. Kalton and sketch how, by adapting the techniques described in our second lecture, we can relate the values of differential games played in this alternate manner to the solutions of certain parabolic equations. Both cases, when the players use constant controls and measurable controls, are discussed and it is of interest that in general the values obtained are different. Full details can be found in the paper [2] written with A. Friedman and N. Kalton.

2. ITERATED TWO PERSON ZERO SUM GAMES.

Suppose we have a continuous real valued function $\phi(y,z)$, where y is chosen by a player J_1 from a compact metric space Y with the object of maximising ϕ and z is chosen by a player J_2 from a compact metric space Z with the object of minimising ϕ. As in the first lecture we see that the best that J_1 can force is

$$\max_y \min_z \phi(y,z) ,$$

and the best that J_1 can force is

$$\min_z \max_y \phi(y,z) .$$

Consider the effect of playing the game several times and taking the average payoff, but that now J_1 and J_2 play alternately in an overlapping manner. Danskin [1] approximated the play in differential games in this way.

Suppose $0 \leq \sigma \leq 1$, and for simplicity first consider a three step game in which J_2 chooses $z_0 \in Z$ at time 0, then J_1 chooses y_0 at time 0^2 and finally, in reply, J_2 chooses $z_1 \in Z$ at time σ. The payoff is defined to be:

$$P(z_0, y_0, z_1) = \sigma \phi(y_0, z_0) + (1-\sigma) \phi (y_0, z_1)$$

and the best the two players can force with this method of play, that is the value of this game, is:

$$\min_{z_0} \max_{y_0} \min_{z_1} P(z_0, y_0, z_1) .$$

More generally consider this game played repeatedly over a time interval of length n : J_2 chooses elements of Z at times $0, \sigma, 1+\sigma, \ldots , n-1+\sigma$ and J_1 chooses elements of Y at times $0, 1, \ldots , n-1$. The value of this game is V_n^σ where V_n^σ is defined as follows:

$$V_0^\sigma (z) = 0 ,$$

$$V_n^\sigma (z) = \max_y \min_\xi \{\sigma\phi(y,z)+(1-\sigma)\phi(y,\xi)+V_{n-1}^\sigma (\xi)\}$$

and $V_n^\sigma = \min_z V_n^\sigma(z)$

Write $B \equiv \max_y \max_z |\phi(y,z)|$.

Then by considering

$$V_{m+n}^\sigma = \min_{z_0} \max_{y_0} \ldots \min_{z_{m+n}} (\sigma\phi(y_0, z_0)+ \ldots +(1-\sigma)\phi(y_{m+n}, z_{m+n}))$$

we see

$$|V_{m+n}^\sigma - V_m^\sigma - V_n^\sigma | \leq 4B \qquad (*)$$

However, we are really interested in the average values of repeated plays of the original game so write:

$$r_n^\sigma = V_n^\sigma/_n$$

From (*) we deduce

$$|r_{mn}^\sigma - r_m^\sigma| \le 4B/_m$$

$$|r_{mn}^\sigma - r_n^\sigma| \le 4B/_n ,$$

so $$|r_m^\sigma - r_n^\sigma| \le 8B\ (^1/_m + {}^1/_n) .$$

Consequently r_n^σ is a Cauchy sequence. Writing $\phi_\sigma^C = \lim_{n\to\infty} r_n^\sigma$ we observe that ϕ_σ^C is the limit of average values of repeated plays of the original game when the moves of the two players overlap in the ratio $\sigma : 1-\sigma$.

Note that

$$\phi_0^C = \max_y \ \min_z \ \phi(y,z)$$

$$\phi_1^C = \min_z \ \max_y \ \phi(y,z)$$

and $$\phi_0^C \le \phi_\sigma^C \le \phi_1^C$$

The C above indicates that J_1 and J_2 use constant controls at each step. If they can use measurable functions a different result is obtained because they can anticipate their opponent's choice.

Write $M_1([a,b])$ (resp. $M_2([a,b])$ for the set of measurable functions on $[a,b]$ with values in Y (resp. Z). For $0\le\sigma\le 1$ and $z\in M_2\ ([0,\sigma])$ define:

$$W_0^\sigma(z) = 0$$

$$W_n^\sigma(z) = \sup_{y\in M_1([0,1])} \ \inf_{\xi\in M_2([\sigma,\ 1+\sigma])} \ \Big(\int_\theta^\sigma \phi(y(t),z(t))dt +$$

$$+\int_\sigma^1 \phi(y(t),\zeta(t))dt + W_{n-1}^\sigma(\zeta)\Big)$$

and $$W_n^\sigma = \inf_{z\in M_2([0,\sigma])} \ W_n^\sigma(z) .$$

We now see that

$$W_n^\sigma = n(\sigma\phi_1^C + (1-\sigma)\phi_0^C).$$

If we write $s_n^\sigma = W_n^\sigma/_n$ then s_n^σ can be considered as the average of repeated plays of the original game in which J_1 and J_2, playing in an overlapping way, can choose measurable functions as controls.

Clearly $s_n^\sigma = \sigma\phi_1^C+(1-\sigma)\phi_0^C$ so $\lim_{n\to\infty} s_n^\sigma = \phi_\sigma^F$ exists and $\phi_\sigma^F = \sigma\phi_1^C+(1-\sigma)\phi_0^C$.

3. ALTERNATE PLAY IN DIFFERENTIAL GAMES

Let us return now to the differential game with dynamics

$$\frac{dx}{dt} = f(t,x,y,z)$$

$$x(o) = 0 \ , \ x\in R^m, \ t\in[0,1] \ ,$$

and payoff: $P = g(x(1))$ as described in the first part of this talk. Suppose $0\leq\sigma\leq 1$ Consider a partition of the time interval $[0,1]$ into N equal subintervals and an approximating game GC_σ^N (C for constant controls), played in the following manner: At time 0 J_2 chooses a control value $z_0\in Z$. Then also at time 0, with the knowledge of z_0, J_1 chooses a control value $y_0\in Y$. At times $i+\sigma/_n$, $i=0\ 1, \ldots , N-1$, with the knowledge of previous control choices by both players, J_2 chooses control values $z_i\in Z$, and similarly J_1 chooses control values $y_i\in Y$ at times $i/_N$, $i=1, \ldots , N-1$, with the knowledge of the previous controls chosen. In this way J_1 and J_2 are playing in an overlapping way. The piecewise constant controls when substituted in the dynamics determine a trajectory and a payoff. The best that J_1 and J_2 can force with this method of play is the value:

$$VC^N(\sigma) = \min_{z_0}\ \max_{y_0} \ldots \min_{z_n}\ g(x(1)).$$

If the players make choices of controls at the same time as described above, but they may now choose measurable functions instead of constant control values, an approximating game

GF_σ^N, (F for functions), is generated. These measurable controls again determine a trajectory and payoff, and the best that J_1 and J_2 can obtain with this method of play is the value:

$$VF^N(\sigma) = \inf_{z_0(t)}\ \sup_{y_0(t)} \ldots \inf_{z_n(t)}\ g(x(1)),$$

where $y_i(t)\in M_1([\frac{i}{N}, \frac{i+1}{N}))$, $i=0, \ldots ,N-1$, $z_0(t)\in M_2([0,\sigma])$,

$z_i(t)\in M_2((\frac{i+\sigma}{N}, \frac{i+1+\sigma}{N}])$, $i=0, \ldots N-2$, and $z_n(t)\in M_2((N-1+\sigma/_n, 1]$.

4. AVERAGED HAMILTONIANS AND STOCHASTIC GAMES.

For any $p\in R^m$, $x\in R^m$ and $t\in[0,1]$ we can consider the Hamiltonian

$H(y,z) = H(t,x,p;y,z) = p.f(t,x,y,z)$ as a function of $y\in Y$ and $z\in Z$.

Consequently, as described in section 2, for $0\leq\sigma\leq 1$, we have both the H^C_σ and the H^F_σ values of $H(y,z)$, and

$$H^C_o = H^F_o = \max_y \min_z p.f(t,x,y,z)$$

$$H^C_1 = H^F_1 = \max_z \min_y p.f(t,x,y,z)$$

Write H_σ for H^C_σ or H^F_σ. Then again by results on parabolic equations due to Friedman [6] and Oleinik and Kruzkov [7] the equation:

$$(\varepsilon^2/_2)\ \nabla^2\phi+\frac{\partial\phi}{\partial t}+H_\sigma(x,t,\nabla\phi)=0$$

$$\phi(1,x) = g(x)$$

has a unique solution ϕ^σ_ε .

As in the second lecture, related stochastic versions of GC^N_σ and GF^N_σ can be introduced in which a discrete "amount ε of noise" enters the dynamics at times i/N, $i=1, \ldots ,N$. In the piecewise constant control case further subdivisions of the time interval into MN subintervals are introduced to discuss the convergence of the Hamiltonian to $H^C_\sigma(t,x,p)$,(but noise only enters at the i/N division points). These stochastic games have values $VC^N_\varepsilon(\sigma)$ and $VF^N_\varepsilon(\sigma)$ respectively. By introducing modified definitions of stochastic controls and strategies and adapting the results described in the second lecture it is shown that $VC^N_\varepsilon(\sigma)$ and $VF^N_\varepsilon(\sigma)$ are approximated by the appropriate unique solution ϕ^σ_ε as $N\to\infty$. Also it is proved that as

$\varepsilon \to o$ $VC_{\varepsilon}^{N}(\sigma)$ approaches $VC^{N}(\sigma)$ and $VF_{\varepsilon}^{N}(\sigma)$ approaches $VF^{N}(\sigma)$.

Consequently we can conclude that

$$\lim_{N\to\infty} VC^{N}(\sigma) = V^{C}(\sigma)$$

and $$\lim_{N\to\infty} VF^{N}(\sigma) = V^{F}(\sigma)$$

both exist.

Further, the dynamic programming identities of [5] can be proved in this alternate play situation and from them it is deduced that, at points of differentiability:

$$\frac{\partial V^{C}}{\partial T} + H_{\sigma}^{C}(t,x,\nabla V^{C}) = 0$$

and $$\frac{\partial V^{F}}{\partial t} + H_{\sigma}^{F}(t,x,\nabla V^{F}) = 0$$

Details of the above can be found in reference [2].

REFERENCES

1. J. Danskin, Values in differential games, Bull. Amer. Math. Soc. 80 (1974) To appear.
2. R.J. Elliott, A Friedman and N. J. Kalton, Alternate play in differential games, J. Diff. Eqns., 15 (1974), 560-588.
3. R.J. Elliott and N.J. Kalton, The existence of value in differential games, Memoir of the Amer. Math. Soc. 126, Providence, R.I. (1972)
4. R.J. Elliott and N.J. Kalton, Upper values of differential games, J. Diff. Eqns. 14 (1973), 89-100.
5. R.J. Elliott and N.J. Kalton, Cauchy problems for certain Isaacs-Bellman equations and games of survival, Trans. Amer. Math. Soc. To appear.
6. A. Friedman, Partial Differential Equations of Parabolic Type, Prentice Hall, Englewood Cliffs, N.J. (1964)
7. O.A. Oleinik and S.N. Kruzkov, Quasi-linear parabolic equations of second order with many independent variables, Usp. Mat. Nauk. 16 (1961), 115-155.

COMPUTATIONAL ASPECTS IN TWO-LEVEL DIFFERENTIAL GAMES

L.F.PAU

Laboratoire d'Automatique,ENS des Télécommunications,
46 rue Barrault,F 75634 Paris Cedex 13, France°

SUMMARY. The basic principle of most planning and budgeting systems is an allocation of resources between a number of subordinate units who prepare a set of requests , and a central decision-maker who eventually approves or rejects the requests . Decentralized planning following primal or dual decomposition schemes are well known {2,3,7 }; however , the lower units are here assumed to have full information at that level , and to act as to pursuing the objectives of the central unit.
The first purpose of this paper is therefore to elaborate upon the way in which the planning problems at the supremal and infimal levels may be treated through constraint coordination , in the case where all units (including the supremal unit) have conflicting goals . Generalizing the approaches in {5,6,7,8,10}, the evolutions of the supremal and infimal units are to be governed by a global differential game of fixed duration,with state constraints and a given initial state (Sections 1.,4.). Stochastic hierarchical sequential games,with discrete states at each level,inter-level and intra-level transition probabilities ,have besides been studied in {4}.
The second purpose of this paper is to show how projection operators from the global state and control spaces into the individual state and control spaces,may define the information structure in a two-level game(Section 1). This allows to describe situations where a large country(supremal player) has only partial control on, and information about, a community of smaller countries (infimal players).
Lastly, a constraint coordination algorithm is studied for the purpose of approximating closed-loop Nash equilibrium strategies in a N-person game, via the computation of such strategies in an infimal (N-1)-person game played in a two-level system with the N'th player at the supremal level (Sections 2,3,5) .

J. D. Grote (ed.), The Theory and Application of Differential Games. 209-220.

1. TWO-LEVEL GAME

Let be given: E^k, euclidean space of dimension k; $\nu,\rho,N,n_k,\rho_k,l_k \in \mathbb{N}^*$, $k=1,..,N$, given integers; $\mathcal{E}_k \subseteq E^{n_k}$, $k=1,..,N$, given convex set; $\mathcal{E}_{N+1} \triangleq (0,T) \subset E$; $P(.,\mathcal{E}_k)$, projection operator, $y_k=P(x,\mathcal{E}_k)\in\mathcal{E}_k$, $x\in E^\nu$, $k=1,..,(N+1)$; U^k, Hilbert space on $\mathbb{R}$ of all real control functionals u, piecewise continuous on $(0,T)$ and with values in E^k; $\mathcal{U}_k \subseteq U^{\rho_k}$, $k=1,..,N$, given open convex set; $Q(.,\mathcal{U}_k): U^\rho \to \mathcal{U}_k$, projection operator, $v_k=Q(u,\mathcal{U}_k)\in\mathcal{U}_k$, $u\in U^\rho$, $k=1,..,N$; $\mathcal{E}_k,\mathcal{U}_k$ are convex in order to have the unicity of the projections y_k,v_k, $k=1,..,N$; $<\cdot,\cdot>_E$, scalar product in E^ν; $<\cdot,\cdot>_U$, Hilbert scalar product in U^ρ; $i(.,A,B)$ canonic injection from A into B, where A and B are two sets, and $A \subseteq B$.
Let be given a N-player differential game with players J_k, $k=1,..,N$ {1} on a fixed horizon $t\in(0,T)$, governed by the evolution equation:

$$\dot{z} = F(z,u) \triangleq (f_o(x,u),f(x,u)) \quad f_o \triangleq (f_{o1},..,f_{oN}) \quad x\in E^\nu \quad z^\circ \text{ given}$$

$$z = (x_o,x) \in E^{N+\nu} \quad x_o \triangleq (x_{o1},..,x_{oN}) \in E^N \quad \dot{x} = f(x,u) \tag{1}$$

In the state space $\mathcal{E}_k$ of player J_k, $k=1,..,N$:

$$\dot{x}_{ok} = f_{ok}(y_k,v_k) \qquad y_k\in Y_k \subseteq \mathcal{E}_k \qquad v_k \in V_k \subseteq \mathcal{U}_k \tag{2}$$

$$y_{N+1} = t \in (0,T) \qquad (\nu-1) \geqslant \mathrm{Sup}\{n_k;k=1,..,N\} \quad \rho \geqslant \mathrm{Sup}\{\rho_k;k=1,..,N\}$$

where Y_k,V_k are closed compact non-empty sets defining the constraints on the state y_k and the control v_k, respectively, $k=1,..,N$; F is C^1 on $E^{N+\nu}\times U^\rho$; f_o and f are C^1 on $E^\nu\times U^\rho$; z is the state, and u is the control in the global game (1).
D1. The N players are organized according to a two-level hierarchical game with levels I and II, J_1 being at the supremal level I, J_k $k=2,..,N$ at the infimal level II, iff one of the conditions (i) or ((i) and (ii)) holds :

(i) $\forall u \in U^\rho\ \forall k=2,..,N \quad <i(v_1,\mathcal{U}_1,U^\rho),i(v_k,\mathcal{U}_k,U^\rho)>_U = 0$
(ii) $\forall x \in E^\nu\ \forall k=2,..,N \quad <i(y_1,\mathcal{E}_1,E^\nu),i(y_k,\mathcal{E}_k,E^\nu)>_E = 0$

If (i), the game is weakly hierarchical; if both (i) and (ii) hold, the game is strongly hierarchical. ∎
In a weakly hierarchical game, the states of the level-II players may depend on the control v_1 of I, and the state y_1 of J_1 may depend on the states of the level-II players. In a strongly hierarchical game, the states of the level-II players may depend on the control v_1 of I, but the state y_1 of J_1 is independent on both controls and states of the level-II players. This latter dependence relation is in the sense of the scalar products $<\cdot,\cdot>_U$, $<\cdot,\cdot>_E$ of D1., and is directly connected to the definitions of the sets $\mathcal{E}_k,\mathcal{U}_k$ and of the operators $P(.,\mathcal{E}_k),Q(.,\mathcal{U}_k)$, $k=1,..,N$.
If D1., then J_k will minimize at level-II a terminal cost $\phi_k:E\to E$ having the value $\phi_k(x_{ok})$, $k=2,..,N$; this will usually require the knwledge of both y_1,v_1 from (1).

In the following , J_1 of I will use controls $v_1 \in V_1$, while the level-II players will use Nash-optimal strategies defined by D2, D3,D4,D5 .

D2. $u \in U^\rho$ is a feasible control functional with respect to $v_1 \in V_1$ and $z^\circ \in E^{N+\nu}$, iff $\dot{z}=F(z,u)$ (1) has a solution $\underline{z}(t) \triangleq (\underline{x}_o(t),\underline{x}(t)) \in E^{N+\nu}$, such that: $\underline{z}(0)=z^\circ$, $v_1=Q(u,\mathcal{U}_1) \in V_1$, $Q(u,\mathcal{U}_k) \in V_k$ for $k=2,.,N$, and $P(\underline{x}(t),\mathcal{E}_k) \in Y_k$ for $k=1,.,(N+1)$; we may then define ψ : $U^\rho \times V_1 \times E^{N+\nu} \times (0,T) \to E^{N+\nu}$ $\underline{z}(t) \triangleq \psi(u,v_1,z^\circ,t)$; $\underline{z}(t)$ is the emission of (1) from z° for the control u . ▮

D3. A strategy p_k of J_k ,k=1,.,N, on X,$X \subseteq E^\nu$,is a function of x (1) defined on $X \ni x$; $p_k : X \to \mathcal{U}_k$; $\forall x \in X: p_k(x) \in V_k$; $y_{N+1}=t \in (0,T)$ ▮

D4. $x \in X$ is attainable relatively to the strategy $p \triangleq (p_2,\ldots,p_N)$, and for given $v_1 \in V_1$ and $z^\circ \in E^{N+\nu}$,iff there exists a control functional $u \in U^\rho$ such that :

(i) $v_1=Q(u,\mathcal{U}_1)$, $p_k(x)=Q(u,\mathcal{U}_k)$, k=2,.,N;

(ii) u is feasible with respect to v_1,z°;

(iii) x belongs to the path $\{(t,\xi): \xi=\underline{x}(t) \in E^\nu, \underline{z}(t)=\psi(u,v_1,z^\circ,t) = (\underline{x}_o(t),\underline{x}(t)),\ t \in (0,T)\} \subset (0,T) \times E^\nu$

We may then define $u \triangleq p_{v1}(x)$ ▮

D5. $p^* =(p_2^*,\ldots,p_N^*)$, defined on X , is called a Nash-optimal (N-1)-tuple in x° for given v_1,z°, iff any $x \in X$ is attainable relatively to the strategies p^* and $p_k^\bullet=(p_2^*,\ldots,p_{k-1}^*,p_k,p_{k+1}^*,\ldots,p_N^*)$ k=2,.,N, and if $\forall$ k=2,.,N :

(i) $\phi_k^*=\underline{x}_{ok}^*(T)$ is defined and its value is given by $\underline{z}^*(t)= \psi(p_{v1}^*(x),v_1,z^\circ,t) = (\underline{x}_o^*(t),\underline{x}^*(t))$

(ii) $\phi_k^\bullet=\underline{x}_{ok}^\bullet(T)$ is defined and its value is given by $\underline{z}^\bullet(t)= \psi(p_{kv1}^\bullet(x),v_1,z^\circ,t)= (\underline{x}_o^\bullet(t),\underline{x}^\bullet(t))$

(iii) $\phi_k^\bullet \geqslant \phi_k^*$ ▮

2. CONSTRAINT COORDINATION : THE LINEAR CASE

Coordination constraints (5) will sequentially be introduced, whereby y_1 from I determines constraints on y_k,k=2,.,N, of II. Moreover, II and y_k,k=2,.,N, determine sequentially a constraint (4) on the state y_1 of I .

D6. Let be given $a^\circ(t) \in L(E^{n1},E^{l1})$; $\forall t \in (0,T)$,we define a recursive relation (3),where η is C^1 , and defined everywhere in $L(E^{n1},E^{l1}) \times E^\nu$:

$$a^{m+1}(t) = \eta(a^m(t),x) \qquad m \in \mathbb{N} \qquad t=y_{N+1} \in (0,T) \quad ▮ \tag{3}$$

D7. Problem Π_I^m of J_1 ,$m \in \mathbb{N}$: Assume the existence of an optimal control {8} $\tilde{u}^m \in U^\rho$ of the following problem called Π_I^m, where $v_1 \in V_1$: find u such that :

$$\tilde{\phi}_1^m \triangleq \operatorname*{Min}_{v_1 \in V_1} \underline{x}_{o1}(T) \qquad y_1= P(x,\mathcal{E}_1) \in Y_1 \qquad v_1=Q(u,\mathcal{U}_1) \quad u \in U^\rho$$

$$\underline{z}(t) \triangleq (\underline{x}_o(t),\underline{x}(t))= \Psi(u,v_1,z^\circ,t) \text{ solution of (1)} \qquad t \in (0,T)$$

$$y_1 \geqslant 0 \quad a^m(t)y_1 \leqslant b(t) \quad , \ b(t) \in E^{l1} \text{ given} , \ m \in \mathbb{N} \tag{4}$$

▮

<u>D8. Problem Π_{II}^{m*} of player J_k; $k=2,..,N$; $m\in\mathbb{N}$</u> : Define $\tilde{v}_1^m=Q(\tilde{u}^m,\mathcal{U}_1)$, $\tilde{z}^m=(\tilde{x}_o^m,\tilde{x}^m)=\psi(\tilde{u}^m,\tilde{v}_1^m,z^\circ,t)$ $\tilde{y}_1^m=P(\tilde{x}^m,\mathcal{E}_1)$ $t=P(\tilde{x}^m,\mathcal{E}_{N+1})\in(0,T)$. Assume the existence {14} of a Nash-optimal (N-1)-tuple $p^{*m}\triangleq(p_2^{*m},..,p_N^{*m})$ at point x° , defined on $X^m\subseteq E^\nu$ for given $v_1=\tilde{v}_1^m$,z° , and solution of the following problem called Π_{II}^{m*}: find a strategy p^*, in accordance with D5, for :

$$X^m = G^m\cap\{x\in E^\nu ; e_{lk}(\tilde{y}_1^m)\leqslant\lambda_{lk}y_k \quad \mu_{lk}y_k\leqslant d_{lk}(\tilde{y}_1^m), l=1,..,n_1 \ k=2,..,N\}$$

$$G^m = \{x\in E^\nu ; y_k= P(x,\mathcal{E}_k)\in Y_k , k=1,..,N\} \tag{5}$$

λ_{lk}, $\mu_{lk}\in L(E^{n_k},E)$; e_{lk}, d_{lk} real continuous functions , strictly increasing, and defined everywhere on E ; e_{lk}^{-1}, d_{lk}^{-1} inverse functions of the preceding ones ; we may then define: $\forall x\in X^m$: $\underline{z}^{*m}(t) = (x_o^{*m}(t),x^{*m}(t))=\psi(p^{*m}_{\tilde{v}}(x),\tilde{v}_1^m,z^\circ,t)\triangleq(x_o^{*m},x^{*m})$ $y_k^{*m}=Q(x^{*m},\mathcal{E}_k)$ $\phi_k^{*m}=\underline{x}_{ok}^{*m}(T)$ $k=2,..,N$ ∎

In this two-level game , when solving Π_{II}^{m*} it is necessary previously to have solved the optimal control problem Π_I^m ,$m\in\mathbb{N}$. The recursive relation (6) obtained from (3) will sequentially modify the coordination constraint (4) of I :

$$a^{m+1}(t) = \eta(a^m(t),x^{*m}) \qquad a^\circ(t) \text{ given} \tag{6}$$

Using (6) , one proceeds from the problem Π_I° to the problems Π_I^m, Π_{II}^{m*} ,$m\in\mathbb{N}$, η being defined for example by L1.

<u>L1.</u> If the function η is defined by:

$$a_{ij}^{n+1}(t)=\frac{1}{N-1}\sum_{k=2}^{N}(a_{ij}^n(t)d_{jk}^{-1}(\mu_{jk}y_k^{*n}))\Big/(e_{jk}^{-1}(\lambda_{jk}y_k^{*n})) \tag{7}$$

$$a^n(t) = a_{ij}^n(t) \qquad i=1,..,l_1 \qquad j=1,..,n_1 \qquad n$$

,and if p^{*n},$\tilde{x}^n$,$\tilde{u}^n$ exist for m=n , then $\tilde{y}_1^n$,$\tilde{x}^n$,$\tilde{u}^n$ will also verify (4) for m=(n+1) . ∎ Proof: {11} .

<u>L2.</u> If $\tilde{u}^\circ$, $p^{*\circ}$ exist , and if the domains $X^\circ,..,X^n$ $n\in\mathbb{N}$, are non-empty , then it is possible to construct by (3) a non-increasing sequence of player J_1's performance index at level I :

$$\tilde{\phi}_1^n\leqslant\cdots\cdots\leqslant\tilde{\phi}_1^\circ \quad . \ ∎ \tag{8}$$

According to L1 , one may show that $X^\circ\subseteq\cdots\subseteq X^n$, because the sequences $[a_{ij}^n(t)]$ are non-increasing $\forall n\in\mathbb{N},\forall t\in(0,T)$.

By modifying $a^m(t)$ sequentially (6) , one mooves the equilibrium among the players $J_2,..,J_N$, and changes the priorities among these (i.e. the resources allocated) (5) , in order to minimize the index of performance of the supremal player J_1 .

3. CONSTRAINT COORDINATION : THE NON-LINEAR CASE

<u>D9.</u> Let be given $\beta_o:E^{n_1}\to E$, $\Psi:E^{n_1}\times E^\nu\to E$; for all $\varepsilon\geqslant 0$,we define the recursive relation (9) , in which Ψ , β_m,$m\in\mathbb{N}$ are C^1 and defined

on $E^{n_1}, E^{n_1} \times E^{\nu}$ respectively:

$$\beta_{m+1}(y_1) = \beta_m(y_1) + \varepsilon\, \Psi(y_1, x^{*m}) \qquad m \in \mathbb{N}, x \in E^{\nu}, y_1 = P(x, \mathcal{E}_1) \quad (9)$$

where: x^{*m} is the Nash-optimal state provided by the problem Π_{II}^{m*} for given $v_1 = \tilde{v}_1^m$ and $z°$ (D8) , $m \in \mathbb{N}$. ∎

Assume the existence of C^1 functions: $\gamma_k : \mathcal{E}_k \to E, \mu_k : \mathcal{U}_k \to E, k=1,..,N$ which may characterize feasible states y_k and feasible controls v_k of player J_k for $k=1,..,N$:

$$\begin{aligned} y_k &= P(x, \mathcal{E}_k) \in Y_k \iff \gamma_k(y_k) \leq 0 \qquad k=1,..,N \\ v_k &= Q(u, \mathcal{U}_k) \in V_k \iff \mu_k(v_k) \leq 0 \qquad k=1,..,N \end{aligned} \quad (10)$$

According to D2 , $u \in U^{\rho}$ will then be a feasible control with respect to $z° \in E^{N+\nu}$,**iff** $\dot{z} = F(z,u)$ (1) has a solution $\underline{z}(t) \triangleq (\underline{x}_o(t), \underline{x}(t)) \in E^{N+\nu}$ $t \in (0,T)$, such that $\underline{z}(0) = z°$, and fulfilling: (10) , $P(\underline{x}(t), \mathcal{E}_{N+1}) \in (0,T)$, $\gamma_k(P(\underline{x}(t), \mathcal{E}_k)) \leq 0$ for $k=1,..,N$.

D10. Problem $\Pi_I^m(\varepsilon)$ of J_1 , $m \in \mathbb{N}$: Assume that there is an $\varepsilon_o > 0$ independent of $m \in \mathbb{N}$, such that there exists {8} an optimal control $\tilde{u}^m(\varepsilon) \in U^{\rho}$, solution of the following problem called $\Pi_I^m(\varepsilon)$ where $v_1 \in V_1$, for any ε $0 \leq \varepsilon < \varepsilon_o$: find a feasible control u relatively to $z°$ such that :

$$\tilde{\phi}_1^m(\varepsilon) \triangleq \operatorname*{Min}_{v_1 \in V_1} \underline{x}_{o1}(T,\varepsilon) \quad y_1 = P(x, \mathcal{E}_1) \quad v_1 = Q(u, \mathcal{U}_1) \quad u \in U^{\rho}$$

$$\gamma_k(y_k) \leq 0 \quad \mu_k(v_k) \leq 0 \quad t = y_{N+1} \in (0,T) \quad k=1,..,N \quad 0 \leq \varepsilon < \varepsilon_o$$

$$\underline{z}(t,\varepsilon) \triangleq (\underline{x}_o(t,\varepsilon), \underline{x}(t,\varepsilon)) = \psi(u, v_1, z°, t) \quad \text{solution of (1)}$$

$$\beta_{m+1}(y_1) = \beta_m(y_1) + \varepsilon\, \Psi(y_1, x^{*m}) \leq 0 \quad , m \in \mathbb{N} \quad (11)$$

We may then define $\tilde{v}_k^m(\varepsilon) \triangleq Q(\tilde{u}^m(\varepsilon), \mathcal{U}_k)$ $k=1,..,N$; $\tilde{z}^m(\varepsilon) = (\tilde{x}_o^m(\varepsilon), \tilde{x}^m(\varepsilon)) \triangleq \psi(\tilde{u}^m(\varepsilon), \tilde{v}_1^m(\varepsilon), z°, t)$; $\tilde{y}_k^m(\varepsilon) \triangleq P(\tilde{x}^m(\varepsilon), \mathcal{E}_k)$ $k=1,..,N$. ∎

According to this latter definition , for any given $\varepsilon \geq 0$ in the recursive relation D9 , $\tilde{u}_1^m(0)$ is an optimal control for the problem $\Pi_I^m(0)$ with the index of performance $\tilde{\phi}_1^m(0)$, while $\tilde{u}^{m+1}(0) = \tilde{u}^m(\varepsilon)$ is an optimal control for the problem $\Pi_I^{m+1}(0) \equiv \Pi_I^m(\varepsilon)$ with the index of performance $\tilde{\phi}_1^{m+1}(0) = \tilde{\phi}_1^m(\varepsilon)$.

Let A,B be two reflexive Banach spaces, and $g: A \to B$ a differentiable functional ; we note $dg(c)/da \in L(A,B)$ the Fréchet derivative of g at the point $c \in A$.

We will now apply to the problem $\Pi_I^m(\varepsilon)$ the necessary conditions of the maximum principle , taking into account the constraints on the control $u \in U^{\rho}$ and on the state $x \in E^{\nu}$, together with the coordination constraint (11) .

T1. Necessary optimality conditions for $\Pi_I^m(\varepsilon)$, $m \in \mathbb{N}$, $0 \leq \varepsilon < \varepsilon_o$:
Assume that , for any ε $0 \leq \varepsilon < \varepsilon_o$:

(i) $p^{*m}, x^{*m}, \tilde{u}^m(\varepsilon) \in U^{\rho}, \tilde{x}^m(\varepsilon)^o, \tilde{z}^m(\varepsilon)$ exist (D10) for given $z°$

(ii) $$\begin{bmatrix} d\beta_{m+1}(\tilde{y}_1^m(\varepsilon))/du & \beta_{m+1}(\tilde{y}_1^m(\varepsilon)) & 0 & \cdots & & & \cdots & 0 \\ 0 & 0 & \gamma_1(\tilde{y}_1^m(\varepsilon)) & & & & & \vdots \\ \vdots & \vdots & & \ddots & & & & \vdots \\ 0 & 0 & \cdots\ 0 & \gamma_N(\tilde{y}_N^m(\varepsilon)) & 0 & \cdots & & 0 \\ d\mu_1(\tilde{v}_1^m(\varepsilon))/du & 0 & \cdots & 0 & \mu_1(\tilde{v}_1^m(\varepsilon)) & 0 & \cdots & 0 \\ \vdots & \vdots & & & & \ddots & & 0 \\ d\mu_N(\tilde{v}_N^m(\varepsilon))/du & 0 & \cdots & & \cdots\ 0 & & & \mu_N(\tilde{v}_N^m(\varepsilon)) \end{bmatrix}$$

$=C(\tilde{x}^m(\varepsilon),\tilde{u}^m(\varepsilon),t)\in L(E^{2N+1}\times U^\rho,E^{2N+1})$ is a linear mapping of rank $(2N+1)$, for all ε $0\leq\varepsilon<\varepsilon_o$ and $t\in(0,T)$.

Then, there exist $(2(N+1)+\nu)$ non-negative scalar multipliers : q_o, $q_i(t,\varepsilon)$ $i=1,.,\nu$, $r_k(t,\varepsilon)$ $k=0,.,2N$, not simultaneously equal to zero, and a function :

$$H(x,u,q,r)= -q_o f_{o1}(y_1,v_1)+q(t,\varepsilon)f(x,u)-r_o(t,\varepsilon)\{\beta_m(y_1)+\varepsilon\Psi(y_1,x^{*m})\} - \sum_{k=1}^{N}(r_k(t,\varepsilon)\gamma_k(y_k)+r_{N+k}(t,\varepsilon)\mu_k(v_k))$$

$$q_o\geq 0 \ ; \ q(t,\varepsilon)=\{q_i(t,\varepsilon),i=1,.,\nu\}\in L(E^\nu,E) \qquad t=y_{N+1}\in(0,T)$$

$$r(t,\varepsilon)=\{r_k(t,\varepsilon),k=0,.,2N\}\in L(E^{2N+1},E)$$

where : $\dot{q}(t,\varepsilon)= - dH(\tilde{x}^m(\varepsilon),\tilde{u}^m(\varepsilon),q(t,\varepsilon),r(t,\varepsilon))/dx$

such that :

(i) $r(t,\varepsilon)$ is a piecewise continuous function of $t\in(0,T)$, and continuous whereever $t\in(0,T)\to\tilde{u}^m(\varepsilon)$ is continuous;

(ii) $r_k(t,\varepsilon)\geq 0$ $k=0,.,2N$;

(iii) $r_o(t,\varepsilon)(\beta_m(\tilde{y}_1^m(\varepsilon))+\varepsilon\Psi(\tilde{y}_1^m(\varepsilon),x^{*m}))=0$ $\quad r_k(t,\varepsilon)\gamma_k(\tilde{y}_k^m(\varepsilon))=0$ $\quad r_{N+k}(t,\varepsilon)\mu_k(\tilde{v}_k^m(\varepsilon))=0$ $k=1,.,N$;

(iv) $q(t,\varepsilon)$ is continuous and has piecewise continuous derivative with respect to $t\in(0,T)$;

(v) the function $t\in(0,T)\to H(\tilde{x}^m(\varepsilon),\tilde{u}^m(\varepsilon),q(t,\varepsilon),r(t,\varepsilon))$ cont whenever t is such that $t\in(0,T)\to\tilde{u}^m(\varepsilon)$ is continuous;

(vi) for any feasible control $u\in U^\rho$ of $\Pi_I^m(\varepsilon)$ relatively to z° and for $t\in(0,T)$:

$$H(\tilde{x}^m(\varepsilon),u,q(t,\varepsilon),0)\leq H(\tilde{x}^m(\varepsilon),\tilde{u}^m(\varepsilon),q(t,\varepsilon),0)\ . \blacksquare$$

Let us now study the sensitivity of J_1's performance index $\tilde{\phi}_1^{m+1}(0)$, or equivalently of the performance $\tilde{\phi}_1^m(\varepsilon)$ of $\Pi_I^m(\varepsilon)$, with respect to the coordination constraint (11) for various values of ε $0\leq\varepsilon<\varepsilon_o$, Ψ being given .

<u>T2. Sensitivity of $\tilde{\phi}_1^{m+1}(0)$ with respect to the coordination constraint (11) :</u> Our notations and basic assumptions will be those of T1 and $\Pi_I^m(\varepsilon)$ for $0\leq\varepsilon<\varepsilon_o$. We make the following supplementary assumptions:

(i) $q_o=1$;

(ii) $\partial r(t,0)/\partial\varepsilon$ exists for all $t\in(0,T)$;

(iii) $t\in(0,T)\to d\tilde{x}^m(0)/d\varepsilon$, and $t\in(0,T)\to d\tilde{u}^m(0)/d\varepsilon$ are piecewise continuous functions ;

(iv) $\overline{(d\dot{\tilde{x}}^m(0)/d\varepsilon)}$ exists and is almost everywhere equal to $d(\overline{\dot{\tilde{x}}^m}(0))/d\varepsilon$ on $(0,T)$.

Then :

(a) $\forall \varepsilon \in (0,\varepsilon_0(: \tilde{\phi}_1^m(\varepsilon)=\tilde{\phi}_1^{m+1}(0) \geqslant \left[\tilde{\phi}_1^m(0)+\int_0^T r_o(t,0)\Psi(\tilde{y}_1^m(0),x^{*m})dt\right]$

(b) $d\tilde{\phi}_1^m(0)/d\varepsilon = \int_0^T r_o(t,0)\,\Psi(\tilde{y}_1^m(0),x^{*m})dt$ ∎

Which confirms the fact that the multiplier $r_o(t,0)$ represents the rate of change of the performance index $\tilde{\phi}_1^m(0)$ with respect to variations in the constraining equation (11).

Using (9),D10, one may proceed from $\Pi_I^o(0)$ to the problems $\Pi_I^m(0)$, Π_{II}^{m*} ,m∈ℕ, the goal being ,just as for lemma L2 , to build a non-increasing sequence of the performance index of player J_1 at level I; in order to achieve this , it is necessary to select an appropriate function Ψ , and to adjust ε , in such a way that the corresponding coordination constraint (9) may contribute to this goal . In this Section 3. , we do assume that the the set X^m of feasible states at level II (D8)(5) , simply depends upon $\tilde{y}_1^m(\varepsilon)$, and may therefore be more general than treated previously .

<u>T3. Updating the coordination constraint (9) , m∈ℕ :</u> Our notations and basic assumptions are those of T2, $\Pi_I^m(\varepsilon)$ for $0 \leqslant \varepsilon < \varepsilon_o$, and namely by definition $\tilde{\phi}_1^{m+1}(0)=\tilde{\phi}_1^m(\varepsilon)$. Then :

(i) $$\left(\tilde{\phi}_1^{m+1}(0)-\tilde{\phi}_1^m(0)\right) \geqslant \int_0^T r_o(t,0)\Psi(\tilde{y}_1^m(0),x^{*m})dt \tag{12}$$

yields a bound on the variations of I's performance, and according to T1 : $\forall\, t \in (0,T)$:

$$r_o(t,0) \geqslant 0 \qquad r_o(t,0)\{\beta_m(\tilde{y}_1^m(\varepsilon))+\varepsilon\Psi(\tilde{y}_1^m(\varepsilon),x^{*m})\} = 0 \tag{13}$$

(ii) if : $$\int_0^T r_o(t,0)\Psi(\tilde{y}_1^m(0),x^{*m})dt \leqslant 0 \text{ , then :} \tag{14}$$

$$\left(\lim_{\varepsilon\to 0_+} \tilde{\phi}_1^{m+1}(0)\right) \leqslant \tilde{\phi}_1^m(0) \quad ∎ \tag{15}$$

In L2, we showed in a special case the existence of a function Ψ leading into a non-increasing sequence $\{\tilde{\phi}_1^m\}$. (15) characterizes other functions Ψ , and especially non-linear functions which may speed-up the convergence .

T3 allows us to select Ψ-functions of two different kinds :

(a) $$\forall y_1 \in Y_1 \quad : \quad \Psi(y_1,x^{*m}) \leqslant 0 \tag{16}$$

According to (13) , (14) is then fulfilled , thus implying (15);if (6) is selected as coordination constraint:

$$\beta_m(y_1)=a^m(t)y_1-b(t) \qquad \varepsilon\Psi(y_1,x^{*m})=(a^{m+1}(t)-a^m(t))y_1 \quad y_1 \geqslant 0$$

Because the sequence $\{a^n(t),n\in\mathbb{N}\}$ is non-increasing(L2), $\Psi(y_1,x^{*m}) \leqslant 0$,and T3 confirms thereby L2 .

(b) $\Psi(y_1,x^{*m})$ of variable sign : there are clearly C^1 functions changing sign on)0,T(, and such that the integral (14) becomes non-positive .

We should stress the fact that $\{\tilde{\phi}_1^m(0)\}$ being a non-increasing sequence does not imply the convergence of the sequences $\{\tilde{v}_1^m, m \in \mathbb{N}\}$, $\{p^{*m}, m \in \mathbb{N}\}$; a convergence theorem has however been obtained in a rather specific case , with strong convexity assumptions : it requires the definition of a topology on the set of Nash-optimal strategies which has to be compatible with the topology in the Hilbert space U^ρ

4. EXTENSIONS

As a consequence of the constraint coordination schemes (6)(9),and of the hierarchy (D1), the level I may ignore the nature of the equilibrium introduced at level II,for all $m \in \mathbb{N}$. It is especially possible to replace at the level II the Nash optimality (D5,D8) by the Pareto optimality {1} :

D10. $p^{**}=(p_2^{**},..,p_N^{**})$,defined on X, is a Pareto-optimal (N-1)-tuple in x° for given $v_1 \in V_1$, $z^\circ \in E^{N+\nu}$,iff $\forall x \in X$, x is attainable in v_1, z° relatively to p^{**} and to the strategy p , and if $\forall k=2,..,N$:

(i) $\phi_k^{**} = \underline{x}_{ok}^{**}(T)$ is defined and its value given by $\underline{z}^{**}(t) = (p_{v1}^{**}(x), v_1, z^\circ, t) \triangleq (\underline{x}_o^{**}(t), \underline{x}^{**}(t))$;

(ii) $\phi_k = \underline{x}_{ok}(T)$ is defined and its value given by $\underline{z}(t)$ (D4);

(iii) $(\forall k=2,..,N: \phi_k \leqslant \phi_k^{**}) \Rightarrow (\phi_k = \phi_k^{**}, \forall k=2,..,N)$ ▮ (17)

T4. Assume that Π_{II}^{m**} is the problem identical to Π_{II}^{m*} , except that the Nash-optimality (D5) of p^{*m} is replaced by the Pareto-optimality (D10) of p^{**m} . Assume moreover that all symbols with one single upper case star * are supplemented by a second star next to the first, in (6),L1,L2,(9),(11),T1,T2,T3. Then : the results of L1,L2,T1,T2,T3 still hold . ▮

Define $J_I = J_1$; $J_{II} = \{J_2,..,J_N\}$,set of all (N-1) level-II players; $K_I: E \to E$ $K_I(\underline{x}_{o1}(T)) \triangleq \underline{x}_{o1}(T)$; $\mathcal{E}_{II} \subseteq E^\nu$,largest subset of E^ν in the sense of the relation of inclusion , and such that $\forall x \in \mathcal{E}_{II}$ $\forall k=1,..,N$: $P(x, \mathcal{E}_k) \in Y_k$; $X_I(.): E^\nu \to \mathcal{E}_1$,continuous application with non-empty values , and such that $\forall x \in E^\nu$: $X_I(x) \subseteq \mathcal{E}_1$,and $\forall \bar{y}_1 \in X_I(x)$: $\bar{y}_1 \in Y_1$; $X_{II}(.)$, continuous application with non-empty values , such that $\forall \bar{y}_1 \in Y_1$: $X_{II}(y_1) \subseteq \mathcal{E}_{II} \subseteq E^\nu$,$X_{II}(.): \mathcal{E}_1 \to E^\nu$.

D11. Intra-level Θ_{II}-equilibrium at the level II : Let Θ_{II} be a given type of equilibrium in the game among the players J_k, k=2,.,N of J_{II}. $p_{II} \triangleq (p_{2II},..,p_{NII})$ is a Θ_{II}-optimal (N-1)-tuple for an intra-level Θ_{II} -equilibrium defined on $X_{II}(y_1)$ for given v_1, y_1, z°, iff $\forall \bar{x} \in X_{II}(y_1) \subseteq \mathcal{E}_{II}$,$\bar{x}$ is attainable relatively to the strategy p_{II} (D4) and if :

(i) $\phi_{IIk} = \underline{x}_{oIIk}(T)$ is defined for k=1,.,N , and its value is given by : $\underline{z}_{II}(t) = (p_{IIv1}(\bar{x}), v_1, z^\circ, t) \triangleq (x_{oII}, x_{II}) \triangleq (\underline{x}_{oII}(t), \underline{x}_{II}(t))$ (D2) , solution of (1) ;

(ii) $\forall t \in (0,T)$ $\underline{x}_{II}(t) \in X_{II}(y_1)$, $t = P(\underline{x}_{II}(t), \mathcal{E}_{N+1})$;

(iii) ϕ_{IIk},k=2,.,N fulfill the conditions for having an equilibrium of type Θ_{II} .

If p_{II} exists , then we may write $\Theta_{II}(v_1,y_1)\triangleq p_{II}$. ▮
The Nash-optimal strategies p^{*m}(D8) , and the Pareto-optimal strategies p^{**m}(D10) , build two examples of such intra-level equilibriums at level II , for $v_1=\tilde{v}_1^m$ $y_1=\tilde{y}_1^m$ $X(\tilde{y}_1^m)=X^m$, $m \in \mathbb{N}$.
We may now give the general formulation of the coordination between I and II , with a minimization of the index of performance ϕ_1 of J_1 (L2,T3,T4) :

$$\begin{array}{llll} \underset{(v_1,p_{II})}{\text{Min}}\ \phi_1 & v_1 \in V_1 \quad y_1 \in X_I(x) & x \in X_{II}(y_1) & \bar{x} \in \mathcal{E}_{II} \\ & p_{II} = \Theta_{II}(v_1,y_1) & y_1 = P(x,\mathcal{E}_1) & \\ (x_o,x) = \psi(p_{IIv_1}(\bar{x}),v_1,z^\circ,t) & & \phi_1 = \underline{x}_{o1}(T) & \end{array} \qquad (18)$$

In other words ,p_{II} is a Θ_{II}-optimal strategy for any $v_1 \in V_1$; but at the same time , because of the state constraints ,v_1 must belong to a domain depending on p_{II} ($X_{II}(.)$), and p_{II} must belong to another domain depending on v_1 ($X_I(.)$) . Notice that the player J_1 at level I may again ignore the nature of the intra-level equilibrium at level II .
In a weakly hierarchical game of the type studied here (1)(2), one may substitute to the latter concept (D11) (approximated by sequential modifications of the constraints (9)) , the notion of a Θ_I-equilibrium between the player J_I at the level I , and the set J_{II} of all (N-1) level-II players behaving collectively according to an equilibrium Θ_{II} which may eventually be of a different type.
An algorithm providing the Θ_I-equilibrium , called inter-level equilibrium , will not necessarily lead to a minimization of ϕ_1 .
<u>D12. Inter-level Θ_I-equilibrium between J_I and J_{II}</u> : Let be given on the fixed time-horizon $t \in (0,T)$ the differential game among the two players J_I and J_{II} , governed by the evolution equation (1).
In the state space $\mathcal{E}_1$ of the player J_I :

$$\dot{x}_{o1} = f_{o1}(y_1,v_1) \quad y_1 \in Y_1 \quad v_1 \in V_1 \qquad y_1 \in X_I(x) \quad y_{N+1}=t \quad (19)$$

In the state space $\mathcal{E}_{II}$ of the player J_{II} ,$\forall k=2,..,N$ $\forall \bar{x} \in X_{II}(y_1)$:

$$\begin{array}{ll} \dot{x}_{ok} = f_{ok}(P(x,\mathcal{E}_k),Q(p_{IIv_1}(\bar{x}),\mathcal{U}_k)) & p_{II} = \Theta_{II}(v_1,y_1) \\ x \in X_{II}(y_1) \quad P(x,\mathcal{E}_k) \in Y_k & Q(p_{IIv_1}(\bar{x}),\mathcal{U}_k) \in V_k \end{array} \qquad (20)$$

J_I minimizes his terminal cost $K_I(\underline{x}_{o1}(T))$, while J_{II} minimizes his terminal cost worth $K_{II}(\underline{x}_{o2}(T),..,\underline{x}_{oN}(T))$, where $K_{II} : E^{N-1} \to E$ is a non-decreasing function . The levels I and II are then said to achieve an inter-level equilibrium of type Θ_I ,iff the terminal costs of the players J_I and J_{II} fulfill a given equilibrium condition Θ_I . ▮
<u>D13.</u> $p \triangleq (p_I,p_{II})$, defined on $\mathcal{E}_{II} \subseteq E^{\nu}$, is a Θ_I- Θ_{II}-optimal N-tuple <u>in z°</u> , iff the following conditions hold simultaneously :
(i) $p_1 \triangleq p_I$ is a strategy of J_1 on $\mathcal{E}_{II}$ (D3) ;

(ii) $\forall \bar{x} \in \mathcal{E}_{II}$, $\exists\, u \in U^{\rho}$ so that $p_1(\bar{x}) = v_1 = Q(u, \mathcal{U}_1) \in V_1$, and $p_{IIk}(\bar{x}) = Q(u, \mathcal{U}_k) \in V_k$ $k=2,..,N$; we may then write $u \triangleq p_{I-II}(\bar{x})$;

(iii) $y_1 = P(x^{\circ}, \mathcal{E}_1) \in X_I(x)$ and $x \in X_{II}(y_1)$ where $(x_o, x) = \psi(p_{I-II}(\bar{x}), p_1(\bar{x}), z^{\circ}, t)$ $t \in (0,T)$;

(iv) $p_{II} = \Theta_{II}(p_1(\bar{x}), y_1)$ is a Θ_{II}-optimal (N-1)-tuple of II in x°, defined on $X_{II}(y_1) \subseteq \mathcal{E}_{II}$ for $p_1(\bar{x}), y_1, z^{\circ}$ given in D12.;

(v) $\forall \bar{x} \in \mathcal{E}_{II}$, $\bar{x}$ is attainable relatively to the strategy p_{II} when $p_1(\bar{x}), z^{\circ}$ are given ;

(vi) $K_I(\underline{x}_{o1}(T))$ and $K_{II}(\underline{x}_{o2}(T),..,\underline{x}_{oN}(T))$ are defined , and their values fulfill the given equilibrium condition Θ_I.

If necessary hereabove , the domains where the strategies p_I, p_{II} are defined should be extended in order to include $(\mathcal{E}_{II} - X_{II}(y_1))$. ∎

The definition D12 for an inter-level equilibrium does not assume I to know the type of equilibrium chosen by II. The notion of Θ_I-Θ_{II}-optimal strategy (D13) is mainly interesting in the case where the game (1) is weakly hierarchical : in that case , the Θ_{II}-optimal strategy p_{II} may act on the evolution of $\underline{x}_{o1}(t)$, $t \in (0,T)$, and act at the same time on the set $X_I(x)$ of feasible states y_1 of J_I (the latter action applying also in a strong hierarchy). An example of inter-level equilibria , of the zero-sum Minmax type , has been solved with an intra-level Nash equilibrium at level II .

5. HIERARCHICAL COMPUTATION OF NASH-OPTIMAL STRATEGIES , AND COORDINATION ALGORITHM

The coordination algorithm $\mathcal{A}$ approximates a point of accumulation (v_1^*, p^*) of the sequence $\{(\tilde{v}_1^m, p^{*m}), m \in \mathbb{N}\}$ through successive approximations and coordination , if such a point of accumulation exists . The assumptions are those of T1,T2,T3 .

$\mathcal{A}$1. Give β_o ;m=0; Ψ fulfilling T3; check the existence of $\tilde{v}_1^{\circ}(0)$, elsewhere : go to $\mathcal{A}$6.

$\mathcal{A}$2:Level I : Solve the optimal control problem $\Pi_I^m(0)$;if $\tilde{v}_1^m(0) = \tilde{v}_1^{m-1}(0)$ and $m \geqslant 1$, go to $\mathcal{A}$5; elsewhere: transfer $\tilde{v}_1^m(0)$, $\tilde{x}_1^m(0)$ to the players J_k ,k=2,.,N of II .

$\mathcal{A}$3:Level II: Solve the Nash game Π_{II}^{m*} ; if there is no p^{*m}, go to $\mathcal{A}$6. ; if $p^{*m} = p^{*(m-1)}$ and $m \geqslant 1$, go to $\mathcal{A}$5.; elsewhere : transfer x^{*m} to J_1 at I .

$\mathcal{A}$4:Coordination: According to T1, T3, solve the scalar optimization problem :

$$\text{Min } \tilde{\phi}_1^m(\varepsilon) = \tilde{\phi}_1^m(\varepsilon_m) \tag{21}$$

If $\tilde{\phi}_1^m(\varepsilon_m) > \tilde{\phi}_1^m(o)$ go to $\mathcal{A}$5.

$\mathcal{A}$5: Convergence: Define $v_1^* = \tilde{v}_1^{m-1}$ $p^* = p^{*(m-1)}$ $m \geqslant 1$

$\mathcal{A}$6: The problem of type (18) has no solution. ∎

$\mathcal{A}$ has been implemented in an economic two-level planning problem with conflicting interests {11} : more specifically the concern was about the budgeting in a university organization with conflicting departments {10} .
It should be noticed that , if Θ_{II} is the Nash equilibrium , then by definition of such an equilibrium (D5) , $\forall \bar{x} \in \mathcal{E}_{II}$, any point of accumulation of the sequence $\{(\tilde{v}_1^m(0), p^{*m}(\bar{x})) \in \pi V_k / k=1,..,N; m \in \mathbb{N}\}$ will build not only a Nash-optimal intra-level equilibrium at the level II , conditionnally with respect to v_1^*, but also a Nash-optimal equilibrium among the N players $J_1,..,J_N$ for strategical N-tuples taking values in $\pi V_k / k=1,..,N$. This is a direct consequence of L2,T3 and (18) and (21) .
T5. Assume that the algorithm $\mathcal{A}$ has locally a point of accumulation (v_1^*, p^*) of the sequence $\{(\tilde{v}_1^m, p^{*m}), m \in \mathbb{N}\}$. Then , $\forall \bar{x} \in (\mathcal{E}_{II} - X_{II}(y_1^*))$, $(v_1^*, p_2^*(\bar{x}),..,p_N^*(\bar{x})) \in V_1 \times V_2 \times \cdots \times V_N$, is a Nash-optimal control of the original global game (1) having the structure of D12. If we then define p_1^* by $p_1^*(\bar{x}) = v_1^*$, then $(p_1^*, p_2^*,..,p_N^*)$ is a Nash-optimal N-tuple on $(\mathcal{E}_{II} - X_{II}(y_1^*))$.
The main result is however that locally Nash equilibrium controls may be computed in closed loop using $\mathcal{A}$, beginning with a two-player intra-level Nash-equilibrium , with branching coordination providing the N-player equilibrium through proper decompositions of the global game into inter-level and intra-level games of the types D11,D12 .

REFERENCES

1. A.Blaquière,L.Juricek,K.E.Wiese,Geometry of Pareto equilibria in N-person differential games ,in:Topics in differential games , A.Blaquière ed., North Holland,Amsterdam,1973,271-309
2. A.Charnes,R.W.Clower,K.O.Kortanek, Effective control through coherent decentralization with preemptive goals, Econometrica, 35(april 1967),no 2
3. G.Dantzig,P.Wolfe, The decomposition algorithm for linear programs , Econometrica , 29(1961),767-778
4. Yu.A.Flerov, Multilevel dynamic games ,J. of cybernetics , 3(no 1,1973),79-107
5. A.M.Geoffrion,W.W.Hogan,Coordination of two-level organizations with multiple objectives, in: Techniques of optimization , A.V.Balakrishnan ed.,Academic Press,N.Y.,1972,455-466
6. G.Jumarie,Towards a theory of multilevel hierarchical games and its perspectives in economic systems , in: IFAC/IFORS Int. Conf. on dynamic modelling and control of national economies , IEE Conf. Publ. no 101,London,1973,270-281
7. J.Kornai,T.Liptak, Two-level planning ,Econometrica , 33(no 1, jan.1965),141-169
8. R.Kulikowski,Decomposition and competition in multilevel environment control systems , in:Differential games and related topics , H.W.Kuhn & G.P.Szegö ed.,North-Holland,Amsterdam, 1971 , 415-428

8. E.B.Lee,L.Markus , Foundations of optimal control theory , John Wiley & sons,N.Y.,1967
9. M.D.Mesarovic,D.Macko,Y.Takahara, Theory of hierarchical multilevel systems , Mathematics in science and engineering no 68 , Academic Press , N.Y.,1970
10. OECD,Center for educational research and innovation , Decision, planning and budgeting (Danish contribution),OECD,Paris,1972
11. L.F.Pau, Coordination algorithm among conflicting subsystems in a two-level hierarchical system,IMSOR Tech.Univ.Denmark, Copenhagen-Lyngby , july 1973
12. L.F.Pau , Coordination par les contraintes dans un jeu différentiel hiérarchisé , C.R.Acad.Sci.Paris,Série A,Paris,1974
13. D.W.Peterson, On sensitivity in optimal control problems , J. of optimization theory and appl.,13(no 1,1974),56-73
14. P.Varaiya,J.Lin , Existence of saddle-points in differential games , J.SIAM on Control , 7(1969),141-157

COMPUTED SOLUTIONS OF DIFFERENTIAL GAMES

Daniel Tabak

Department of Electrical Engineering, Ben-Gurion
University of the Negev, Beer-Sheva, Israel

ABSTRACT. Computational methods, applicable to the numerical solution of differential games, are described. Both zero- and non-zero sum differential games are considered. The advantages and difficulties, involved in the methods presented, are discussed.

1. INTRODUCTION

The theory of differential games [1-3] has reached an advanced stage in its development. The numerical solution of many classes of differential games is a very difficult and in many cases - still an unsolved problem. Computer solutions of several classes of differencial games have been reported in the literature. A part of the available information in this area will be surveyed in the following paragraphs, covering both zero- and non-zero sum differential games. Some alternatives for obtaining solutions of certain classes of differential games will be outlined.

2. ZERO-SUM GAMES

2.1 Formulation of the problem

We will assume that we have an n-dimensional dynamic system, characterized by the state vector $\underline{x}$, satisfying the state equations:

$$\dot{\underline{x}} = \underline{f}(\underline{x}, \underline{u}, \underline{v}, t) \quad ; \qquad \underline{x}(t_0) = \underline{x}_0 \tag{1}$$

where $\underline{u}$ is the control vector of the minimizing and $\underline{v}$ that of the maximizing player. The performance criterion is

J. D. Grote (ed.), The Theory and Application of Differential Games. 221-227.

$$J = h[\underline{x}(t_f), t_f] + \int_{t_0}^{t_f} g(\underline{x}, \underline{u}, \underline{v}, t)dt \tag{2}$$

where the final time t_f may be fixed or variable. The problem is to find $\underline{u}^0$, $\underline{v}^0$ so that

$$J(\underline{u}^0, \underline{v}) \leq J(\underline{u}^0, \underline{v}^0) \leq J(\underline{u}, \underline{v}^0) \tag{3}$$

In addition, we may have a set of terminal conditions

$$\underline{s}[\underline{x}(t_f), t_f] = \underline{0} \tag{4}$$

Define a Hamiltonian

$$H(\underline{x}, \underline{u}, \underline{v}, \underline{p}, t) = g(\underline{x}, \underline{u}, \underline{v}, t) + \underline{p}^T \underline{f}(\underline{x}, \underline{u}, \underline{v}, t) \tag{5}$$

The necessary conditions for the solution are [2]:

$$\dot{\underline{p}} = -\frac{\partial H}{\partial \underline{x}} \quad ; \quad \underline{p}(t_f) = \frac{\partial h[\underline{x}(t_f), t_f]}{\partial \underline{x}} + \left(\frac{\partial \underline{s}[\underline{x}(t_f), t_f]}{\partial \underline{x}}\right)^T \underline{\lambda} \tag{6}$$

where $\underline{\lambda}$ is a vector of Lagrange multipliers and f is the conjugate vector.

$$\frac{\partial H}{\partial \underline{u}} = \underline{0} \quad ; \quad \frac{\partial H}{\partial \underline{v}} = \underline{0} \tag{7}$$

$$\frac{\partial^2 H}{\partial \underline{u}^2} \geq \underline{0} \quad ; \quad \frac{\partial^2 H}{\partial \underline{v}^2} \leq \underline{0} \tag{8}$$

Naturally, eqs. (1) and (4) constitute a part of the necessary conditions. If t_f is not specified we also have the condition [4]:

$$H(t_f) - \frac{\partial h(t_f)}{\partial t} - \underline{\lambda}^T \frac{\partial \underline{s}(t_f)}{\partial t} = 0 \tag{9}$$

Conditions (7), (8) can be expressed as:

$$H^0(\underline{x}, \underline{p}, t) = \min_{\underline{u}} \max_{\underline{v}} H(\underline{x}, \underline{u}, \underline{v}, \underline{p}, t) \tag{10}$$

A more detailed formulation, involving open and closed-loop solutions is given in [2].

2.2 Solution approaches

Applying conditions (7), (8), we arrive to a two-point-boundary-value-problem (TPBVP), [4,5], of the following form:

$$\left.\begin{aligned} \underline{\dot{x}} &= \underline{f}_1(\underline{x}, \underline{p}, t) \quad ; \quad \underline{x}(0) = \underline{x}_0 \\ \underline{\dot{p}} &= \underline{f}_2(\underline{x}, \underline{p}, t) \quad ; \quad \underline{p}(t_f) = \frac{\partial h(t_f)}{\partial \underline{x}} + \left(\frac{\partial \underline{s}(t_f)}{\partial \underline{x}}\right)^T \underline{\lambda} \end{aligned}\right\} \quad (11)$$

with eq. (9) added if t_f is not specified. A near optimal solution for this case was proposed by Anderson [4,5]. It uses an initial reference solution and linearization of the equations. The linearized set of equations is solved using the Riccati equation approach, as described in [2]. The solution serves as the reference solution for the next iteration and so on. The convergence of this procedure has been demonstrated on some pursuit-evasion examples [4,5]. Quintana and Davison [6] solved a similar problem with a quadratic performance index J and a fixed t_f using a Newton-Raphson and a gradient approach.

In the preceding formulation $\underline{f}$, g and h could be nonlinear functions. If the system dynamics (or $\underline{f}$) are linear and the functions g and h are quadratic, a feedback strategy can be worked out leading to a Riccati-type equation [2]. Detailed solutions of problems of this type have been worked out by Krikelis and Rekasius [7] and by Medanic and Andjelic [8,9]. The solution in [7] is also applicable to nonzero-sum two-player games. It uses Newton's method for the numerical solution of the matrix equations involved. The method in [8,9] is applicable to cases where a saddle-point does not exist. It also allows for more than one situation defining the performance index J. The numerical solution of the matrix equation is done by a gradient projection procedure. A mathematical programming formulation (without a numerical solution) of a pursuit-evasion problem, with linear dynamics, was given by Jacob and Polak [10]. Numerical solutions of algebraic minimax problems were proposed by Salmon [11] and Heller et. al. [12].

2.3 Approximate iterative solutions

For differential games where the dynamics of the system are such that the state variables incur relatively slow changes, during the time interval considered, we can use the following approximate solution [13]. We divide the time interval into a finite number of subintervals and assume that $\underline{u}$ and $\underline{v}$ retain constant values in each one of them. We then convert all the differential equations involved into difference equations, using finite points approximations. The integral part in J is converted into a finite sum. We then solve a sequence of nonlinear programming (NLP) problems, minimizing J with respect to $\underline{u}$ at one step and maximizing with respect to $\underline{v}$ at the other. At each step we use the solution obtained at the previous

one. The approximation involved in this method may be somewhat crude in many cases, however it is very efficient in handling state and control variables constraints.
If the system dynamics are changing rapidly, the approach described above is not applicable, since we need to use very small time sub-intervals, which would render the dimensionality of the NLP problem untreatable. In this case we can use the approach proposed by Brusch [14]. A modified integral performance index is formulated in analogy with the SUMT formulation using a parameter r, which decreases from iteration to iteration [15]. Both the integral as well as the differential equations are solved numerically at each iteration using small steps. Another alternative would be to use the cubic splines approximation, recently implemented in an optimal control problem [16]. However, in general, this approach could lead to a highly complicated NLP problem. Nevertheless it is worth investigating in future research. The Differential Dynamic Programming (DDP) method, originated by Jacobson and Mayne [17] can also be applied to the solution of zero-sum differential games.

3. NONZERO-SUM GAMES

3.1 Formulation of the problem

The interest in nonzero-sum differential games arose in the late sixties. Many details in problem formulation, necessary conditions, examples and practical solution considerations can be found in the work of Starr and Ho [18-21]. In this case we have N players whose corresponding control vectors are $\underline{u}_1, \underline{u}_2, \ldots, \underline{u}_N$. The system dynamics are given by the state equations

$$\dot{\underline{x}} = \underline{f}(\underline{x}, t, \underline{u}_1, \ldots, \underline{u}_N) \quad ; \qquad \underline{x}(t_o) = \underline{x}_o \tag{12}$$

Each player has a different performance index to minimize

$$J_i = h_i[\underline{x}(t_f), t_f] + \int_{t_o}^{t_f} g_i(\underline{x}, t, \underline{u}_1, \ldots, \underline{u}_N)dt \qquad i = 1, \ldots, N \tag{13}$$

We can distinguish between several types of equilibria or solutions, such as Nash, minimax and noninferior [18-20]. Particular attention was given to the numerical solution attaining the Nash equilibrium $(\underline{u}_1^*, \ldots, \underline{u}_N^*)$, defined as follows [18]:

$$J_i(\underline{u}_1^*, \ldots, \underline{u}_{i-1}^*, \underline{u}_i, \underline{u}_{i+1}^*, \ldots, \underline{u}_N^*) \geq J_i(\underline{u}_1^*, \ldots, \underline{u}_N^*) \qquad i = 1, \ldots, N \tag{14}$$

where $\underline{u}_i$ is any admissible control vector, or strategy, of player i. If we define a Hamiltonian for each player:

$$H_i = g_i + \underline{p}_i^T \underline{f} \quad ; \qquad i = 1, \ldots., N \tag{15}$$

we can write the necessary conditions for the Nash equilibrium [21], to which eq. (12) is added:

(1) Open-loop solution

$$\dot{\underline{p}}_i = - \frac{\partial H_i}{\partial \underline{x}} \; ; \qquad i = 1, \ldots., N \; ; \qquad \underline{p}_i(t_f) = \frac{\partial h_i(t_f)}{\partial \underline{x}} \tag{16}$$

$$\frac{\partial H}{\partial \underline{u}_i} = \underline{0} \quad ; \qquad i = 1, \ldots., N \tag{17}$$

$\dfrac{\partial^2 H}{\partial \underline{u}_i^{\,2}}$ is positive semidefinite for all i

(2) Closed-loop solution

$$\underline{u}_j = \underline{\psi}_j(\underline{x}, t) \quad ; \qquad j = 1, \ldots., N \tag{18}$$

$$\dot{\underline{p}}_i = - \frac{\partial H_i}{\partial \underline{x}} - \sum_{j \neq i}^{N} \frac{\partial \underline{\psi}_j}{\partial \underline{x}} \frac{\partial H_i}{\partial \underline{\psi}_j} \quad ; \qquad \underline{p}_i(t_f) = \frac{\partial h_i(t_f)}{\partial \underline{x}} \tag{19}$$

$$i = 1, \ldots., N$$

$$\underline{u}_j \text{ minimizes } H_j(\underline{x}, t, \underline{\psi}_1, \ldots., \underline{\psi}_{j-1}, \underline{u}_j, \underline{\psi}_{j+1}, \ldots, \underline{\psi}_N) \tag{20}$$

3.2 Solutions

A detailed solution for linear-quadratic games, involving a Riccati type matrix equation has been worked out in [18, 20, 22].
A comprehensive feasibility analysis of obtaining the numerical solution of Nash equilibria, using Dynamic Programming and Differential Dynamic Programming, was performed by Starr [20, 21]. Many difficulties, involved in obtaining a numerical solution, were demonstrated. Since then, numerical solutions for some classes of non-zero-sum differential games have been worked out.
Holt and Mukundan [23] proposed the following iterative procedure. An initial set of control vectors is guessed at a set of discrete times in the interval $[t_0, t_f]$. The state eqs. (12) are integrated forward in time, obtaining $\underline{x}(t_f)$, which in turn permits the integration of the co-state eqs. (16), backwards in time. The control vectors are now calculated using eq. (17) and the procedure is re-

started. The convergence of the algorithm was demonstrated on an example for a linear-quadratic two-players problem. A linearization scheme with respect to the controls $\underline{u}_1, \ldots, \underline{u}_N$ of the $\underline{f}$ and g_i functions, was proposed by Krikelis [24]. No computational experience was reported.

A considerable amount of work in obtaining solutions for the Stackelberg strategy, was performed by Cruz and associates [25-29]. It involves a case where a player has to announce his strategy first. If he follows the Stackelberg strategy he will do no worse than the corresponding Nash solution [25]. Solutions for two-players nonzero-sum games were obtained. In [26] the solution was extended to a many players case. However the players were divided into two main groups: leaders and followers.

Approximate methods, mentioned in paragraph 2.3, could be extended to nonzero-sum games as well. Naturally, the actual implementation of the algorithms would be much more complicated.

4. CONCLUSION

Practically all numerical algorithms, applicable to problems other than linear-quadratic, require an initial guess of a reference solution. This is of course a serious drawback. Moreover, in most cases, convergence was demonstrated experimentally in some particular cases, but not proved rigorously.

In view of the existing situation there are two main avenues of research which are open in this field:

(a) Mathematical justification of existing methods, including proof of convergence;

(b) Continuation of the experimental development of new computational algorithms and their implementation in the solution of real-life problems.

A great amount of work along these lines, yet unpublished, is going on in various institutions. Hopefully, new results will soon be presented to the scientific community.

REFERENCES

1. R. Isaacs, Differential Games, Wiley, N.Y., 1965.
2. A.E. Bryson and Y.C. Ho, Applied Optimal Control, Blaisdell, Waltham, Mass., 1969, Chapter 9.
3. A. Friedman, Differential Games, Wiley, N.Y., 1971.
4. G.M. Anderson, J. Optimization Theory and Appl., 13, pp.303-318, 1974.
5. G.M. Anderson, IEEE Trans. on Automatic Control, AC-17, pp.576-577, 1972.
6. V.H. Quintana and E.J. Davison, Int. J. Control, 16, pp.465-474, 1972.

7. N.J. Krikelis and Z.V. Rekasius, IEEE Trans. on Automatic Control AC-16, pp.140-147, 1971.
8. J. Medanic and M. Andjelic, J. Optimization Theory and Appl., 8, pp.413-430, 1971.
9. J. Medanic and M. Andjelic, IEEE Trans. on Automatic Control, AC-17, pp.597-604, 1972.
10. J.P. Jacob and E. Polak, IEEE Trans. on Automatic Control, AC-12, pp.752-755, 1972.
11. D.M. Salmon, IEEE Trans. on Automatic Control, AC-13, pp.369-376, 1968.
12. J.E. Heller, J.B. Cruz and J. Medanic, Proc. 1st Int. Conf. on the Theory and Appl. of Diff. Games, Amherst, Mass., Sept. 29 - Oct. 1, 1969 (VII-26).
13. D. Tabak and B.C. Kuo, Optimal Control by Mathematical Programming, Prentice-Hall, Englewood Cliffs, N.J., 1971.
14. R.G. Brusch, J. Optimization Theory and Appl. 13, pp.94-118, 1974
15. A.V. Fiacco and G.P. McCormick, Nonlinear Programming: SUMT, Wiley, N.Y., 1968.
16. C.P. Neuman and A. Sen, Automatica, 9, pp.601-613, 1973.
17. D.H. Jacobson and D.Q. Mayne, Differential Dynamic Programming, American Elsevier, N.Y., 1970.
18. A.W. Starr and Y.C. Ho, J. Optimization Theory and Appl., 3, pp.184-206, 1969.
19. A.W. Starr and Y.C. Ho, J. Optimization Theory and Appl., 3, pp.207-219, 1969.
20. A.W. Starr, Nonzero-Sum Differential Games: Concepts and Models, Tech. Report, No. 590, Harvard Univ., 1969.
21. A.W. Starr, Proc. 1st Int. Conf. on the Theory and Appl. of Diff. Games, Amherst, Mass., Sept. 29 - Oct. 1, 1969 (IV-13).
22. M.H. Foley and W.E. Schmitendorf, J. Optimization Theory and Appl., 7, pp.357-377, 1971.
23. D. Holt and R. Mukundan, Int. J. Systems Sc., 2, pp.379-387, 1972.
24. N.J. Krikelis, J. Optimization Theory and Appl., 9, pp.359-363, 1972.
25. M. Simaan and J.B. Cruz, J. Optimization Theory and Appl., 11, pp.533-555; pp.613-626, 1973.
26. M. Simaan and J.B. Cruz, IEEE Trans. on Automatic Control, AC-18, pp.322-324, 1973.
27. M. Simaan and J.B. Cruz, Int. J. Control, 17, pp.1201-1209; 18, pp.57-63, 1973.
28. J.B. Cruz and C.I. Chen, J. Optimization Theory and Appl., 7, pp.240-257, 1971.
29. C.I. Chen and J.B. Cruz, IEEE Trans. on Automatic Control, AC-17, pp.791-798, 1972.

OPTIMALITY AND TRANSITION SURFACES USING THE REPRISAL CONCEPT

J. Bradley* and P.L. Yu**

*Department of Mathematics, Roberts Wesleyan College, Rochester, New York.

**Department of General Business, University of Texas, Austin, Texas

1. INTRODUCTION

One major difficulty in differential games is that the solution path of a pair of admissible strategies may not meet the terminal surfaces (or sets) [1]. One approach to this difficulty is to introduce the concept of "playable" strategies [2-5]. Although this has been a helpful approach, there are difficulties with it. First, (μ_1,ν_1) and (μ_2,ν_2) being playable does not necessarily imply that (μ_1,ν_2) and (μ_2,ν_1) are. Thus the value of the game may not be well-defined when there is more than one pair of optimal strategies. A second problem is that the playability requirement involves the implicit assumption of the presence of a third party or "umpire" who can rule pairs of strategies "admissible" or not. This is counter to the hypothesis (which prevails in most games) that each player should be free to select any strategy under his command independently of his opponent's choice. A third problem is that in many games (such as pursuit-evasion), it may be to one player's advantage to seek strategies which can prevent termination. Thus insisting that all pairs of strategies under consideration be playable may result in omitting crucial aspects of these games.

As an alternative to the playability assumption, we assume that the payoff is constant whenever no finite termination occurs. Thus the payoff is defined for both terminating and non-terminating cases. With this assumption, we introduce a new concept of optimal strategies which consists of a "local semipermeable condition" and a "global reprisal condition."

J. D. Grote (ed.), The Theory and Application of Differential Games. 229-242.

(See section 3, after the preliminary discussion of section 2.) Also in section 3 we report that subject to some assumptions the value function is well-defined and the optimal strategies are interchangeable. We also report that optimal strategies must satisfy Isaacs' main equation. Whenever all admissible strategy pairs satisfy a condition similar to playability, the reprisal condition is redundant, and our formulation reduces to the saddle-point formulation. As a consequence, in fixed-duration or all-terminating games, there is no need for the reprisal conditions.

In the rest of the paper we report some workable sufficiency conditions for optimality as well as some necessary conditions for transition surfaces to occur. In section 4, under a closure and terminating assumption, we state that a pair of idealized strategies are optimal if and only if they are semipermeable and a transition surface occurs only when at least one of a set of switching functions vanishes. In Section 5, we focus on games which do not terminate for some pair of admissible strategies. Sufficiency conditions for optimality are stated along with three alternative necessary conditions for transition surfaces to occur. These conditions can be imbedded in backward integration for solving the game. Finally, we supply an example to illustrate the results derived in Section 5. All results are reported here without proof. For the proofs see [12-13].

2. DEFINITION OF THE GAME AND OTHER BASIC DEFINITIONS

Definition 2.1. By a differential game we shall mean the following:

(i) a region $E \subset \mathbb{R}^{n+1}$ called the playing space;

(ii) two sets of functions; †

$$\{\mu \,|\, E \to \Phi \text{ such that } \mu \text{ is of class } C^{(2)}\}$$
$$\{\nu \,|\, E \to \Psi \text{ such that } \nu \text{ is of class } C^{(2)}\}$$

where $\Phi \subset R^m$ and $\Psi \subset R^{\ell}$ are compact and convex. The elements of these sets are called admissible strategies.

(iii) a system of ordinary differential equations:

$$\frac{dx}{dt} = f(x,\mu,\nu) \tag{1}$$

where $x = (x^0,\ldots,x^n)$. We ordinarily denote $\frac{dx}{dt}$ by $\dot{x}$ and call these equations the kinematic equations. f is defined over $E \times \Phi \times \Psi$ and is of class $C^{(2)}$ in each of its arguments.

(iv) a set $C \subset E$ called the terminal surface. We assume that C can be written as $C(s^1,\ldots,s^n) = (C^0(s),\ldots,C^n(s))$, where $s = (s^1,\ldots,s^n)$ is a member of S, an open connected subset

† We will introduce discontinuous strategies under the name "idealized strategies" in Definition 2.3.

of $\mathbb{R}^n$. We use the symbol C to denote both the surface S and the mapping $C: \mathbb{R}^n \to \mathbb{R}^{n+1}$ such that $C[S] = C$.

C is one-to-one and of class $C^{(2)}$, and the Jacobian matrix of C has rank n for each $s_0 \in S$. We also assume that C separates E; i.e., that $E \setminus C$ consists of two disjoint components E_1 and E_2 such that any continuous curve in E which meets both E_1 and E_2 also meets C. For convenience, any surface satisfying all these properties we shall call a $C^{(2)}$-separating surface.

(v) a function, H(x), called the payoff function defined on a neighborhood of C in $\mathbb{R}^{n+1}$. We assume H maps this neighborhood into $\mathbb{R}$ such that H is of class C^2 and $\nabla H(x_0) \neq 0$ and is not normal to C for each x_0 in C.

If (μ,ν) is admissible, x(t) is a solution to $\dot{x} = f(x,\mu,\nu)$, $x(0) = x_0 \in E$, and t_1 is the least t such that $x(t_1)$ is a member of C, we say (μ,ν) yield payoff $H(x(t_1))$ at x_0 and t_1 is the terminating time resulting from (μ,ν) and x_0. If x(t) is not in C for all t, we say (μ,ν) yield payoff $+\infty$†. We assume μ is under the control of a player who wants to minimize this payoff while ν is under the control of a player who wants to maximize it.

As long as the appropriate continuity conditions are satisfied, we can use the standard transformations to include integral payoff games and games in which the kinematic equation is not time-autonomous within this formulation.

Definition 2.2. By a linear differential game we shall mean the game of Definition 2.1 with the following modifications: $\Phi = [-1,1]^m$, $\Psi = [-1,1]^\ell$ and

$$\begin{aligned} f(x,\phi,\psi) &= A(x) + \sum_{i=1}^{m} B^i(x)\phi^i + \sum_{j=1}^{\ell} C^j(x)\psi^j \\ &= A(x) + B(x)\phi + C(x)\psi \end{aligned} \tag{2}$$

where $B = (B^1,\ldots,B^m)$, $C = (C^1,\ldots,C^\ell)$, $\phi = (\phi^1,\ldots,\phi^m) \in \Phi$ and $\psi = (\psi^1,\ldots,\psi^\ell) \in \Psi$.

Given a pair of admissible strategies (μ,ν), we suppose that $f(x,\mu(x),\nu(x))$ is not tangent to C at each point x of C. Then it can be shown that there exists a region†† R which contains C such that for each point x in R the payoff, denoted by $W(x;\mu,\nu)$,

---

†Alternatively we could have assigned any extended real number as payoff. This would require us, however, to decompose the playing space according to which player would benefit by avoiding termination. We choose $+\infty$ to eliminate this complication.

††More specifically $R = N(C;\mu,\nu)$ in terms of [7].

and the terminating time, denoted by $T(x;\mu,\nu)$, resulting from (μ,ν) are uniquely determined. Furthermore, both $W(x;\mu,\nu)$ and $T(x;\mu,\nu)$ are $C^{(2)}$ and ∇W and ∇T satisfy the same adjoint equation associated with $\dot{x} = f(x,\mu,\nu)$, although their initial transversality conditions are different.

Definition 2.3. Let M be a $C^{(2)}$-separating surface which separates E into E_1 and E_2. By (μ_*,ν_*) is a pair of idealized strategies with respect to M, we mean there exists two pairs of admissible strategies (μ_i,ν_i), $i = 1,2$ such that

$$(\mu_*(x),\nu_*(x)) = \begin{cases} (\mu_1(x),\nu_1(x)) & \text{if } x \in E_1 \\ (\mu_2(x),\nu_2(x)) & \text{if } x \in E_2 \cup M \end{cases}$$

and (μ_*,ν_*) are discontinuous at each point of M. A similar definition for individual μ_* or ν_* to be idealized holds.

Under suitable non-tangency and other conditions (Assumption 4.1 of 7), one can show that there exists a region R_* which contains C and M such that for each point x in R_* the payoff, $W_*(x;\mu_*,\nu_*)$ and the terminating time, $T_*(x;\mu_*,\nu_*)$ resulting from (μ_*,ν_*) are uniquely defined. Furthermore, both $W_*(x;\mu_*,\nu_*)$, $T(x;\mu_*,\nu_*)$ are absolutely continuous in R_* if x denotes a solution curve to $\dot{x} = f(x,\mu_*,\nu_*)$ and $C^{(2)}$ in R_* except possibly on M. Also ∇W_* and ∇T_* satisfy the same adjoint equation associated with $\dot{x} = f(x;\mu_*,\nu_*)$ although their transversality conditions on both C and M are different.

Notation 2.1. We shall use the subscript * only when denoting idealized strategies or functions determined by them. Whenever we use the symbols R_1, R_*, T_1, T_*, W, or W_* etc., it will be as above. Also $\gamma(t,x;\mu,\nu)$ or $\gamma(t,x;\mu_*,\nu_*)$ will denote points of the solution path resulting from (μ,ν) or (μ_*,ν_*) whenever this path exists and such that $\gamma(0,x;\mu,\nu) = x$ or $\gamma(0,x;\mu_*,\nu_*) = x$.

We let
$$\begin{aligned} \lambda(x;\mu,\nu) &= \nabla W(x;\mu,\nu) \\ \lambda_*(x;\mu_*,\nu_*) &= \nabla W_*(x;\mu_*,\nu_*) \\ \eta(x;\mu,\nu) &= -\nabla T(x;\mu,\nu) \\ \eta_*(x;\mu_*\nu_*) &= -\nabla T_*(x;\mu_*\nu_*) \end{aligned}$$

where λ_* and η_* on M are defined in the forward time limit sense. For more detailed construction see Definition 4.1 of [7].

In some situations, it will be convenient to treat M as the terminal surface. In such cases, $T_M(x;\mu,\nu)$ will denote the terminating time from x and (μ,ν) and η_M will denote $-\nabla T_M$ etc. Note we may allow T_M to be negative. However if $T_M(x;\mu,\nu) < 0$ then M can never be reached in a positive interval of time from x by (μ,ν) unless the path leaves R_* first. Under a mild assumption, it can be shown that $T_M(x;\mu,\nu)$ is well defined in a neighborhood of M (See [7]).

Definition 2.4. We say a set $R_0 \subset R$ is terminal if for each admissible (μ,ν) and any $x_0 \in R_0$, the path $\gamma(t,x_0;\mu,\nu)$ meets $\mathcal{C}$ at some $t(x_0;\mu,\nu) < \infty$. We say it is closed and terminal if, in addition $\gamma(t,x_0,\mu,\nu)$ is in R_0 for $0 \leq t \leq t(x_0,\mu,\nu)$.

Definition 2.5. Let N be a neighborhood of R_0 in R. We say that R_0 is α-closed with respect to $\eta(x)$ and N if for each admissible (μ,ν) such that $\eta(x)\cdot f(x,\mu,\nu) \geq \alpha > 0$ at each $x \in N$, the path $\gamma(t,x_0;\mu,\nu)$ starting at any point x_0 of R_0 meets $\mathcal{C}$ at some $t(x_0;\mu,\nu) < \infty$ and is in R_0 for $0 \leq t \leq t(x_0;\mu,\nu)$. If R_0 is α-closed with respect to $\eta(x)$ and any neighborhood of R_0, we say that R_0 is α-closed with respect to $\eta(x)$.

Remark 2.1. The following are evident.

(i) If R_0 is α_0-closed with respect to $\eta(x)$ and N_0, then R_0 is α-closed with respect to any $\alpha > \alpha_0$ and any $N \supset N_0$.

(ii) If R_0 is closed and terminal then it is α-closed with respect to $\eta(x)$ for any $\alpha > 0$.

(iii) The union of two sets α-closed with respect to $\eta(x)$ and N is again α-closed with respect to $\eta(x)$ and N.

3. A CONCEPT OF OPTIMAL STRATEGIES

Definition 3.1. Let R be the nonempty region induced by (μ_0,ν_0) which is either an admissible or idealized strategy pair. Let W(x) denote $W(x;\mu_0,\nu_0)$. We say (μ_0,ν_0) is optimal over R if for each $x_0 \in R$ and each $\varepsilon > 0$:

(i) for any admissible ν such that $\gamma(t,x_0;\mu_0,\nu)$ exists and is in R for $0 \leq t < t_1$, $W(\gamma(t,x_0;\mu_0,\nu)$ is non-increasing in t on $[0,t_1)$.

(ii) for any admissible μ such that $\gamma(t,x_0;\mu,\nu_0)$ exists and is in R for $0 \leq t < t_2$, $W(\gamma(t,x_0;\mu,\nu_0))$ is non-decreasing on $[0,t_2)$.

(iii) for any admissible ν', there is a strategy μ', (called a reprisal strategy) admissible or idealized, such that $P(x_0;\mu',\nu') \leq W(x_0) + \varepsilon$.

(iv) for any admissible μ'', there is a reprisal strategy ν'', admissible or idealized, such that $P(x_0;\mu'',\nu'') \geq W(x_0) - \varepsilon$.

If $R' \subset R$, we say (μ_0,ν_0) are optimal on R' if and only if conditions (i) - (iv) hold for R'.

The first two conditions of the definition will be called the semipermeable conditions, the last two the reprisal conditions. The former roughly means that no player can be better off by a "temporary" deviation from (μ_0,ν_0). In Theorem 3.5, we report

(i) - (ii) are satisfied iff (μ_0, ν_0) are semipermeable with respect to $\nabla W(x)$ in R.

Remark 3.1. Because of the difficulties that arise with the existence of solutions of the kinematic equations, we assume that each player uses admissible strategies for the game and regards idealized strategies as limiting strategies which can be approximated by admissible strategies. Note that optimal strategies do not often exist in the class of admissible strategies. By including idealized strategies in the formulation of optimal strategies, we can find optimal strategies in a broader class of games and make it easier to treat singular surfaces.

Definition 3.2. A strategy μ (or ν'), admissible or idealized, is subject to terminating reprisal within R if there is a reprisal strategy ν(or μ'), admissible or idealized, such that for each $x \in R$, $\gamma(t,x;\mu,\nu)$ (or $\gamma(t,x;\mu',\nu')$) exists and stays in R until it meets C in finite time; furthermore, $P(x;\mu,\nu) \geqq W(x) - \varepsilon$ (or $P(x;\mu',\nu') \leqq W(x) + \varepsilon$) for any $\varepsilon > 0$, where $W(x)$ is specified in Definition 3.1. We shall call ν(or μ') a terminating reprisal strategy. Clearly, if each admissible ν(or μ) is subject to terminating reprisal within R, then (iii) (or (iv)) of Definition 3.1. is satisfied. In many cases, it is much easier to work with terminating reprisal rather than with the general reprisal condition (iii) - (iv).

Theorem 3.1. Suppose that (μ_1,ν_1) and (μ_2,ν_2) both satisfy (i) - (ii) of Definition 3.1 over R and that either both μ_1 and μ_2 or both ν_1 and ν_2 are subject to terminating reprisal within R. Then $W(x;\mu_1,\nu_1) = W(x;\mu_2,\nu_2)$ for all $x \in R$.

Definition 3.3. If (μ,ν) are optimal over R, we call $W(x;\mu,\nu)$ the value of the differential game for each $x \in R$.

Note that Theorem 3.1 implies that the value is uniquely defined over R whenever a suitable terminating reprisal condition is satisfied.

Definition 3.4. R is closed and terminal with respect to (μ,ν) if for each $x_0 \in R$, $\gamma(t,x_0;\mu,\nu)$ exists and meets C at some $t(x_0) < \infty$ and $\gamma(t,x_0;\mu,\nu)$ is contained in R for $0 \leqq t \leqq t(x_0)$.

Theorem 3.2. Suppose that (μ_1,ν_1) and (μ_2,ν_2) are optimal over R, that the terminating reprisal condition of Theorem 3.1 is satisfied, and suppose that R is closed and terminal with respect to (μ_1,ν_2). Then (μ_1,ν_2) is also optimal over R.

The following result is a simple modification of Theorems 4.2 and 5.2 of [7].

Definition 3.5 Let $\lambda(x)$ be defined over a region R. We say that (μ,ν) are semipermeable with respect to $\lambda(x)$ over R if and only if the following inequalities hold:

$$\lambda(x) \cdot f(x,\mu(x),\psi) \leq 0 \quad (3)$$

$$\lambda(x) \cdot f(x,\phi,\nu(x)) \geq 0 \quad (4)$$

for each $x \in R$, $\phi \in \Phi$ and $\psi \in \Psi$.

Theorem 3.3 Let (μ_0,ν_0) be a pair of admissible or idealized strategies such that its associated R is nonempty. Then (μ_0,ν_0) satisfies optimality condition (i)-(ii) of Definition 3.1 if and only if (μ_0,ν_0) are semipermeable over R with respect to $\nabla W(x;\mu_0,\nu_0)$.

Theorem 3.4. Let (μ_0,ν_0) be a pair of admissible or idealized strategies with nonempty R such that R is closed and terminal with respect to all (μ_0,ν) and (μ,ν_0), where ν,μ are arbitrary admissible strategies. Suppose that (μ_0,ν_0) satisfies optimality conditions (i) - (ii) of Definition 3.1. Then conditions (iii) - (iv) are also satisfied with μ_0 and ν_0 as reprisal strategies against any ν and μ respectively. Thus (μ_0,ν_0) is optimal over R.

Theorem 3.5. Let (μ_0,ν_0) be a pair of admissible strategies such that its associated R is nonempty and such that $R_0 \subset R$ is closed and terminal with respect to all admissible strategies. Then (μ_0,ν_0) are optimal over R_0 if and only if for each admissible strategy μ,ν and each $x \in R_0$, the following inequalities hold:

$$P(x;\mu_0,\nu) \leq P(x;\mu_0,\nu_0) \leq P(x;\mu,\nu_0). \quad (5)$$

Hence also, (μ_0,ν_0) is optimal over R_0 if and only if (μ_0,ν_0) are semipermeable with respect to $\lambda(x;\mu_0,\nu_0)$ over R_0.

The following result can aid in determining terminating reprisal strategies:

Theorem 3.6. Let (μ_0,ν_0) be a pair of admissible or idealized strategies such that its associated R is nonempty. Let μ (or ν') be an arbitrary admissible or idealized strategy. Suppose there is ν (or μ'), an admissible or idealized strategy, such that for each $x \in R$, $\gamma(t,x;\mu,\nu)$ (or $\gamma(t,x;\mu',\nu')$) exists and stays in R until termination in finite time. Suppose also that $\lambda(x;\mu_0,\nu_0)\cdot f(x,\mu,\nu) \geq 0$ (or $\lambda(x;\mu_0,\nu_0)\cdot f(x,\mu',\nu') \leq 0$). Then ν (or μ') is a terminating reprisal strategy against μ (or ν').

4. OPTIMALITY AND TRANSITION SURFACES IN CLOSED AND TERMINAL REGIONS

In this section we present necessary and/or sufficient conditions for optimality and necessary conditions for transition surfaces.

Definition 4.1. Suppose that (μ_*,ν_*) is optimal over R_*. Then M is called a transition surface of our differential game.

We shall extend Theorem 3.6 for the case of idealized strategies in:

Theorem 4.1. Let R_* be terminal and closed. Suppose that there exist $\alpha,\sigma > 0$ such that $\eta_2(x)\cdot f(x,\phi,\psi) \geq \alpha$ for all x such that $|T_M(x;\mu_2,\nu_2)| < \sigma$ and $(\phi,\psi) \in \Phi \times \Psi$. Then (μ_*,ν_*) are optimal over R_* iff (μ_*,ν_*) are semipermeable with respect to $\lambda_*(x)$ over R_*. $(\eta_2(x) = -\nabla T_M(x;\mu_2,\nu_2))$.

Corollary 4.1. Suppose that the assumptions of Theorem 4.1 hold. Then in linear differential games (μ_*,ν_*) is optimal over R_* iff

(i) $\lambda_*(x) \cdot [A(x) + B(x)\mu_*(x) + C(x)\nu_*(x)] = 0$
(ii) $\mu_*^i(x) = -\text{sgn}[\lambda_*(x)\cdot B^i(x)]$ if $\lambda_*(x)\cdot B^i(x) \neq 0$
(iii) $\nu_*^j(x) = \text{sgn}[\lambda_*(x)\cdot C^j(x)]$ if $\lambda_*(x)\cdot {}^j(x) \neq 0$

where

$$\text{sgn}(a) = \begin{cases} 1 \text{ if } a > 0 \\ -1 \text{ if } a < 0 \\ 0 \text{ otherwise} \end{cases}$$

Remark 4.1. The following Theorem enables us to determine optimal strategies near $\mathcal{C}$ in order to initialize the backward integration process.

Theorem 4.2. Let $x_0 \in C \cap R$ be such that $\eta(x_0)\cdot f(x_0,\phi,\psi) \geq \alpha_0 > 0$ for all $(\phi,\psi) \in \Phi \times \Psi$. Suppose that (μ_0,ν_0) are admissible strategies and semipermeable with respect to $\lambda(x)$ over a neighborhood containing x_0. Then there is a set $\Omega \subset R$ of (n+1) dimensions which contains x_0 and in which (μ_0,ν_0) are optimal. (Here $\eta(x)=\eta(x;\mu_0,\nu_0)$ and $\lambda(x)=\lambda(x;\mu_0,\nu_0)$.)

Corollary 4.2. Suppose that the first assumption of Theorem 4.2 holds. Then in linear differential games, the conditions that (μ_0,ν_0) are semipermeable at x_0 with respect to $\lambda(x_0)$ and that $\lambda(x_0)\cdot B^i(x_0) \neq 0$ and $\lambda(x_0)\cdot C^j(x_0) \neq 0$ for each $i = 1,\ldots,m$ and $j = 1,\ldots,\ell$; will assure the existence of a set $\Omega \subset R$ of (n + 1) dimensions which contains x_0 and in which (μ_0, ν_0) are optimal.

Remark 4.2. As a consequence of Corollary 2.1 and 2.2 of [11], $(\mu_0, \nu_0) = (-\text{sgn}\ \lambda(x_0)\cdot B(x_0),\ \text{sgn}\ \lambda(x_0)\cdot\ (x_0))$, is a constant strategy (with each component being -1 or 1) over Ω. Furthermore (μ_0, ν_0) are the unique pair of optimal strategies.

Theorem 4.3. Let (μ_*, ν_*) be optimal over R_* and $\eta_2(x)\cdot f(x,\phi,\psi) \geqq \alpha > 0$ for each $x \in M$ and $(\phi,\psi) \in \Phi \times \Psi$. (Thus M is the transition surface in R_*). Then in linear differential games, at each $x_0 \in M$ at least one of the following equations holds:

$$\lambda_1(x_0) \cdot B^i(x_0) = 0,\ i = 1,\ldots,m$$

$$\lambda_1(x_0) \cdot C^j(x_0) = 0,\ j = 1,\ldots,\ell$$

(Here $\lambda_1 = \lambda(x;\mu_1,\nu_1)$, $\eta_2(x) = \eta(x;\mu_2,\nu_2)$ and with C and M respectively as the terminal surface.)

Remark 4.3. Note that each $\lambda_1(x)\cdot B^i(x)$ and each $\lambda_1(x)\cdot C^j(x)$ is known as a " switching function." Theorem 4.3 is significant because it states that under a simple condition transition surfaces can occur only when one of the switching functions vanishes. The conditions can be used in backward integration as part of the usual techniques for solving differential games. Note that the terminal and closed condition is needed; in the next section we shall point out that without this condition the switching functions are not adequate to locate transition surfaces.

5. OPTIMALITY AND TRANSITION SURFACES IN THE NON-CLOSED AND TERMINAL CASE.

5.1 Conditions for Optimal Admissible Strategies

Theorem 5.1. Let $R_0 \subset R$ be induced by (μ_0, ν_0). Suppose that the following hold; (with $\lambda(x) \equiv \lambda(x;\mu_0,\nu_0)$,)

(i) (μ_0, ν_0) are semipermeable with respect to $\lambda(x)$ over R_0;

(ii) there exists $\alpha > 0$ and a neighborhood N of R_0 in R such that for each admissible ν' there exists admissible μ' such that for each $x \in N$, $\eta(x)\cdot f(x,\mu',\nu') \geqq \alpha$ and $\lambda(x)\cdot f(x,\mu',\nu') \leqq 0$ hold;

(iii) for each admissible μ'' there is admissible ν'' such that for each $x_0 \in R_0$, $\lambda(x)\cdot f(x,\mu'',\nu'') \geqq 0$ and $\gamma(t,x_0;\mu'',\nu'')$ is in R_0 until termination or is such that, with respect to the same α and N of (ii), $\eta(x)\cdot f(x,\mu'',\nu'') \geqq \alpha$ and $\lambda(x)\cdot f(x,\mu'',\nu'') \geqq 0$ hold for each $x \in N$;

(iv) R_0 is α-closed with respect to $\eta(x)$ and N.

Then (μ_0, ν_0) are optimal over R_0.

Theorem 5.2. Let $x_0 \in C \cap R$ where R is induced by $(\mu_0,\nu_0) = (\phi_0,\psi_0)$, a pair of constant strategies. Let $\lambda(x) = \lambda(x;\mu_0,\nu_0)$ and $\eta(x) = \eta(x;\mu_0,\nu_0)$. In linear differential games, suppose that

(i) $\eta(x)\cdot f(x,\mu_0,\nu_0) \geqq \alpha > 0$ in a neighborhood containing x_0,

(ii) (ϕ_0,ψ_0) is semipermeable with respect to $\lambda(x_0)$ at x_0

(iii) each $\lambda(x_0)\cdot B^i(x_0) \neq 0$ and each $\lambda(x_0)\cdot C^j(x_0) \neq 0$, $i = 1,\ldots,m$; $j = 1,\ldots,\ell$.

(iv) for each extreme point $\psi_k \neq \psi_0$ of Ψ [resp. $\phi_k' \neq \phi_0$ of Φ], there is $\phi_k \in \Phi$ [resp. $\psi_k' \in \Psi$] such that $\eta(x_0)\cdot f(x_0,\phi_k,\psi_k) \geqq \alpha > 0$ and $\lambda(x_0)\cdot f(x_0,\phi_k,\psi_k) < 0$ [resp. $\eta(x_0)\cdot f(x_0,\phi_k',\psi_k') \geqq \alpha$ and $\lambda(x_0)\cdot f(x_0,\phi_k',\psi_k') > 0$] hold. Then there is a set $\Omega \subset R$ of (n+1) dimensions which contains x_0 and in which (μ_0,ν_0) are optimal.

5.2 A Necessary Condition for Transition Surfaces

Definition 5.1. Given $x \in R$ and $\psi \in \Psi$, we define

$$\Phi_\lambda(x,\psi) = \{\phi \in \Phi \mid \lambda(x)\cdot f(x,\phi,\psi) \leqq 0\} \tag{6}$$

and

$$\Phi_\eta(x,\psi) = \{\phi \in \Phi \mid \eta(x)\cdot f(x,\phi,\psi) \geqq 0\} \tag{7}$$

The following simplified notation will be used:

$$\sup_{\Phi_\lambda} \eta(x)\cdot f(x,\phi,\psi) \equiv \sup_{\phi\in\Phi_\lambda(x,\psi)} \eta(x)\cdot f(x,\phi,\psi) \tag{8}$$

and

$$\inf_{\Phi_\eta} \lambda(x)\cdot f(x,\phi,\psi) \equiv \inf_{\phi\in\Phi_\eta(x,\psi)} \lambda(x)\cdot f(x,\phi,\psi) \tag{9}$$

Remark 5.1. Whenever $\Phi_\lambda(x,\psi)$ and $\Phi_\eta(x,\psi)$ are nonempty, (8) - (9) are well defined. By continuity of f in ϕ and compactness of Φ_λ and Φ_η, in fact one can replace "sup" and "inf" by "max" and "min" respectively in (8) - (9).

Lemma 5.1. Let x and ψ be given. Suppose that there is a $\phi \in \Phi$ so that $\eta(x)\cdot f(x,\phi,\psi) > 0$. Then

$$\sup_{\Phi_\lambda} \eta(x)\cdot f(x,\phi,\psi) = 0 \text{ implies that}$$

$$\inf_{\Phi_\eta} \lambda(x)\cdot f(x,\phi,\psi) = 0.$$

Definition 5.2 Given (μ_*,ν_*) we say that M satisfies the η_i-condition if for each $x \in M$, $\eta_1(x)\cdot f(x,\mu_1, \nu_1) \geqq 0$ and for each $\alpha > 0$ and any two neighbourhoods $N_i(x)$, $i = 1,2$, there is a set $\Omega \subset N_1(x) \cap N_2(x) \cap R_2$ of n+1 dimensions such that Ω is α-closed with respect to $\eta_i(x)$ and $N_i(x)$ for both $i = 1, 2$ when M is regarded as the terminating surface. (Refer to Def. 2.3. Note that $\eta_i(x) \equiv \eta\ (x;\mu_i, \nu_i)$).

Theorem 5.3 In a linear differential game, suppose that $(\mu_*\ \nu_*)$ is optimal over R_* with M satisfying the η_i-condition. Assume that for each extreme point ψ_k of Ψ and each $x \in M$ there is $\phi \in \Phi_{\lambda_1}(x, \psi_k)$ such that $\eta_1(x)\cdot f(x,\phi,\psi_k) \geqq 0$. Assume also that at $x_0 \in M$, there is a neighbourhood $N_2(x_0)$ and $\alpha > 0$ such that for each admissible ν there exists an admissible μ so that $\lambda_2(x)\cdot f(x,\mu,\nu) \leqq 0$ and $\eta_2(x)\cdot f(x,\mu,\nu) \geqq \alpha$ hold at each $x \in N_2(x_0)$. Then at x_0 at least one of the following conditions holds:

(i) $\lambda_1(x_0)\cdot B^i(x_0) = 0$ for some i, $1 \leqq i \leqq m$;

(ii) $\lambda_1(x_0)\cdot C^j(x_0) = 0$ for some j, $1 \leqq j \leqq \ell$;

(iii) $\sup_{\Phi_{\lambda_1}} \eta_1(x_0)\cdot f(x_0,\phi,\psi_k) = 0$ for some extreme point ψ_k of Ψ.

Remark 5.2. Theorem 5.3 allows us to locate transition surfaces by backward integration. We shall supply an example in the next section to illustrate its application.

5.3 A Sufficient Condition for Idealized Strategies to be Optimal

We shall use the following notation for Theorem 5.4. (See Figure 1)

$$B_\sigma = \{x \in R \mid |T_M(x)| \leqq \sigma\}$$

$$B_{\sigma\delta} = \{x \in R \mid -\delta-\sigma < T_M(x) < -\sigma.\}$$

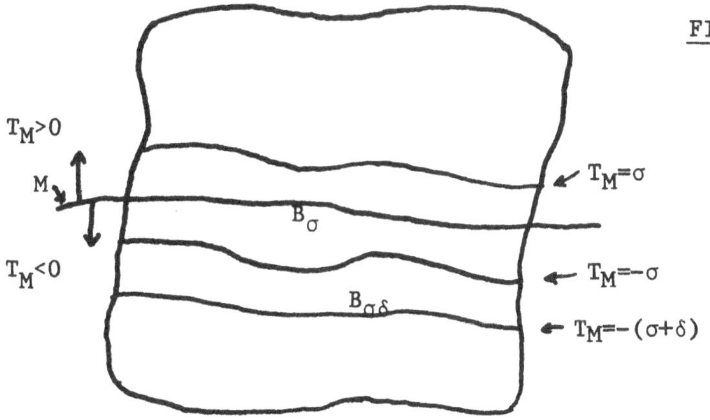

FIGURE 1

Theorem 5.4. (μ_*, ν_*) are optimal in R_* if the following hold:

(i) (μ_*, ν_*) is semipermeable with respect to $\lambda_*(x)$ at each $x \in R$; (Note, $\lambda_*(x) \equiv \lambda_*(x;\mu_*,\nu_*)$);

(ii) for each admissible ν there exists an admissible μ and two positive numbers α_0 and σ_0 such that for each $\sigma \in (0,\sigma_0)$ there is $\delta(\sigma) > 0$ so that for each $\delta \in (0,\delta(\sigma))$ and σ the following hold:

(A) for each $x \in B_\sigma$, $\eta_2(x)\cdot f(x,\mu,\nu) \geqq \alpha_0$

(B) for each $x \in B_{\sigma\delta}$, $\eta_2(x)\cdot f(x,\mu,\nu) \geqq \alpha_1(\sigma) > 0$

(C) for each $x \in R_* \setminus (B_\sigma \cup B_{\sigma\delta})$, $\eta_*(x)\cdot f(x,\mu,\nu) \geqq \alpha_2(\sigma,\delta) > 0$ and $\lambda_*(x)\cdot f(x,\mu,\nu) \leqq 0$

(iii) $\lambda_*(x)\cdot f(x,\phi,\psi)$ is bounded on $(B_\sigma \cup B_{\sigma\delta}) \times \Phi \times \Psi$;

(iv) for each admissible μ' there is an admissible or idealized ν' so that $\lambda_*(x)\cdot f(x,\mu',\nu') \geqq 0$ and $\gamma(t,x_0;\mu',\nu')$ stays in R_* until termination for each $x_0 \in R_*$.

(v) R_* is α-closed with respect to $\eta_*(x)$ for each $\alpha > 0$.

Remark 5.3. Condition (iv) clearly can be replaced by similar conditions as (ii) - (iii) (with $\lambda_*\cdot f(x,\mu',\nu') \geqq 0$ replacing $\lambda_*\cdot f(x,\mu,\nu) \leqq 0$).

Remark 5.4. We use the results of this section as follows: by applying Theorem 5.2, we can frequently find strategies which are optimal near C. Using these strategies, we then locate the zero sets specified by the conditions of Theorem 5.3. Reapplying Theorem 5.2, we treat such a zero set as a new terminal surface and again determine optimal strategies near it. If the conditions of Theorem 5.4 are satisfied, we can then use it to verify that the strategies we have constructed by this process are in fact optimal. (See the example of the next section for an illustration.)

6. AN EXAMPLE

Let us consider the following linear differential game:

(i) The playing space is $E = \{x = (x^0,x^1,x^2) \mid x^2 \geqq 0\}$;

(ii) The dynamic system is

$$\dot{x} = \begin{pmatrix} 8 \\ 2x^2 \\ -10 \end{pmatrix} + \begin{pmatrix} 5 \\ 1 \\ 15 \end{pmatrix}\phi + \begin{pmatrix} -3x^2 \\ 0 \\ 7 \end{pmatrix}\psi;$$

(iii) The terminal surface is $C = \{x \mid x^2 = 0\}$;

(iv) The payoff is $H(x) = -x^0$; if no termination occurs the payoff is $+\infty$.

Note that in this game, for each initial point, there exist strategy pairs which do not lead to termination.

With the help of Theorem 5.2, we can show that $(\mu_1,\nu_1) = (1,-1)$ is optimal in a set containing C.

We can use the conditions of Theorem 5.3 to locate candidates for transition surfaces. First, it is readily verified that $\lambda_1(x)\cdot B(x)$ and $\lambda_1(x)\cdot C(x)$ are never vanishing in E. Thus we obtain no condidates for transition surfaces from the switching functions. Using (iii) of Theorem 5.3, we have, with $\psi=1$, $M = \{x \in E | x^2 = 3\}$. Also, we find that

$$\sup_{\Phi_{\lambda_1}} \eta_1(x)\cdot f(x,\phi,-1) = 2$$

which is never vanishing in E. Thus we conclude that M as specified above is the only candidate for a transition surface in E.

Now let $(\mu_2,\nu_2) = (-1,1)$ be a pair of constant strategies and

$$(\mu_*,\nu_*) = \begin{cases} (1,-1) & \text{if } 0 \leqq x^2 < 3 \\ (-1,1) & x^2 \geqq 3. \end{cases}$$

We can verify (see [10] or [13]) that the conditions of Theorem 5.4 are satisfied. Thus (μ_*,ν_*) are optimal over E and M is indeed a double transition surface. (i.e., both μ_* and ν_* switch simultaneously).

7. CONCLUSION

In each of the examples the authors have studied, (iii) of Theorem 5.3 has located double transition surfaces. We conjecture that in general the switching functions (i) and (ii) locate single transition surfaces (unless more than one of them should become zero simultaneously) and (iii) locates double transition surfaces.

A heuristic explanation of the double transition surface phenomenon is as follows: "below" the surface the minimizing player can, in response to any ψ, everywhere choose a ϕ which meets both his objectives; i.e. such that the payoff is non-increasing and such that the path is approaching termination. "Above" the surface, he is no longer able to meet both these objectives simultaneously and is forced to choose a strategy with the sole objective of reachingthe terminal surface. The transition surface is precisely the set of points where the minimizing player "loses his advantage"; i.e., if he chooses to aim for termination, the payoff relative to that pair of strategies is non-decreasing; if he chooses to attempt to decrease the payoff determined by that pair, he cannot approach termination. The maximizing player is able to take advantage of this forced shift in the minimizing player's strategy and also shift to a strategy more profitable to him.

In a sense, then, the double transition surface separates subregions in which different players have the advantage of being able to force the other player to choose his strategy first, whereas for the single transition surface, it is only to a single player's advantage in achieving his own objectives to switch. The double transition surface, then, reveals more of the conflict aspects of differential games while the single transition surface reveals more of the optimal control aspects.

REFERENCES

1. Isaacs, R., Differential Games, John Wiley and Sons, New York, 1965.
2. Leitmann, G., Two-Person Zero-Sum Games, Paper presented at the 36th National Meeting of the Operations Research Society of America, Miami, Florida, 1969.
3. Leitmann, G., Introduction to Optimal Control, McGraw-Hill, New York, 1966.
4. Blaquiere, A., Gerard, F., and Leitmann, G., Quantitative and Qualitative Games, Academic Press, New York, 1969.
5. Berkovitz, L.D., A Variational Approach to Differential Games, Advances in Game Theory, Annals of Mathematics, Study No. 52, Princeton University Press, Princeton, New Jersey, 1964.
6. Stalford, H. and Leitmann, G., "Sufficient Conditions for Optimality in Two-Person Zero-Sum Differential Games with State and Strategy Constraints" Journal of Mathematical Analysis and Applications 33; pp. 650-654, 1971
7. Bradley, J., and Yu, P.L., "Some Basic Properties of the Payoffs Defined by Closed-Loop Strategies." CS 178, U.T. (See Ref. 12).
8. Friedman, A., Differential Games, Wiley-Interscience, 1971.
9. Yu, P.L., "Some Fundamental Qualifications of Optional Strategies and Transition Surfaces in Differential Games," Journal of Optionization Theory and Applications, Vol. 9, No. 6, June 1972.
10. Bradley, J.,"Transition Surfaces and Optimal Stretegies in a Class of Differential Games," Ph.D. Dissertation, Dept. of Mathematics, University of Rochester, 1974.
11. Bradley, J. and Yu, P.L. "Semipermeable Directions in Linear Differential Games," Univ. of Rochester, Graduate School of Management F7316 (To appear in Journal of Mathematical Analysis and Applications).
12. Bradley,J. and Yu, P.L. "A Concept of Optimality in Differential Games," CS179, Center for Cybernetic Studies, Univ. of Texas,Austin.
13. Bradley,J. and Yu, P.L. "Conditions for OPtimality and Transition Surfaces Using the Reprisal Concept," CS180, Center for Cybernetic Studies, Univ. of Texas, Austin, Texas.

PURSUIT OF A FASTER EVADER

John V. Breakwell

Department of Aeronautics and Astronautics,
Stanford University, Stanford, California, U.S.A.
94305

INTRODUCTION. This paper is concerned with coplanar pursuit-evasion problems with constant speeds, in which the evader is faster than the pursuer or pursuers, but is handicapped by the requirement either to escape from more than one pursuer or to move within a specified boundary. The "payoff" is in every case the distance of closest approach between evader and pursuer (or pursuers).

A typical problem of this kind is Isaacs' DEADLINE GAME ([1], p.260) in which the evader must not cross a straight boundary. In the "one-sided" deadline game ([1], p.265) Isaacs poses this question: For what initial positions of pursuer P (speed 1) and evader E(speed $w > 1$) can E pass between P and the x-axis (see Fig. 1), moving in the positive x-direction without coming closer to P than a specified distance ℓ ? The answer to this "game of kind" is obviously obtainable if we can solve the following equivalent "game of terminal payoff": What is the distance ℓ of closest approach from any given starting position, assuming that P minimizes ℓ and E maximizes ℓ while passing

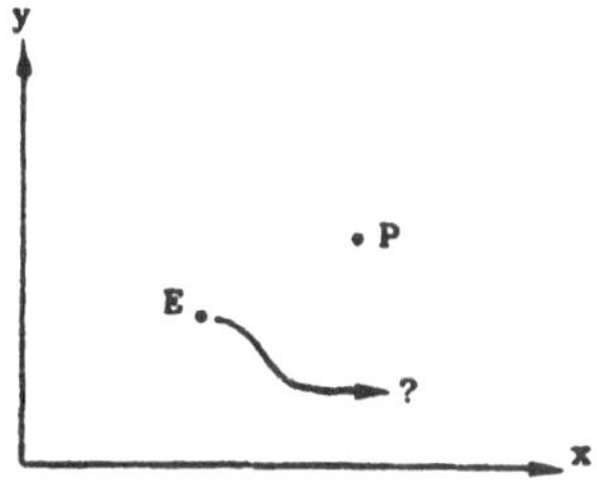

Figure 1.

J. D. Grote (ed.), The Theory and Application of Differential Games. 243-256.

between E and the x-axis? [The answer to this question may, of course, be $\ell = 0$.]

Isaacs analyzes the optimal paths in this game and concludes that they have two phases: a first, straight-line phase during which the distance PE is decreasing, and a second, curved, phase during which the distance PE remains equal to ℓ. The second phase terminates with E's path tangent to the x-axis at E_f and P, now at P_f, moving directly towards E_f (see Figure 2) after which the distance PE increases. P's curved path is determined by the

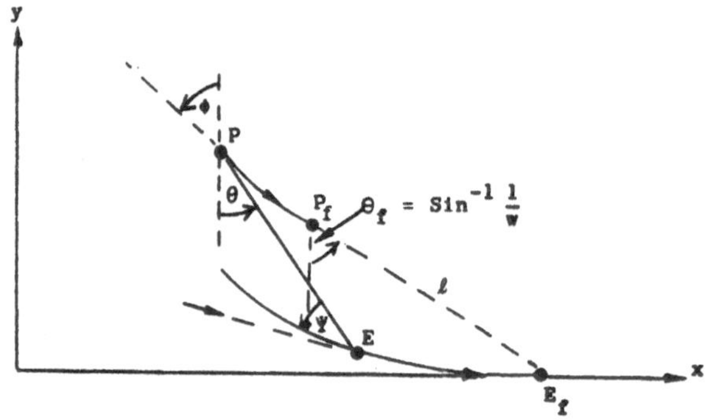

Figure 2.

maximization of $-\frac{dy_E}{d\theta}$ subject to PE = ℓ , which determines, in turn, E's curved path. The first phase straight paths are tangent to the second phase curved paths.

ANALYSIS: We start the analysis by proceeding to verify these features of the optimal paths in the one-sided deadline game.

Let $V(\vec{r}_E, \vec{r}_P)$ denote the closest approach distance ℓ corresponding to initial positions $\vec{r}_E$ and $\vec{r}_P$ of E and P, respectively. If $\hat{\beta}_E$ and $\hat{\beta}_P$ denote unit-vectors parallel to the velocities of E and P respectively, the "main equation" is:

$$0 = \underset{\hat{\beta}_E}{\text{Max}}\ \underset{\hat{\beta}_P}{\text{Min}}\ (V_{\vec{r}_E} \cdot w\hat{\beta}_E + V_{\vec{r}_P} \cdot \hat{\beta}_P)$$

$$= w|V_{\vec{r}_E}| - |V_{\vec{r}_P}| , \tag{1}$$

with $\hat{\beta}_E$, $\hat{\beta}_P$ parallel to $V_{\vec{r}_E}$, $-V_{\vec{r}_P}$ respectively.

The gradient vectors $V_{\vec{r}_E}$, $V_{\vec{r}_P}$ are, furthermore, constant along unconstrained paths, implying straight-line motion. It is clear, moreover, that here $V(\vec{r}_E, \vec{r}_P)$ is a function $V(x, y_P, y_E)$, where $x = X_P - X_E$, so that $V_{X_E} = -V_{X_P}$. It now follows from (1) that the "controls" Ψ, ϕ (see Fig. 2) satisfy:

$$\frac{w}{\sin \Psi} = \frac{1}{\sin \phi} \tag{2}$$

During the final phase, however, E must observe the "state constraint" that EP not be allowed to decrease further. This requires that E's control Ψ be a function $\tilde{\Psi}(\phi, \theta)$ defined by

$$w \cos(\Psi - \theta) = \cos(\phi - \theta) . \tag{3}$$

[If an objection is raised to this implicit assumption that E knows P's present control ϕ, E can achieve as small a change in EP as he pleases by utilizing knowledge of P's control in the recent past, as close as necessary to the present.]

The constrained main equation, using coordinates r, θ, y_E in place of x, y_P, y_E, where r = EP, is:

$$\underset{\substack{\phi \\ \Psi=\tilde{\Psi}(\phi,\theta)}}{\text{Min}} \left(V_\theta \dot{\theta}(\theta, \phi, \tilde{\Psi}) + V_{y_E} \dot{y}_E(\theta, \phi, \tilde{\Psi})\right) = 0 ,$$

which implies the stationarity of $\frac{dy_E}{d\theta}$ w.r.t. ϕ. [Obviously, $-\frac{dy_E}{d\theta}$ should be maximized.] It is also easily verified, by comparing constrained and unconstrained main equations in coordinates r, θ, y_E, that Ψ and ϕ must be continuous at junctions of the unconstrained and constrained paths. This proves that the unconstrained paths are straight tangents to the second phase curved paths.

The controls Ψ, ϕ along the curved paths must satisfy both (2) and (3). We note immediately that both conditions are met at E_f and P_f in Fig. 2, where $\theta = \phi = \text{Sin}^{-1} \frac{1}{w}$, $\Psi = \frac{\pi}{2}$. Note, further, that (3) is a consequence of (2) if

$$\phi + \Psi + \left(\frac{\pi}{2} - \theta\right) = \pi \tag{4}$$

so that, if $0 < \frac{\pi}{2} - \theta \leq \pi$, Ψ and ϕ can be expressed as functions of θ with the aid of Fig. 3.

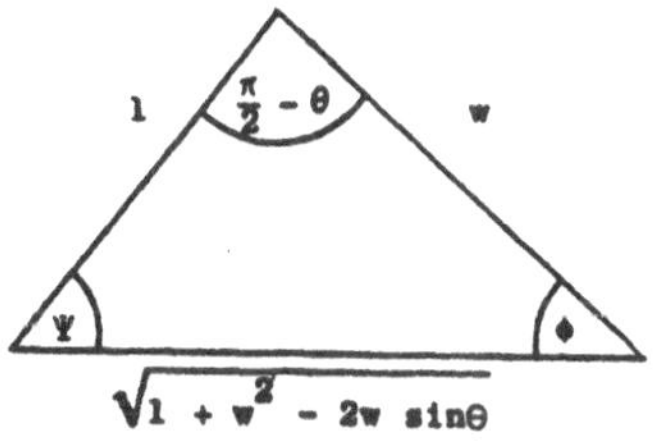

Figure 3.

This yields:

$$\left.\begin{aligned} \cos\Psi &= \frac{1 - w\sin\Theta}{\sqrt{1 + w^2 - 2w\sin\Theta}}, \quad \sin\Psi = \frac{w\cos\Theta}{\sqrt{1 + w^2 - 2w\sin\Theta}} \\ \cos\phi &= \frac{w - \sin\Theta}{\sqrt{1 + w^2 - 2w\sin\Theta}}, \quad \sin\phi = \frac{\cos\Theta}{\sqrt{1 + w^2 - 2w\sin\Theta}} \end{aligned}\right\} \quad (5)$$

(cf [1], p. 262, wherein Θ is denoted by s). These formulas extend, moreover, to values of Θ less than $-\frac{\pi}{2}$.

The integration of the curved paths back from the positions E_f, P_f, is elementary (cf [1], eqn. (9.5.9)).

$$\left.\begin{aligned} y_E &= \frac{\ell}{w^2 - 1}(kw - w\Theta - w^2\cos\Theta) \\ y_P &= \frac{\ell}{w^2 - 1}(kw - w\Theta - \cos\Theta) \\ x_E &= \frac{\ell}{w^2 - 1}(w^2\sin\Theta - w) + x_{E_f} \\ x_P &= \frac{\ell}{w^2 - 1}(\sin\Theta - w) + x_{E_f} \end{aligned}\right\} \quad (6)$$

where $k = \sqrt{w^2 - 1} + \mathrm{Sin}^{-1}\frac{1}{w}$.

These paths may be constructed by rolling a wheel of radius $\frac{w\ell}{w^2-1}$ along a line through E_f parallel to the y-axis, with E, outside, and P, inside, rigidly attached to the wheel (see Fig. 4).

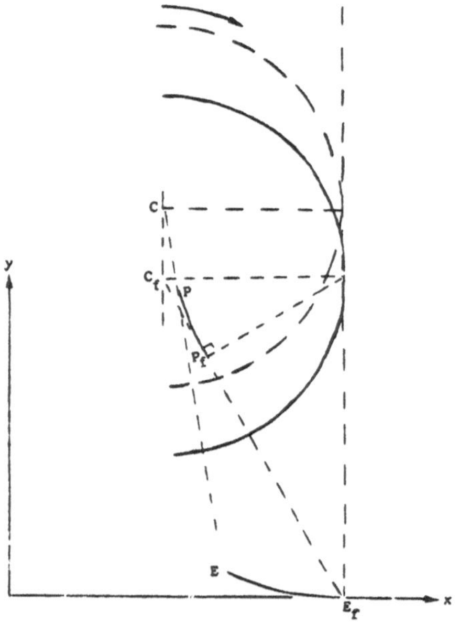

Figure 4.

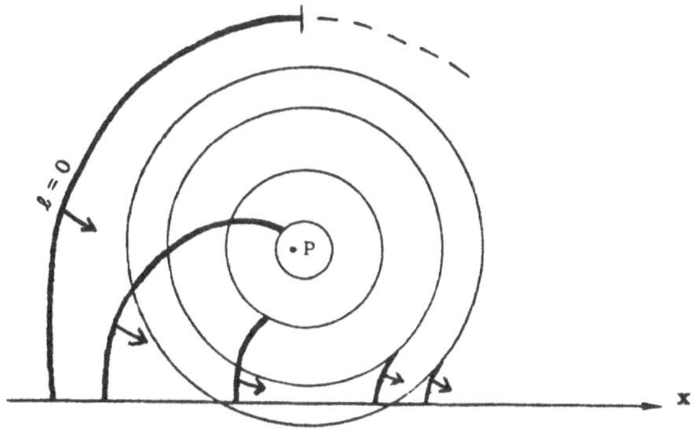

Figure 5.

The inclusion of the first phase straight tangents gives $y_P(\theta, \tau) = y_P(\theta) - \tau y'_P(\theta)$, etc., as in [1], eqn. (9.5.10). This expresses y_E, y_P, $x_E - x_{E_f}$, $x_P - x_{P_f}$ in terms of ℓ, θ, τ:

$$\left.\begin{aligned} y_E &= \frac{\ell}{w^2 - 1}\left\{kw - w\theta - w^2\cos\theta + w\tau(1 - w\sin\theta)\right\} \\ y_P &= \frac{\ell}{w^2 - 1}\left\{kw - w\theta - \cos\theta + \tau(w - \sin\theta)\right\} \\ x_E &= \frac{\ell}{w^2 - 1}\left\{w^2\sin\theta - w - w^2\tau\cos\theta\right\} + x_{E_f} \\ x_P &= \frac{\ell}{w^2 - 1}\left\{\sin\theta - w - \tau\cos\theta\right\} + x_{E_f} \end{aligned}\right\} \quad (7)$$

A section, for fixed y_P, of the surfaces corresponding to various ℓ, is sketched in Fig. 5, E's optimal direction, indicated by arrows, being perpendicular to the local ℓ - contour. The contour corresponding to $\ell = 0$ is easily verified to be an allipse with center on the x-axis, major axis vertical, focus at P and eccentricity 1/w.

If E is permitted to escape in either direction we have the two-sided DEADLINE GAME discussed by Isaacs. Fig. 6 shows a section of the surfaces S_1, S_2 corresponding to a fixed y_P and a single value of ℓ , and to escape to the right, left respectively. The surfaces S_1, S_2 in the three-dimensional state-space (x, y_P, y_E) intersect along a "dispersive edge," designated by (D) in Fig. 6, with the following property: If E at (D) chooses, for example, to escape to the right, he will remain on S_1 but will move off of surface S_2 to the "capture" side, i.e., he will no longer be in a position to escape to the left.

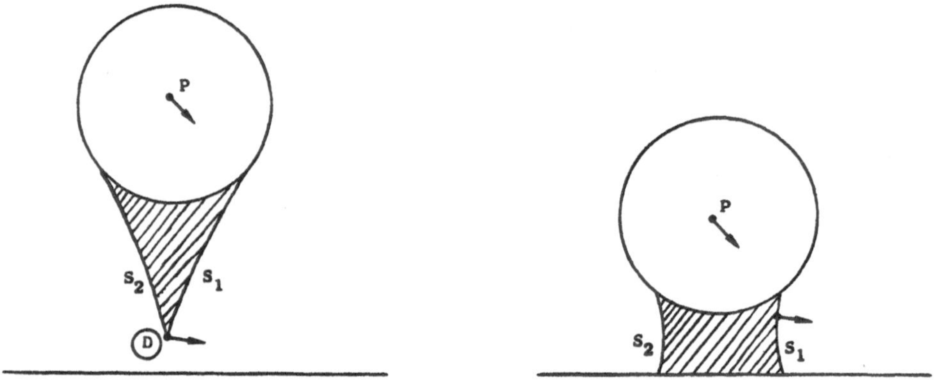

Figure 6.

The surfaces S_1, S_2, together with their junction Ⓓ, form a composite semi-permeable surface (see also [2]), enclosing a region from which E cannot escape (without coming closer to P than ℓ). Analytically this is expressed by

$V_{\vec{r}_E}^{(2)} \cdot w\hat{\beta}_E^{(1)} + V_{\vec{r}_P}^{(2)} \cdot \hat{\beta}_P^{(1)} < 0$ at Ⓓ, which, because of (1), may be expressed in the symmetric form:

$$\hat{\beta}_E^{(1)} \cdot \hat{\beta}_E^{(2)} - \hat{\beta}_P^{(1)} \cdot \hat{\beta}_P^{(2)} < 0 \text{ at Ⓓ}. \tag{8}$$

PATROLLING A CHANNEL

An interesting extension of the one-sided deadline game is the game: "patrolling a channel," also discussed in [1]. Here E can choose whether to pass along the channel (to the right in Fig. 7) above or below P. If the width L of the channel is less than a critical width:

$$L_c = \frac{2\ell w}{w^2-1}\left(\frac{\pi}{2} + k\right), \tag{9}$$

the surfaces S_1 and S_2, corresponding to passage below or above P respectively, form with the ℓ-circle around P the boundary of a capture region (shaded).

If, however, $L > L_c$, the surfaces S_1 and S_2 intersect outside the ℓ-circle in an "attractive edge," designated by Ⓐ (see Fig. 8), which is such that if, for example, E chooses to pass above P, E remains on S_2 but moves off of S_1 to the escape side, i.e., $\hat{\beta}_E^{(1)} \cdot \hat{\beta}_E^{(2)} - \hat{\beta}_P^{(1)} \cdot \hat{\beta}_P^{(2)} > 0$ at Ⓐ. The surfaces S_1 and S_2

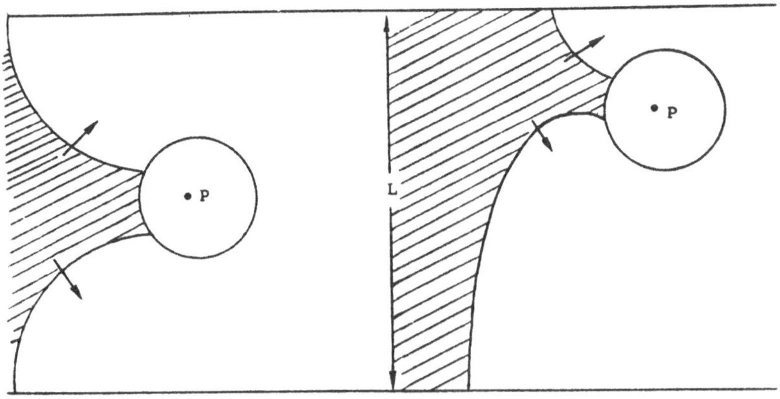

Figure 7.

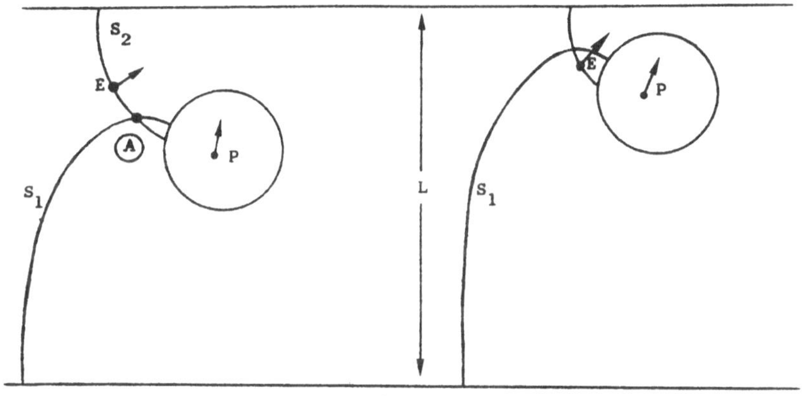

Figure 8.

thus clearly fail to form a composite semi-permeable surface -- if P guards against passage above P, E can pass below P.

THE CORNERED RAT

A further extension is the "cornered rat" game, also mentioned by Isaacs in [1]. The state-space is now essentially four-dimensional and the locus of E's positions for fixed P, corresponding to given ℓ and to escape along the x or y directions, is sketched in Figs. 9a, b, and c. For P sufficiently close to the corner, as in Fig. 9a, the hypersurfaces S_1 and S_2 do not intersect. For P somewhat further from the corner they intersect in two "dispersal hyperedges," denoted by (D) in Fig. 9b.

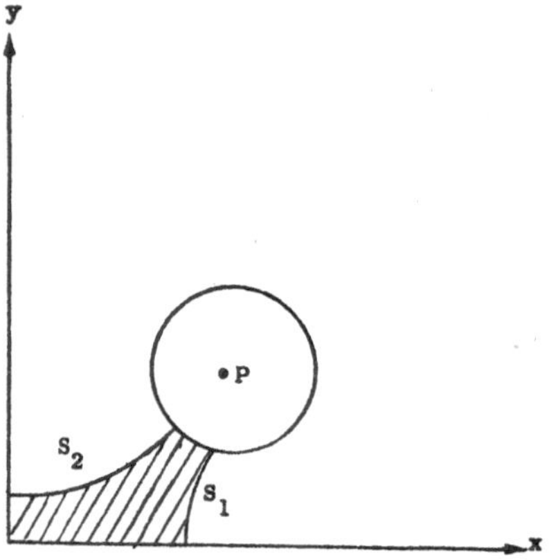

Figure 9a

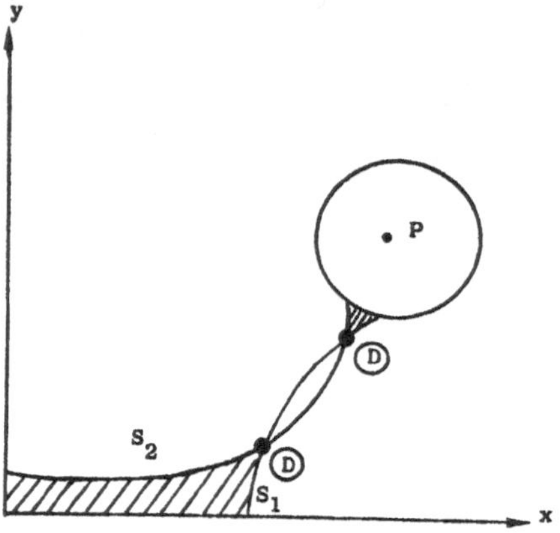

Figure 9b

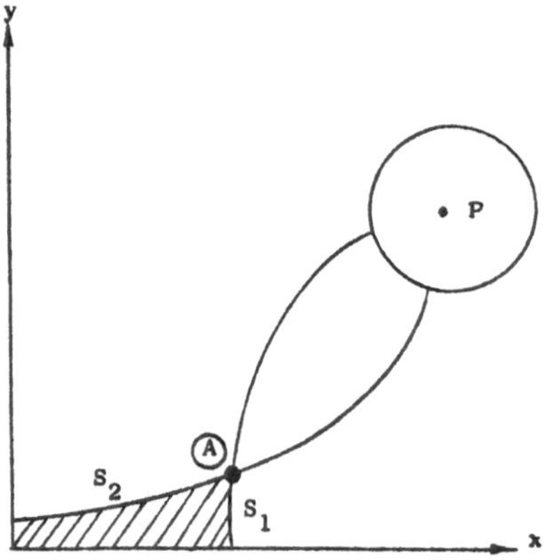

Figure 9c

But for P sufficiently far from the corner, as in Fig. 9c, S_1 and S_2 intersect in an "attractive hyperedge" Ⓐ . The hypersurfaces thus again fail to combine into a composite semi-permeable hypersurface.

The critical positions P^* and E^*, corresponding to a change from dispersive hyperedge Ⓓ to attractive hyperedge Ⓐ , satisfy:

$$\hat{\beta}_E^{(1)} \cdot \hat{\beta}_E^{(2)} - \hat{\beta}_P^{(1)} \cdot \hat{\beta}_P^{(2)} = 0 . \tag{10}$$

It is straightforward to verify that this implies:
$\tau_1 = \tau_2 = 1$ and $\theta_2 = -\theta_1$ (θ_2 being measured from the x-axis

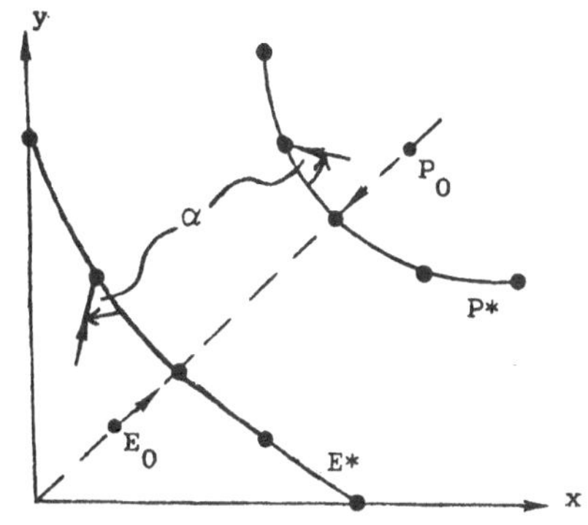

Figure 10.

rather than the y-axis), and that the P*- and E*-loci, sketched in Fig. 10, are given by:

$$\left.\begin{aligned}
\frac{w^2-1}{\ell}\, x_P^* &= kw + w\theta_1 - \cos\theta_1 + w + \sin\theta_1 \\
\frac{w^2-1}{\ell}\, y_P^* &= kw - w\theta_1 - \cos\theta_1 + w - \sin\theta_1 \\
\frac{w^2-1}{\ell}\, x_E^* &= kw + w\theta_1 - w^2\cos\theta_1 + w + w^2\sin\theta_1 \\
\frac{w^2-1}{\ell}\, y_E^* &= kw - w\theta_1 - w^2\cos\theta_1 + w - w^2\sin\theta_1
\end{aligned}\right\} \qquad (11)$$

It may further be easily verified that the arc-lengths of these loci satisfy:

$$|d\vec{r}_E^{\,*}| = w\,|d\vec{r}_P^{\,*}| \qquad (12)$$

It is hereby <u>conjectured</u> that for positions P outside of the P*-locus there is a locus of positions E inside the E*-locus from which P and E move in straight lines toward appropriate positions on the P*- and E*-loci, at which time E chooses which side of P to pass. If so, the directions during the "delayed option" phase of the game are determined by:

$$0 = \frac{dV^*}{d\theta_1} = V^{(3)}_{\vec{r}_E} \cdot \vec{r}_E^{\,*'}(\theta_1) + V^{(3)}_{\vec{r}_P} \cdot \vec{r}_P^{\,*'}(\theta_1)\,, \qquad (13)$$

which may be rewritten:

$$\hat{\beta}^{(3)}_{\vec{r}_E} \cdot \vec{r}_E^{\,*'}(\theta_1) - w\hat{\beta}^{(3)}_{\vec{r}_P} \cdot \vec{r}_P^{\,*'}(\theta_1) = 0\,, \qquad (13a)$$

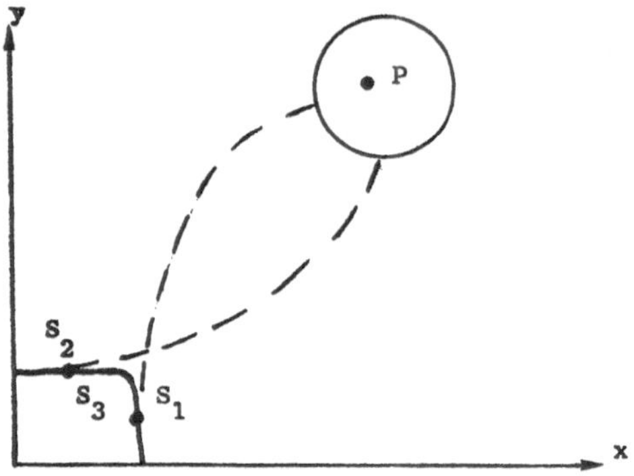

Figure 11.

the superscript (3) denoting the delayed-option phase. Together with (12) this implies that the delayed-option phase straight paths make equal angles, say α, with the E*- and P*-loci, as indicated in Fig. 10. In particular, for the symmetric positions E_0 and P_0 the two players move directly toward each other before E decides which way to pass. The locus of positions E corresponding to a delayed option strategy, when P is outside the P*-locus, can now be computed. Fig. 11 shows a sketch of how this locus should look. Note that equation (13a) is satisfied automatically by $(\hat{\beta}_E^{(3)}, \hat{\beta}_P^{(3)}) = (\hat{\beta}_E^{(i)}, \hat{\beta}_P^{(i)})$, $i = 1$ or 2, since

$(\vec{r}_E^{\,*}(\theta_1), \vec{r}_P^{\,*}(\theta_1))$ belongs to both hypersurfaces S_1 and S_2 for all θ_1. The surface S_3 thus joins smoothly onto the surfaces S_1 and S_2. Fig. 11 replaces Fig. 9c and together with Figs. 9a and 9b defines a composite hypersurface which, it is conjectured, is semi-permeable. To lend further plausibility to this conjecture, it may be noted that on the E^*- and P^*-loci the quantities $V_{\vec{r}_E}^{(i)} \cdot w\hat{\beta}_E^{(3)}(\alpha) + V_{\vec{r}_P}^{(i)} \cdot \hat{\beta}_P^{(3)}(\alpha)$, $i = 1$ and 2, which are homogeneous linear in $\sin\alpha$ and $\cos\alpha$, vanish for two distinct α's not differing by π , and therefore vanish identically. The delayed-option paths thus arrive tangentially to both hypersurfaces S_1 and S_2, thus fulfilling necessary conditions described in [2].

PASSAGE BETWEEN TWO PURSUERS

We now take up the game in which E must pass between two pursuers P_1 and P_2, the payoff being again the distance of closest approach to a pursuer. It is clear that E's "final" position E_f

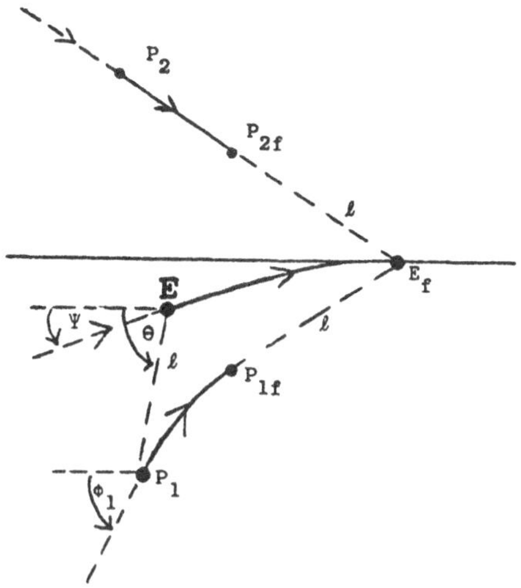

Figure 12.

must be equidistant (ℓ) from the two pursuers, and that prior to this position E will in general have remained for a while at distance ℓ from just one of the two pursuers, e.g., P_1. The situation is illustrated in Fig. 12.

Since V must be a function of just the two vector differences $\vec{r}_E - \vec{r}_{P_1}$ and $\vec{r}_E - \vec{r}_{P_2}$, it follows that $-V_{\vec{r}_E} = V_{\vec{r}_{P_1}} + V_{\vec{r}_{P_2}}$,

so that the main equation is:

$$w\left|V_{\vec{r}_{P_1}} + V_{\vec{r}_{P_2}}\right| - \left|V_{\vec{r}_{P_1}}\right| - \left|V_{\vec{r}_{P_2}}\right| = 0 . \tag{14}$$

Applying the sine law to the triangle in Fig. 13, the main equation becomes:

$$w \sin(\phi_1 - \varphi_2) = \sin(\varphi_1 - \Psi) + \sin(\Psi - \varphi_2) , \tag{15}$$

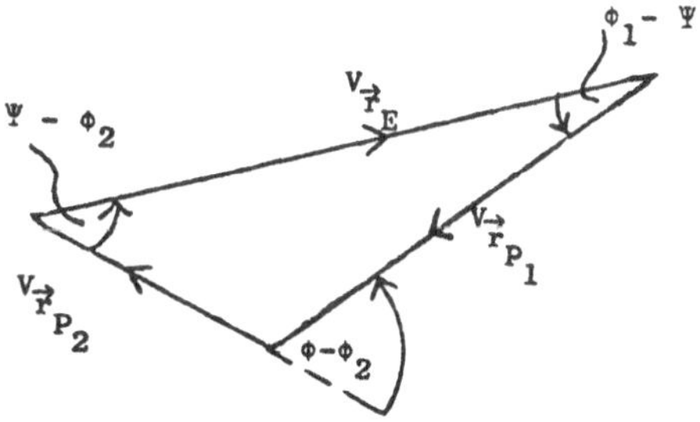

Figure 13.

$$\text{i.e., } w \cos \frac{\varphi_1 - \varphi_2}{2} = \cos\left(\Psi - \frac{\varphi_1 + \varphi_2}{2}\right) \tag{15a}$$

If the control directions Ψ, ϕ_i are measured as in Fig. 12 from E's direction at E_f, the final values are:

$\phi_{2f} = -\mathrm{Cos}^{-1}\frac{1}{w}$, $\phi_{1f} = \mathrm{Cos}^{-1}\frac{1}{w} = \Theta_f$; $\Psi_f = 0$. During the constrained motion by E and P_1, moreover, $\cos(\Theta - \varphi_1) = w\cos(\Theta - \Psi)$. The integration of the constrained paths, with ϕ_1 as independent variable, leads to elliptic integrals (see [3]). After inclusion of the straight-line tangents to the curved paths of E and P_1, the ℓ-surface in the essentially three-dimensional state-space can be computed. Fig. 14 shows E's position contours corresponding to various values of ℓ/P_1P_2, E's optimal direction being, as usual, perpendicular to those contours. The contour corresponding to $\ell = 0$ is again an ellipse with eccentricity $\frac{1}{w}$, the foci being at P_1 and P_2.

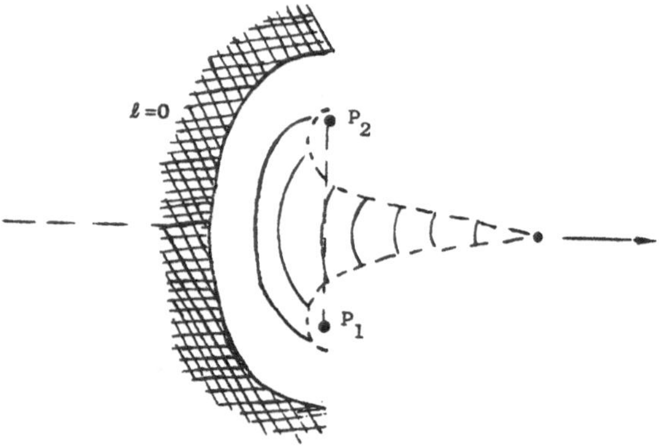

Figure 14.

ESCAPE FROM SEVERAL PURSUERS IN SEQUENCE

We take up, finally, that class of problems in which E passes close to one pursuer after another before passing either (i) between the last pursuer and a straight boundary, or (ii) between the last two pursuers.

In case (ii) V is expressible as a function of the vector differences $\vec{r}_E - \vec{r}_{P_1}, \ldots, \vec{r}_E - \vec{r}_{P_N}$, and in case (i) of these vector differences as well as $y_E = \vec{r}_E \cdot \hat{j}$, $\hat{j}$ being a unit-vector in the y-direction. Hence $-V_{\vec{r}_E} = \sum_j V_{\vec{r}_{P_j}}$ in case (ii), and $-V_{\vec{r}_E} = \sum_j V_{\vec{r}_{P_j}} + V_{y_E}\hat{j}$ in case (i).

In passing a particular pursuer P_k the corresponding $\left|V_{\vec{r}_{P_k}}\right| \to 0$ and has no further influence on the game. Just prior to the instant t_k at which $\left|V_{\vec{r}_{P_k}}\right| = 0$, the distance EP_k remains equal to ℓ, while each $V_{\vec{r}_{P_j}}$, $j > k$, remains constant, as does $V_{y_E}\hat{j}$ in case (i). The main equation: $w\left|V_{\vec{r}_E}\right| - \sum_j \left|V_{r_{P_j}}\right| = 0$, can thus be rewritten:

$$w\left|V_{\vec{r}_E}\right| - V_{\vec{r}_{P_k}} - w\left|V_{\vec{r}_E}(t_k)\right| = 0 , \tag{16}$$

where

$$V_{\vec{r}_E} = - V_{\vec{r}_{P_k}} + V_{\vec{r}_E}(t_k) \tag{17}$$

If directions are measured from E's direction at t_k, the main equation (16) can be rewritten by applying the sine law to Fig. 15.

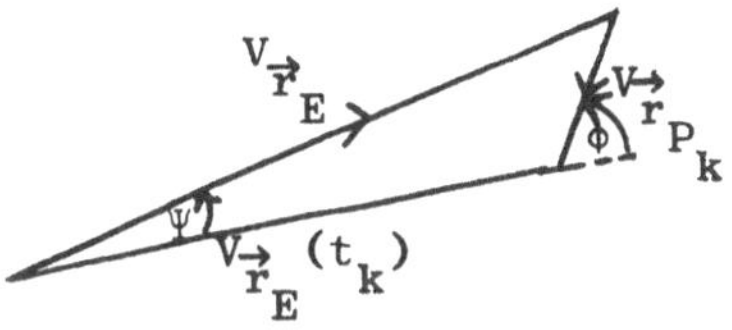

Figure 15.

$$w\left\{\sin\phi - \sin(\phi - \Psi)\right\} = \sin\Psi \tag{18}$$

i.e.,

$$w\cos\left(\phi - \frac{\Psi}{2}\right) = \cos\frac{\Psi}{2} \tag{18a}$$

Again the curved paths can be integrated in terms of elliptic integrals in this case with Ψ as independent variable, with $\Psi(t_k) = 0$. A typical result of such integration is sketched in Fig. 16 for the case of passage past P_1 followed by passage between P_2 and the x-axis. Note that the duration of the two

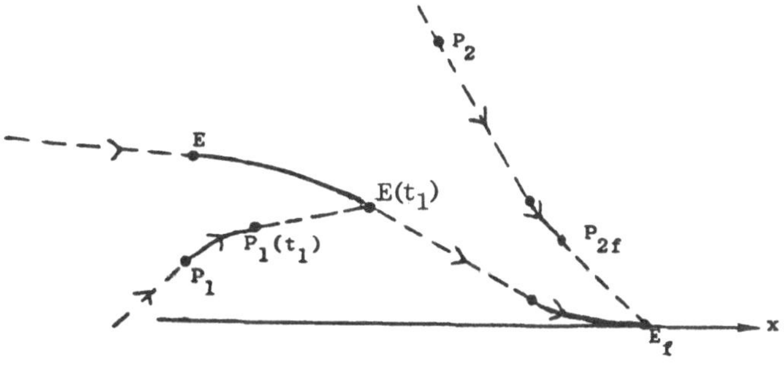

Figure 16.

curved and two straight segments of E's path prior to E_f, together with $\mathcal{L}$ and the non-essential parameter x_{E_f}, constitute the six parameters necessary to match the general initial positions of E, P_1 and P_2.

From some initial positions, E may find it advantageous to pass below rather than above P_1. As in the Cornered Rat game, we may expect a delayed-option phase from some initial positions, terminating on a critical three-dimensional manifold in the (essentially) five-dimensional state-space.

REFERENCES

[1] R. Isaacs, Differential Games, Wiley, 1964.

[2] P. Bernhard, "Conditions de coin pour les jeux differentiels," 1971, Seminaire sur les Jeux Differentiels, Centre D'Automatique, Paris.

[3] P. Hagedorn and J.V. Breakwell, "A differential game with two pursuers and one evader," to be submitted to J.O.T.A.

AERIAL TURNING CHASES

Henry J. Kelley and Leon Lefton

Analytical Mechanics Associates, Inc.
Jericho, New York, U.S.A.

ABSTRACT

A brief review is first given of earlier published results on aerial turning chases, including: vehicle modelling in so-called "energy" approximation, formulation as a differential game with horizontal-plane kinematics simplified to turn-angle difference, capture criteria, a sufficient condition for capture, and a necessary condition for capture in an extended-duration chase. The computation of families of solutions is then discussed with emphasis on trajectory pairs defining the boundary in the space of initial states separating successful pursuit from successful evasion. Some computational results for an example are presented.

INTRODUCTION

Vehicle modelling approximations for 3-D flight have been studied in Ref. 1 via an order-reduction (singular perturbation) approach leading to so-called "energy" models and to even simpler models of lower order. Refs. 2-4 report studies of turning competition in air-to-air combat using energy-modelled vehicles, simplified horizontal-plane kinematics, and a differential-gaming formulation of the pursuit/evasion problem. It is intended to review briefly the main results of these studies, then to discuss computation of solutions using the approach of piecing together Euler-solution segments, concluding with some example computations.

MODELLING OF TURNING CHASES

Each aircraft in the turning competition is described by its turn angle χ (heading) and specific energy (the sum of potential and kinetic energies

J. D. Grote (ed.), The Theory and Application of Differential Games. 257-266.

per unit mass, $E = h + V^2/2g$). Since only the turn angle difference is important, this and the energies of pursuer and evader describe the state of the system. Time to capture is the performance index, to be minimized by the pursuer and maximized by the evader. Capture consists of closing angularly with enough energy to stay with the evader in a pull-up. In 'energy approximation', this is a requirement for sufficient energy to match 'loft-ceilings' with the evader. (Loft-ceiling h_L at a given energy is the upper altitude bound for vertical equilibrium, as limited by $C_{L_{max}}$.) This requirement is developed analytically in Ref. 3 and illustrated with loft-ceiling versus energy computations for two example aircraft. Also illustrated are "target" and "capture" sets for one aircraft pursuing the other, and with the rôles reversed. The target set is that portion of the state subspace $\Delta\chi = 0$ (energy space) for which the pursuer's energy is sufficient for his loft-ceiling to match or exceed that of the evader, while the capture set is a subset of this, characterized by pursuer superiority in maximum instantaneous turn-rate.

In the case of angular closure with the loft-ceiling-match requirement not met, the question of instantaneous altitude-match must receive special attention since the altitudes of both energy-modelled vehicles are control variables, hence can jump instantaneously. This is dealt with by considering, like Marchal (Ref. 5), two extremes of idealization and an intermediate one. One extreme of altitude-dynamics treatment is the assumption that the pursuer is faster than the evader in the sense that the sum of his perception and actuation delay times is small compared to the evader's actuation delay; this results in a pursuer with free choice of altitude between declared limits defining some band within his attainable range of altitudes from his loft-ceiling downward to whatever lower bound may exist, but an evader obliged to remain outside the pursuer's range of altitude choice. The other, "fast evader", extreme leads to free choice of altitude within their respective bounds for both participants. The intermediate idealization is that the pursuer may drive the evader upward to the pursuer's declared limit, but must remain at that altitude himself if he chooses to do this; otherwise, both have free choice of altitude within their respective bounds. This intermediate or neutral case (Marchal's term for a different but somewhat similar case) is really one of a pursuer comparatively fast "in the small", i.e., when near the evader. It appears to be the idealization of main interest.

A SUFFICIENT CONDITION FOR CAPTURE

A comparison of interest may be made of the "hodograph figures" or "domains of maneuverability" with the evader's superimposed on the pursuer's and the energies of the two chosen so that the loft-ceilings are equal. (These figures, nonconvex in the setting of the original problem, are assumed to have been made convex by "relaxation".) An argument is given in Ref. 3 that, if the pursuer's figure completely envelops the evader's, without contact, for all loft-ceilings attainable by the evader, eventual capture is guaranteed. This sufficient condition is obviously quite conservative, i.e., considerably less restrictive ones would seem possible.

TANDEM MOTION

Motion with the pursuer's heading angle χ equal to the evader's but with the loft-ceiling inequality requirement for capture not met, may be termed tandem motion. The only such type of motion emerging from the Euler system for the differential game corresponds to symmetric flight, $\dot{\chi} = 0$ (Ref. 4). Attention will accordingly be confined to this case in the following.

With angular closure effected, the pursuer has two options in the case of neutral altitude dynamics, and chooses between them so as to maximize the Hamiltonian. One is to follow angularly to maintain closure without threatening altitude-match; in this case, both participants adjust their altitude controls for maximum energy rate, each without regard to the other's choice. A second pursuer option is to press the attack, driving the evader outside a declared range of altitudes. With the neutral modelling, the pursuer must maintain his altitude control against the declared limit to do this; thus both craft may fly along at the pursuer's loft-ceiling altitude.

The first type of symmetric tandem motion, with unmatched altitudes, consists of minimum-time climbs to higher energy done independently by the two combatants. The second usually will consist of motion with altitudes matched at the pursuer's loft-ceiling value; this has all controls determined and can be calculated once and for all by numerical integration of the energy-rate differential equations, the specific energy values at initiation of the tandem symmetric motion being parameters of the family.

Subarcs of tandem motion have a character foreign to the general run of pursuit/evasion differential games, in which the Hamiltonian function is separable in pursuer and evader controls. While the Hamiltonian for the present problem is technically separable, the vanishing of the turn angle difference $\Delta\chi$ and its time derivatives along tandem motion arcs effectively couples the controls and the order in which the min and max are taken assumes importance, this being related to the question of which player is assumed faster (Ref. 5).

Tandem motions with $\Delta\chi = 0$ lie in the state subspace of the two specific-energy variables. Regions of the two types of tandem motion in the subspace are shown in Fig. 1 (from Ref. 4) for the case of aircraft B as pursuer and A as evader. The region labelled tandem/trail finds the pursuer opting for unmatched altitudes and both participants operating at maximum energy rate. Maximum energy rates are compared in Fig. 2 (from Ref. 3). In the tandem/loft region, trajectory computations are performed with the common altitude taken as the pursuer's loft-ceiling.

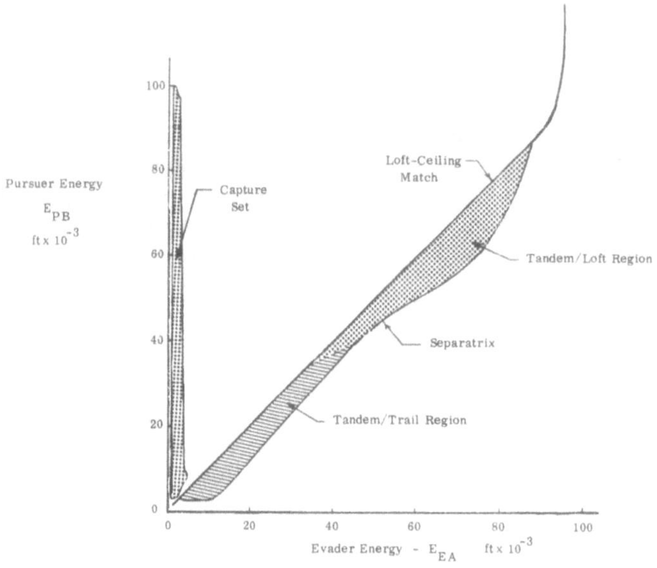

Figure 1 - Tandem-Motion Regions in Energy Space.

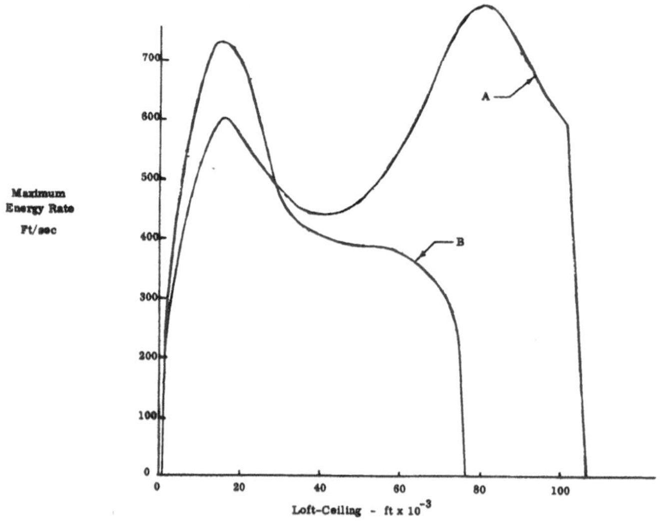

Figure 2 - Maximum Energy Rate versus Loft-Ceiling.

One boundary of the tandem/loft region, the lower right one, is the separatrix dividing those trajectories that eventually cross into the target set from those that do not. It is of interest to compare time-rates-of-change of loft-ceilings for the two aircraft in 2-D flight at their loft-ceiling altitudes with the energies chosen for equal loft-ceilings. Such a comparison is given for the data of aircraft A and B in Fig. 3 (from Ref. 4). 2-D tandem motion at the pursuer's loft-ceiling can be ruled out a priori if the evader's curve lies completely above the pursuer's. Thus with A chasing B, no high 2-D tandem trajectories appear, i. e., A in the röle of pursuer could not contrive to match loft-ceilings, starting from a near match, by driving his opponent aloft.

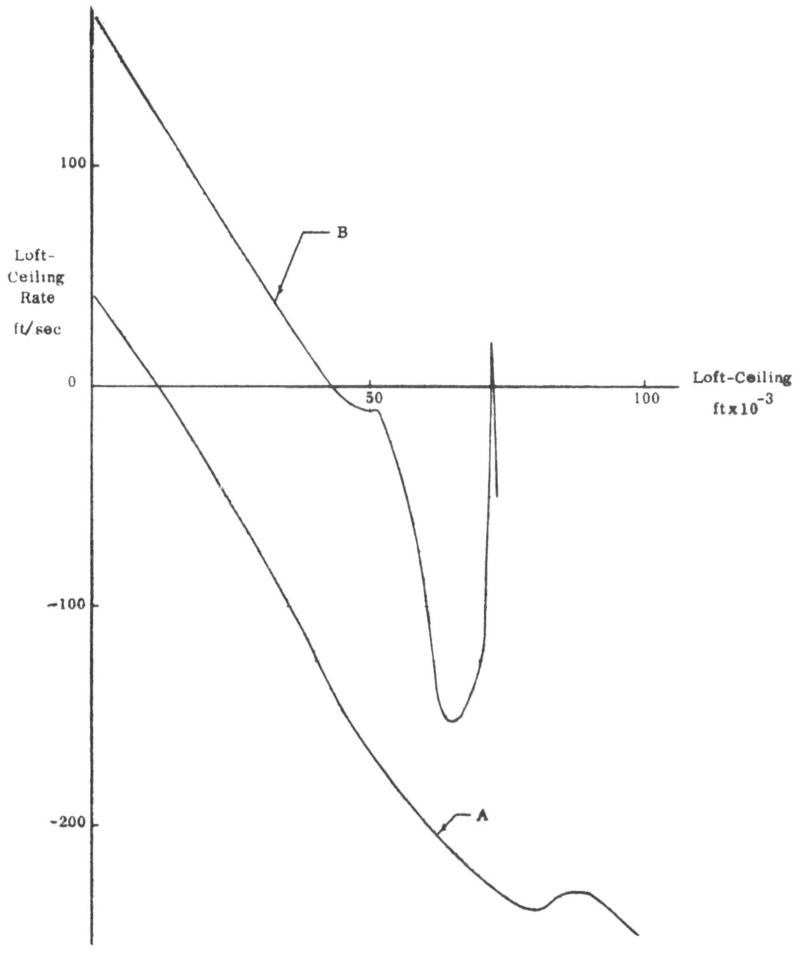

Figure 3
Loft-Ceiling Rate at Loft-Ceiling Altitude versus Loft-Ceiling.

The separation between the tandem/loft region and the tandem/trail region has not yet been studied. If it is a single curve (as drawn) serving as a channel of trajectories, motion along it would have the character of chattering between the two types of tandem motion. As the vehicle energies decrease in the course of tandem motion at the pursuer's loft-ceiling, the curve may be reached and the tandem/trail region penetrated; but then the trajectories would gain energy and cross again into the tandem/loft region. The limit of a sequence of such transitions is a chattering motion along the curve separating the two regions. As the motion reaches the loft-ceiling-match curve, either capture occurs or else a transition to a high turn-rate open-loop trajectory pair, depending upon whether the pursuer's maximum instantaneous turn rate exceeds the evader's or not, respectively. (In the example, the evader happens to have higher instantaneous turn rate at matched loft-ceilings.) It seems likely that the preceding description will turn out to be an oversimplification, as various tandem trajectories passing through the same point of energy space will have slightly different energy-multiplier ratios and, hence, different transitions from one type of tandem motion to the other.

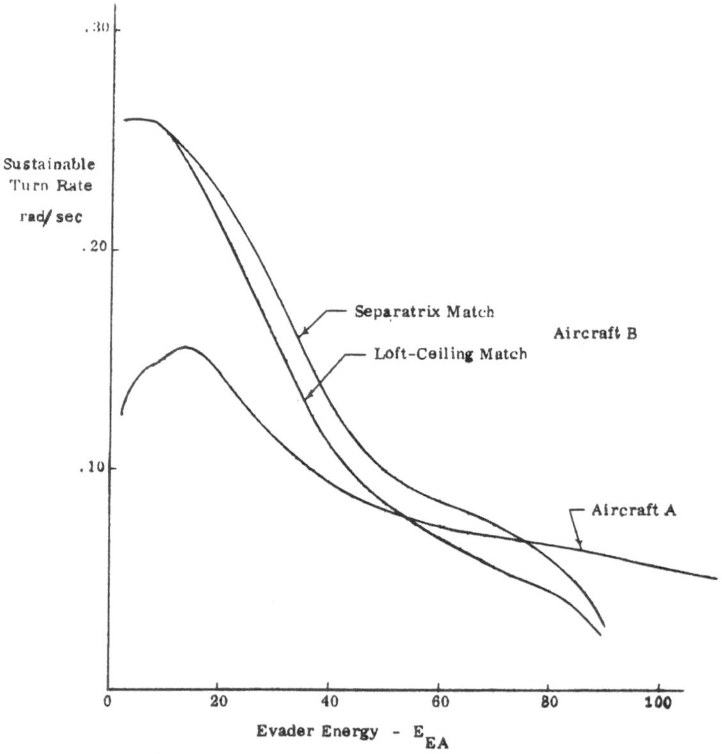

Figure 4 - Sustainable Turn-Rate Comparison
Pursuer/Evader B/A

A NECESSARY CONDITION FOR CAPTURE

A necessary condition for capture in an important subclass of solutions is obtained in Ref. 4 by considering constant-turn-rate evasion, allowing the evader choice of his own energy level, and restricting attention to chases with low initial pursuer energy. This restriction screens out early capture attributable to high energy of pursuer in relation to evader initially, narrowing the range of situations encompassed to those for which capture occurs as a result of resources inherent in the design of the pursuing vehicle. Thus the criterion may be characterized as a necessary condition for capture in an extended-duration turning chase.

Sustainable-turn-rate data for aircraft A and B are given in Refs. 2 and 3. These are combined with the loft-ceiling and separatrix computational results to produce the turn-rate comparison required for the necessary condition for capture in an extended-duration chase in Fig. 4. The necessary condition for extended-duration capture is that the pursuer's maximum sustainable turn-rate equal or exceed the evader's over the attainable range of evader energies, with the pursuer's energy chosen in the comparison to equal or exceed the separatrix value, or the loft-ceiling-match value, if there is no separatrix. If B is designated pursuer (Fig. 4), the necessary condition can be stated as a requirement that the curve for aircraft A lie below a derived curve for B which is the composite of a horizontal segment to the left of the peak and the B-curve to the right. The necessary condition is not met, either for B chasing A or A chasing B in a similar comparison. As suggested by view of the envelope and sustainable turn-rate data (as by overlay in Ref. 2), the outcome of a turning duel will depend strongly upon initial energies and angular separation.

One cannot expect that satisfying the condition with a slight margin would provide a guarantee of eventual capture, i.e., that the condition becomes sufficient when the inequalities are strengthened, for it includes no restriction on maximum instantaneous turn rates or energy rates.

TRAJECTORY COMPUTATIONS

Pursuit/evasion control policies (tactics) are discussed in Ref. 4, where candidates are developed as combinations of open-loop optimal tactics (simple pairing of individually time-optimal energy-turn trajectories) and closed-loop tactics suggested by study of an analogous problem with vehicle models further reduced in order. Attention will be directed in the following to computation of differential-turning solutions for a pursuer superior in the low-energy range (aircraft B) chasing an evader possessing a haven of superiority at high energy (aircraft A), with particular regard to the boundary surface that separates successful pursuits from successful evasions. A qualitative description of the maneuvering will first be given.

The evader spirals upwards and the pursuer closes upon him angularly with slightly more energy than needed to match loft-ceilings, if possible. If he succeeds, both pursuer and evader go into a transition to high-turn-rate, open-loop-optimal trajectories, both losing energy rapidly thereafter and capture occurring at low energy in the shaded capture region of Fig. 1. If, near closure, the evader's maximum instantaneous turn rate at loft-ceiling-matched energies exceeds that of the pursuer, the transition will generally be delayed until an instant before closure (this is the case in the example computations to be presented); otherwise, it will take place somewhat earlier.

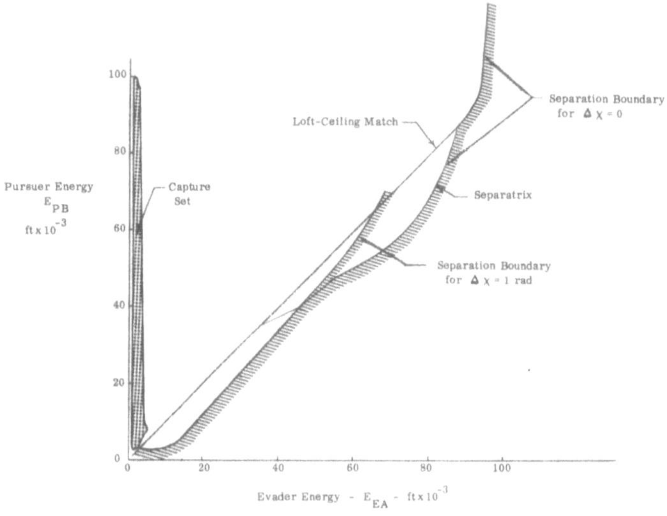

Figure 5 - Boundaries in State Space.

If the pursuer closes angularly and presses the attack with insufficient energy to match loft-ceilings, the evader ascends and the motion becomes of the 2-D tandem/loft type discussed earlier, a matter of relative energy attrition. The limit of energy deficiency for which tandem motion eventually produces loft-ceiling-match is defined by the separatrix curve, mentioned earlier. In the particular example of B on A, a transition (of as-yet-unknown character) to motion of the tandem/trail type may occur before loft-ceiling-match is approached. As and when it is approached, a further transition to high-turn-rate, open-loop-optimal trajectory pairs takes place.

The synthesis is in the spirit of the indirect method of the calculus of variations: it pieces together various fragments with glue containing much conjecture. The computational approach adopted accepts the control policies and seeks the boundary trajectories according to a scheme somewhat as sketched in Ref. 2.

The terminal value of the Hamiltonian function H is -1 for minimax final time by transversality, and it is constant in time along an Euler solution since there is no explicit time dependence in the state system. Trajectories defining the boundary of the region of successful pursuit are characterized by $H = 0$ and are exceptional in that no amount of scaling of the multipliers can produce $H \neq 0$. Such solutions are analogous to those abnormal solutions in optimal control whose endpoints lie on the boundary of the reachable set.

In the subspace $\Delta\chi = 0$, the tandem trajectories defining the separatrix (Fig. 1) are the symmetric members of the $H=0$ family. Turning trajectories of the family may be obtained by backward numerical integration of the Euler system starting from various points on the separatrix between the loft-trail boundary point and the highest point on the separatrix at which the pursuer's sustainable turn-rate equals the evader's. From the data employed to produce Figs. 3 and 4, this point corresponds to pursuer and evader specific energies of 62.5K and 76K, respectively. Although the turn-angle-difference multiplier λ_χ jumps to zero as the singular surface $\Delta\chi = 0$ is entered, the energy multipliers are continuous as is the function H (Refs. 4 and 6). The energy rates on the separatrix together with $\lambda_\chi = 0$ and $H=0$ determine the ratio of the energy multipliers, and this provides the end conditions needed for backward integrations. A one-parameter family of trajectory pairs can be so generated from points along the separatrix.

The procedure just sketched is special to the case in which the evading craft is generally superior in maximum instantaneous turn-rate and for which, as a consequence, the singular surface is defined by $\Delta\chi = 0$; some analogous but more complex procedure will apply to other cases. Constant-turn-rate evasion can be shown to frustrate pursuit starting from points higher on the separatrix than the equal-sustainable-turn-rate point.

Some results obtained according to the procedure are presented in Fig. 5. The boundary for $\Delta\chi = 0$ is a composite of separatrix and loft-ceiling-match curves. That for $\Delta\chi = 1$ radian was obtained by backward integration of energy-turn trajectory pairs. The survey is incomplete, as of the present writing. The results obtained are consistent with the characteristics of the pursuing and evading aircraft in that greater energy margins are required of the pursuer to close an angular gap in the range of high energies where the evader has superior performance.

CONCLUDING REMARKS

The preceding survey of earlier and recent work on the turning-chase problem has been applications-oriented and no effort made toward generalizations beyond the particular system model. The use of two functions of the state to define capture, leading to effective nonseparability of the Hamiltonian, offers a novel mathematical aspect and requires consideration of "control dynamics" idealizations along the lines of Ref. 5.

REFERENCES

1. Kelley, H. J.; "Aircraft Maneuver Optimization by Reduced-Order Approximation," in Vol. X of Controls and Dynamic Systems: Advances in Theory and Applications, C. T. Leondes, ed., Academic Press, New York, 1973.

2. Kelley, H. J. and Lefton, L.; "Differential Turns," AIAA Atmospheric Flight Mechanics Specialists Conference, Palo Alto, California, September 11-13, 1972; also AIAA Journal, Vol. 11, No. 6, June 1973.

3. Kelley, H. J.; "Differential-Turning Optimality Criteria," presented at the AIAA 12th Aerospace Sciences Meeting, Washington, D. C., January 30-February 1, 1974.

4. Kelley, H. J.; "Differential-Turning Tactics," presented at the AIAA Mechanics and Control of Flight Conference, Anaheim, California, August 5-9, 1974.

5. Marchal, C.; "Generalization of the Optimality Theory of Pontryagin to Deterministic Two-Player Zero-Sum Differential Games," Fifth IFIP Conference on Optimization Techniques, Rome, Italy, May 7-11, 1973.

6. Bernhard, P.; "Corner Conditions for Differential Games," Fifth IFAC Congress, Paris, France, June 12-17, 1972.

AVOIDANCE OF GUIDED PROJECTILES*

Thomas L. Vincent

University of Arizona, Tucson

ABSTRACT. The game of two cars defined by Rufus Isaacs is visualized here in terms of missile avoidance by aircraft and torpedo avoidance by surface ships. Having the system dynamics specified and confined to the horizontal plane, the distinctiveness between the "problems" lies not only in the choice of parameters contained in Isaacs' original model, but in the choice of target configuration and certain assigned player rationale as well. The guided projectile may or may not be using optimal guidance.

In particular, the aircraft missile avoidance problem is examined in terms of a circular target. A capture zone associated with the missile is obtained for both optimal missile guidance and proportional navigation missile guidance.

The torpedo avoidance problem is examined assuming the target to have a ship shape. This leads to last minute maneuvers by the target ship to avoid contact.

INTRODUCTION

We have two vehicles I and II. For point of reference, we will assume that we are on vehicle I and are thus interested in the motion of II with respect to I. The system dynamics is written with respect to such a relative coordinate system.

---

* The author is indebted to Dr Willy Peng and Douglas Sticht for their contributions to this project.

J. D. Grote (ed.), The Theory and Application of Differential Games. 267-279.

The dynamical system for Isaacs' game of two cars consists of two vehicles each moving in the horizontal plane with constant velocity. The vehicles can maneuver by turning, but the turning rate of each vehicle is limited. This situation is illustrated in figure 1 with all quantities shown positive.

Let x_1 and x_2 be non-dimensional relative coordinates in the plane of motion and t be a non-dimensional time. Vehicle I is located at the origin and the x_2 axis is taken to be in the direction of I's velocity vector. Let x_3 be the angular displacement of II's velocity vector with respect to the x_2 axis. In terms of such coordinates, the kinematical motion of II with respect to I is given by

$$dx_1/dt = \alpha \sin x_3 + u_1 x_2 \tag{1}$$

$$dx_2/dt = \alpha \cos x_3 - u_1 x_1 - 1 \tag{2}$$

$$dx_3/dt = u_1 - u_2 \tag{3}$$

where $u_1 \leq 1$, $u_2 \leq \delta$ and u_1 is I's turning rate (control) and, u_2 is II's turning rate (control) and α is the ratio of II's speed divided by I's speed. The equations are non-dimensionalized with respect to I's performance. Thus if s_1 is I's speed and u_{1max} is I's maximum turning rate then the coordinates x_1 and x_2 must be multiplied by s_1/u_{1max} to obtain length dimensions and t must be multiplied by $1/u_{1max}$ to obtain time. The quantity δ is the ratio of II's maximum turning rate to I's maximum turning rate.

Each problem is governed by the same system (1)-(3). They are distinguished from one another in terms of targets, parameter values and player rationale. Figure 2 illustrates the relationship between the problems.

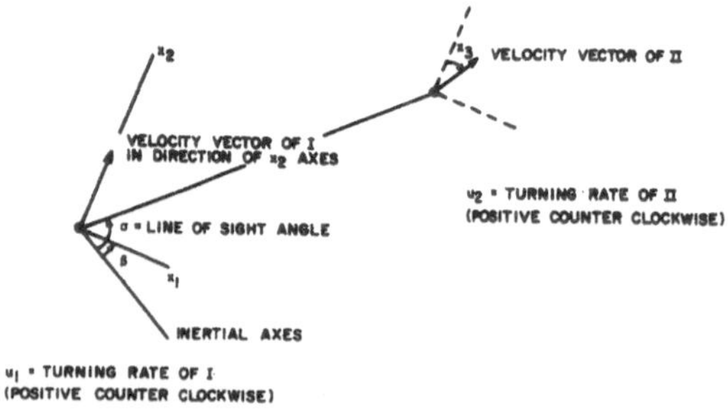

FIGURE I. MOTION OF II WITH RESPECT TO I

Aircraft missile avoidance problem is depicted in Figure 2a. It is assumed that the aircraft will be destroyed if the missile is detonated anywhere in a circular region of radius R = .006 about the aircraft. For the case to be considered the aircraft is faster but has a larger turning radius than the missile (α = .75 δ = 10). Two aspects of this problem will be investigated. We will determine capture zones for the missile (that region of $x_1x_2x_3$space from which the missile may reach the circular target) by first assuming an optimal strategy for both the aircraft and the missile. We will then compare this result by assuming optimal evasion for the aircraft with the familiar proportional navigation strategy for the missile. Ship torpedo avoidance is depicted in figure 2b. In this case, it is assumed that the torpedo must hit the ship in order to inflict damage. Hence a ship shaped target is used to determine if the target, by virtue of its shape, can take maneuvers to avoid a hit, as proposed by Vincent and Peng (1973). Again the ship is assumed faster with a smaller turning radius than the torpedo. We will examine only the case where both the ship and torpedo are using optimal guidance.

PRELIMINARY ANALYSIS

For each of the problems mentioned above, we will seek the corresponding capture set by determining its boundary. Only necessary conditions will be used. Hence trajectories thus obtained, are said to lie on a barrier. If game surfaces or controllability surfaces exist, they must be composed of these barriers. Either one or both vehicles will be maneuvering optimally. With both vehicles maneuvering optimally, the problem becomes a game of kind [Isaacs (1965)]. With one vehicle operating under a given guidance law and the other using optimal guidance, we then have a problem in controllability [Grantham and Vincent (1974]. See also Vincent, et al. (1972) for a more detailed discussion on this point.

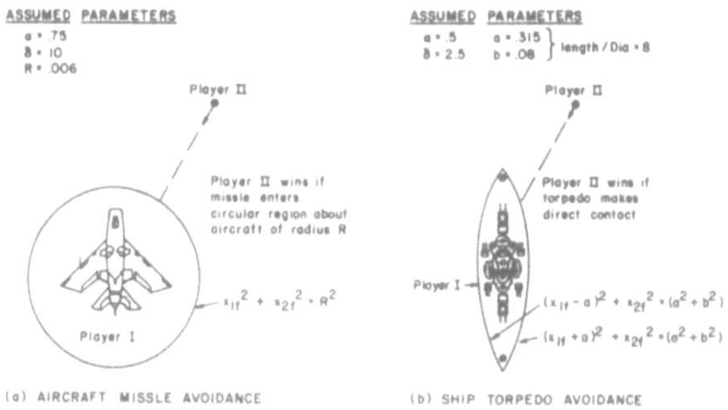

FIGURE 2. PROJECTILE AVOIDANCE PROBLEMS

Theorem 6.1, p. 131 of Blaquiere, et al. (1967) will be used for the determination of trajectories on a game barrier and theorem 1 of Grantham and Vincent (1974) will be used for the determination of trajectories on a controllability barrier. In either case, we have the same Hamiltonian function. Optimal strategy for player I maximizes the Hamiltonian function with respect to u_1 and optimal strategy for player II minimizes the Hamiltonian function with respect to u_2.

The Hamiltonian function is linear in the controls u_1 and u_2 and for convenience is written as

$$H = \lambda_1 \alpha \sin x_3 + \lambda_2 (\alpha \cos x_3 - 1) + \sigma_1 u_1 + \sigma_u u_2 \tag{4}$$

where $\sigma_1 = \lambda_1 x_2 - \lambda_2 x_1 + \lambda_3$ and $\sigma_2 = - \lambda_3$. The optimal control strategies thus consist of maximum and minimum control arcs plus singular control arcs. Singular control arcs for both I and II correspond to null control. In what follows, let the subscript "f" designate quantities evaluated at the final time, and a "dot" denote differentiation with respect to time t.

For both problems the terminal manifold is x_3 cylindrical so that from the transversality condition we have $\lambda_{3f} = 0$. Using this information we may now tentatively conclude the following

$$u_{1f} = \begin{cases} +1 \text{ if } \sigma_{1f} > 0 \\ 0 \text{ if } \sigma_{if} \equiv 0 \\ -1 \text{ if } \sigma_{if} < 0 \end{cases} \qquad u_{2f} = \begin{cases} +\delta \text{ if } \dot{\sigma}_{2f} \geqslant 0 \\ 0 \text{ if } \dot{\sigma}_{2f} \equiv 0 \\ -\delta \text{ if } \dot{\sigma}_{2f} \leqslant 0 \end{cases} \tag{5}$$

Terminal singular control for u_1 is possible provided

$$\cos x_{3f} = 1/\alpha \text{ and } x_{1f} = 0 . \tag{6}$$

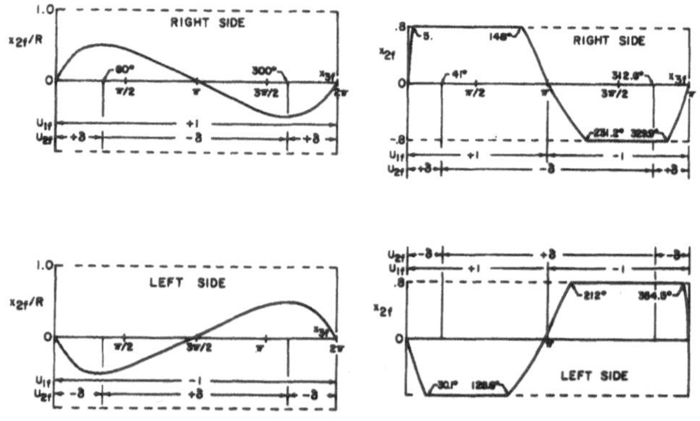

(a) AIRCRAFT MISSLE AVOIDANCE (b) SHIP TORPEDO AVOIDANCE

FIGURE 3. TERMINAL CONDITIONS AS PROJECTED ONTO $x_2 - x_3$ SPACE

In what follows, $\alpha < 1$. We thus conclude that u_{1f} can not be singular.

Singular Control for player II requires that $\sigma_{2f} = \dot{\sigma}_{2f} = 0$. This leads us to the condition

$$\cos x_{3f} = \alpha + u_{1f}(x_{2f} \sin x_{3f} - x_{1f} \cos x_{3f}) \ . \tag{7}$$

Points on the terminal manifold satisfying this requirement are candidates for singular terminal control for player II .

MISSILE AVOIDANCE BY AIRCRAFT

Assuming that the performance characteristics for both vehicles is fixed, then the size of the capture zone for player II will depend on the guidance law used by this player. Optimal guidance will yield the largest capture zone. We can then compare optimal guidance with any other guidance law by simply comparing the corresponding capture zones.

A common guidance law is proportional navigation. Under proportional navigation, the missile control is chosen to satisfy

$$u_2 = k\dot{\sigma} \tag{8}$$

where σ is the line of sight angle (Figure 1) and k the navigation constant. This constant usually has a value between 3 and 5 (Ramo and Puckett 1959).

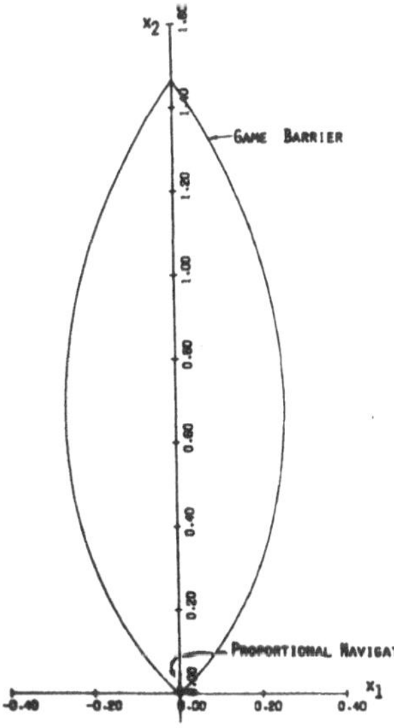

FIGURE 4 BARRIER CROSS SECTION AT x_3 = 0°

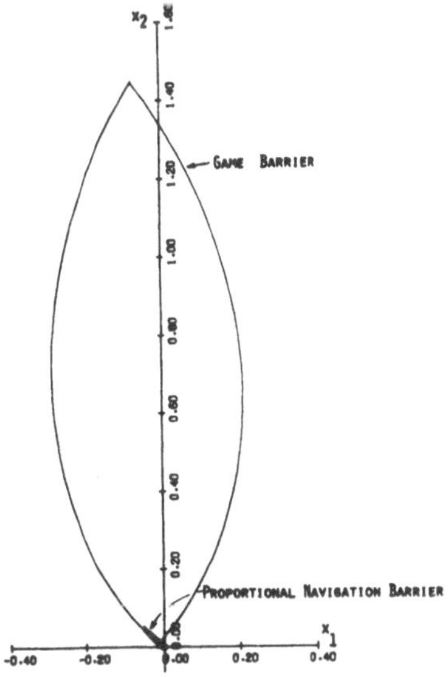

FIGURE 5 BARRIER CROSS SECTION AT x_3 = 30°

It is interesting to note that while proportional navigation is not optimal (as will be shown) it is often derived on the basis of minimizing a quadratic performance index [Bryson and Ho (1969), Cottrell (1971)]. There is certainly no a priori justification for believing that a guidance law derived on such a basis would satisfy the requirements imposed by game theory however.

We will employ the methods of qualitative game theory to determine the largest possible capture set. These results will then be compared with the capture set obtained by requiring that the pursuer use proportional navigation, while the evader uses optimal guidance. The methods of Controllability Theory are used to obtain the capture set in the latter case. For the circular target (Figure 2) we have the additional transversality conditions $\lambda_{1f} = 2\mu\, x_{1f}$ and $\lambda_{2f} = 2\mu\, x_{2f}$ where μ is a positive constant. The location of the terminal points must satisfy the requirement $H_f = 0$. This requirement in conjunction with the transversality conditions yield

$$x_{2f}/x_{1f} = \alpha \sin x_{3f}/(1 - \alpha \cos x_{3f}) \; . \tag{9}$$

We see that for a given value of x_{3f} barrier trajectories are tangent to the circle on both the right and left sides. Terminal points for a right and left barrier as obtained from (9) are shown in Figure 3a.

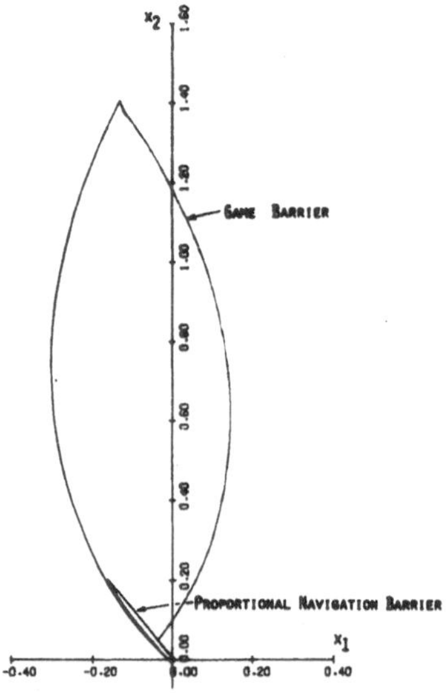

Figure 6 Barrier Cross Section at $x_3 = 60°$

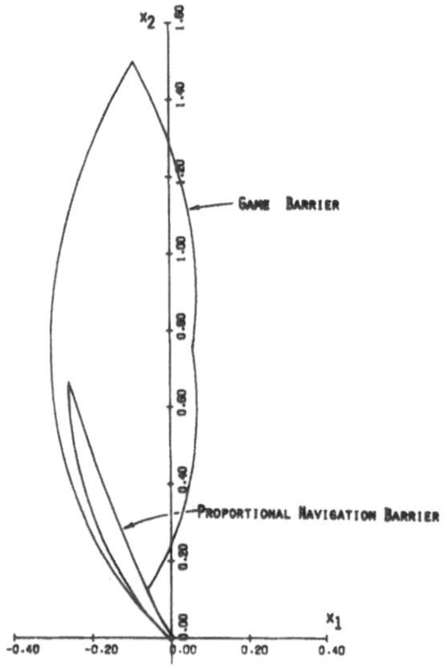

Figure 7 Barrier Cross Section at $x_3 = 90°$

Game Barriers

For the circular target, the transversality conditions require $\sigma_{1f} = 0$ so that the terminal control for u_{1f} at these terminal points are obtained by examining $\dot{\sigma}_{1f}$. Since terminal singular control for u_{1f} is not possible and since $\dot{\sigma}_{1f} = \lambda_{1f} = 2\mu x_{1f}$

$$u_{1f} = \begin{cases} +1 & \text{if } x_{1f} > 0 \quad \text{(right side)} \\ -1 & \text{if } x_{1f} < 0 \quad \text{(left side)} \end{cases} \tag{10}$$

We have already shown that $x_{1f} \neq 0$.

The transversality conditions in conjunction with (9) yields

$$\dot{\sigma}_{2f} = \frac{2\mu\alpha x_{1f}}{1-\alpha \cos x_{3f}} (\cos x_{3f} - \alpha) \ . \tag{11}$$

Since $2\mu\alpha/(1 - \alpha \cos x_{3f}) > 0$ we conclude

$$u_{2f} = \begin{cases} +\delta & \text{if } x_1(\alpha - \cos x_3) \geqslant 0 \\ 0 & \text{if } \cos x_3 = \alpha \\ -\delta & \text{if } x_1(\alpha - \cos x_3) \leqslant 0 \end{cases} \tag{12}$$

These results are also summarized in figure (3a). We note that terminal control is uniquely specified except when $\cos x_{3f} = \alpha$. At such a point II's control $u_{2f} = \pm\,\delta$ and $u_{2f} = 0$ all satisfy the necessary conditions.

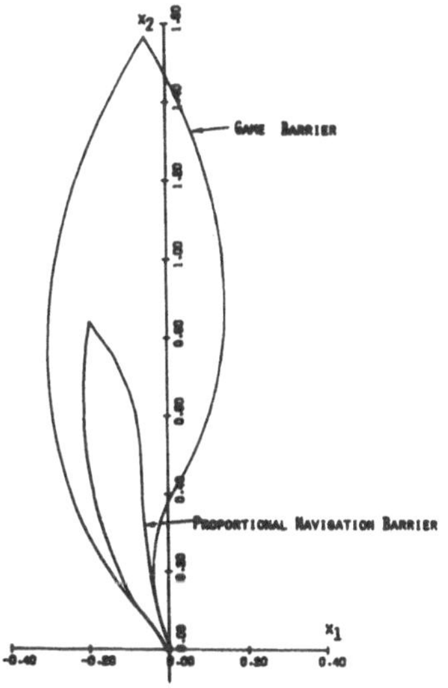

FIGURE 8 BARRIER CROSS SECTION AT x_3 = 120°

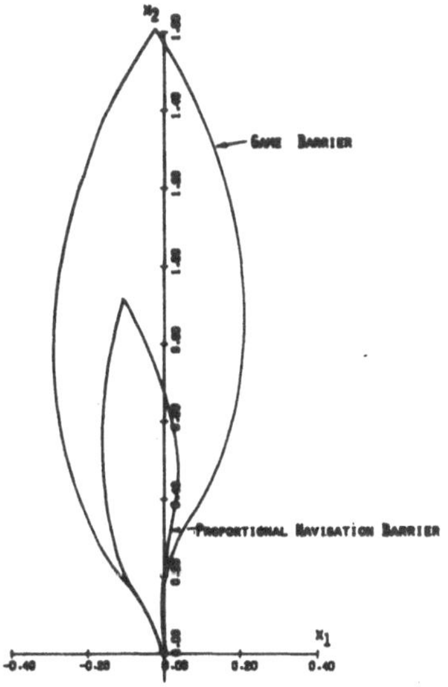

FIGURE 9 BARRIER CROSS SECTION AT x_3 = 150°

Proportional navigation barriers

Consider now the case where player I continues to use optimal control but player II uses proportional navigation. The guidance law for II is given by

$$u_2(t) = \begin{cases} +\delta & \text{if} \quad k\dot{\sigma} \geq \delta \\ k\dot{\sigma} & \text{if} \quad -\delta \leq k\dot{\sigma} \leq \delta \\ -\delta & \text{if} \quad -\delta \leq k\dot{\sigma} \end{cases} \tag{13}$$

where σ is the line of sight angle previously defined and k is the navigation constant. The rate of change of σ may be expressed in terms of the coordinates $x_1 x_2$ and x_3. From Figure 1, $\sigma = \theta + \beta$. Noting that $\dot{\beta} = u_1$, $\tan\theta = x_2/x_1$, and using (1) and (2) to substitute for $\dot{x}_1$ and $\dot{x}_2$ yields

$$\dot{\sigma} = \frac{\alpha(x_1 \cos x_3 - x_2 \sin x_3) - x_1}{x_1^2 + x_2^2} \tag{14}$$

The equations of motion and the H function are the same as before except now $u_2 = k\dot{\sigma}(x_1 x_2 x_3)$. Since $\lambda_{3f} = 0$ on the terminal manifold we arrive at exactly the same terminal conditions for u_{1f} as before. That is on a circular terminal manifold u_{1f} is given by (10).

Let us now examine the terminal control for u_{2f} as given by proportional navigation. We have from (14)

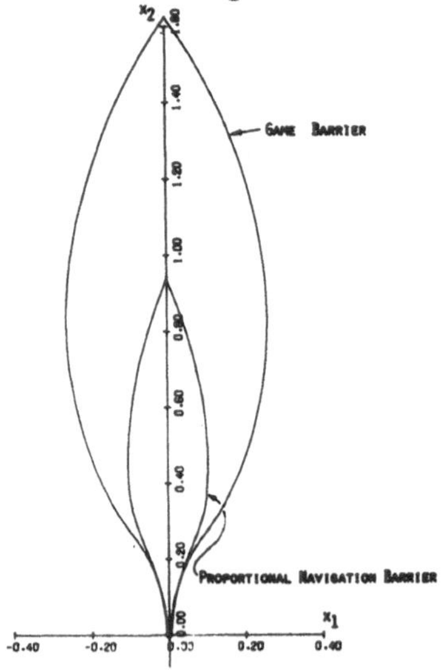

FIGURE 10 BARRIER CROSS SECTION AT $x_3 = 180°$

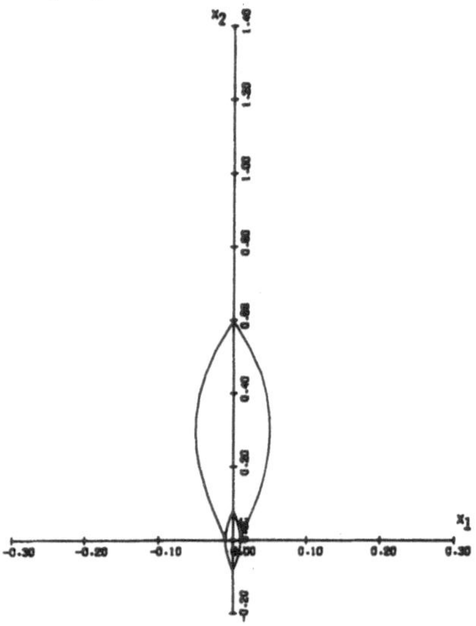

FIGURE 11 BARRIER CROSS SECTION AT $x_3 = 0°$

$$u_{2f} = \frac{k}{R} x_{1f} \left[\alpha \cos x_{3f} - (x_{2f}/x_{1f}) \alpha \sin x_{3f} - 1 \right] \tag{15}$$

Equation (9) is again valid on the terminal manifold. Substituting this expression into (15) yields

$$u_{2f} = \frac{k}{R} x_{1f} \left[2\alpha \cos x_{3f} - \alpha^2 - 1/(1 - \alpha \cos x_{3f}) \right] \tag{16}$$

With $\alpha < 1$, both the numerator and denominator of the term in brackets are positive. Thus under proportional navigation

$$u_{2f} \text{ is } \begin{cases} \text{positive when } x_{1f} > 0 \\ \text{negative when } x_{1f} < 0 \end{cases} \tag{17}$$

It is clear upon comparing (17) with (12) that for some values of x_{3f}, the proportional navigation law has the missile turning the wrong way.

Barrier Cross Sections

Trajectories which lie on a game barrier are obtained by integrating equations (1)-(3) with the controls u_1 and u_2 determined from a mini-max principle [Theorem 6.1, Blaquiere, et al. (1967)]. Trajectories which lie on a proportional navigation barrier are obtained by integrating equations (1)-(3) with the control u_2 determined from (13) and the control u_1 determined from a minimum principle [Theorem 1, Grantham and Vincent (1974)].

Barrier cross sections may then be obtained from these results. There are two routes one can take. The approach used by Vincent, et al. (1972) is to obtain the complete structure of the barrier trajectories requiring construction of all singular surfaces which dominate the solution. The barriers are then obtained by simply taking cross sections. The approach used by Olsder and Breakwell (1973) is to obtain portions of the barrier directly by simply allowing for all candidate trajectories. The barrier is then

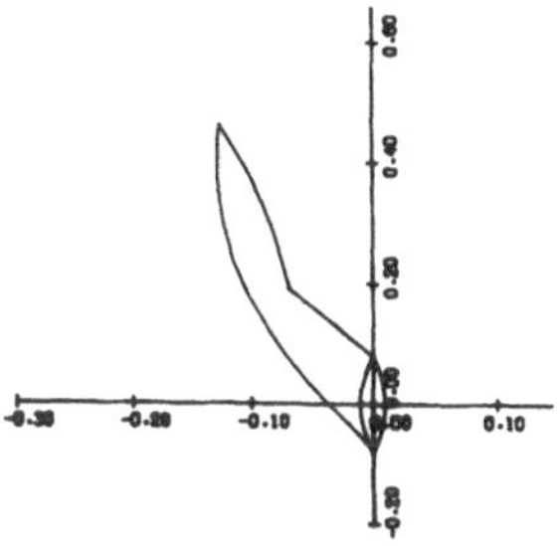

Figure 12 Barrier Cross Section at $x_3 = 30°$

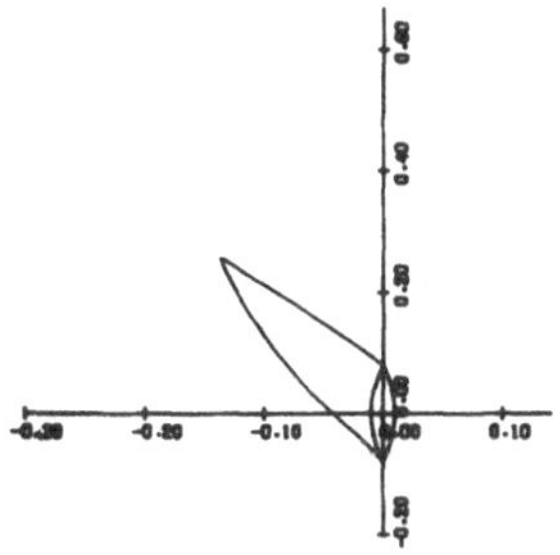

Figure 13 Barrier Cross Section at $x_3 = 60°$

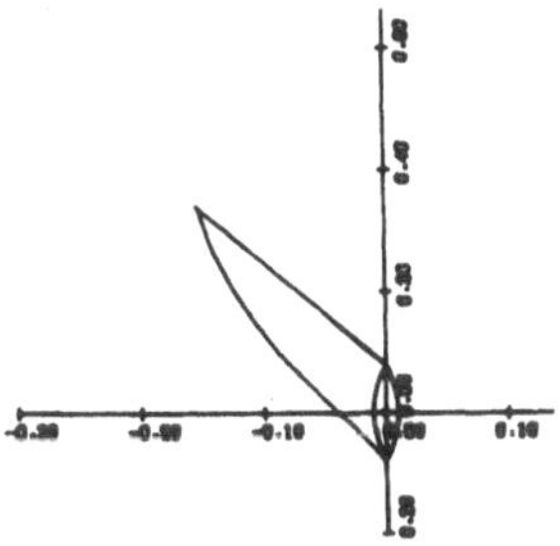

Figure 14 Barrier Cross Section at $x_3 = 90°$

constructed by selecting appropriate segments. Once this is done, the structure of the trajectory solution can then be determined.

A judicious combination of both approaches was used here to obtain the results shown in figures (4)-(10) [α = .75, δ = 10, k = 4]. While only barrier cross section data is presented, it does reflect the structure of the trajectory solutions as well. The cross sections are presented for values of x_3 ranging from $x_3 = 0$ to $x_3 - 180^\circ$ in 30° increments. Any cross section at $x_3 = 180^\circ + \gamma$ can be obtained from the cross section at $x_3 = 180^\circ - \gamma$ by replacing x_1 by $- x_1$.

It is seen that for every cross section the capture set under proportional navigation for II is less than the game set under optimal control for II. The fact that the pursuer turns the wrong way on the terminal manifold for $0 \leq x_{3f} \leq 41.5^\circ$ and $318.5^\circ \leq x_{3f} \leq 360^\circ$ is reflected in the small capture sets for proportional navigation in this cross section range (Figures 4 and 5). Figures (6 and 7) show that while the correct terminal control is used, proportional navigation does not give the correct right barrier control near termination. Proportional navigation yields the correct control direction in Figures (8)-(10), but because of the value of navigation constant used, the control is unsaturated during much of the trajectory. Increasing the navigation constant will increase the proportional navigation cross section in Figures (8)-(10) but will have little effect on the cross sections of Figures (4) and (5).

A number of "corners" are observed in the <u>right</u> hand game barrier of Figures (6)-(7). They result from the fact that there is a dispersal arc associated with the barrier trajectory solutions. Wherever a barrier cross section cuts this dispersal arc there will be an associated corner. Control for the pursuer differs in sign on either side of the dispersal arc.

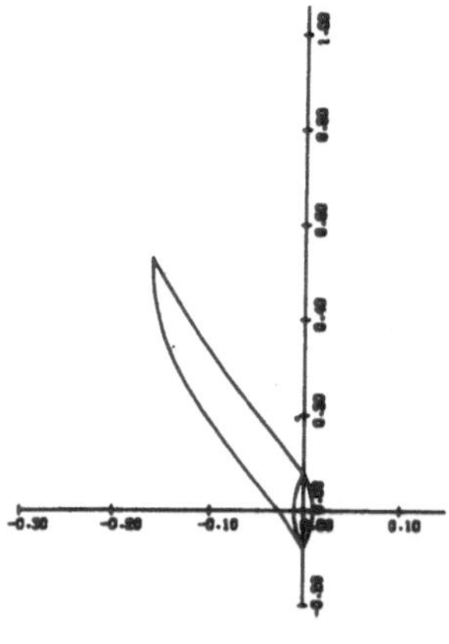

FIGURE 15 BARRIER CROSS SECTION AT x_3 = 120°

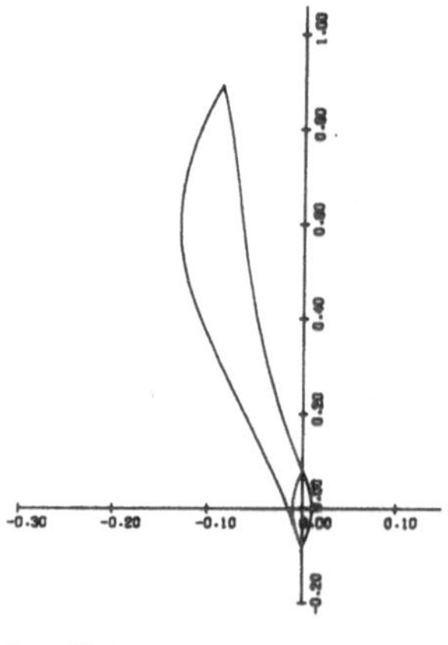

FIGURE 16 BARRIER CROSS SECTION AT x_3 = 150°

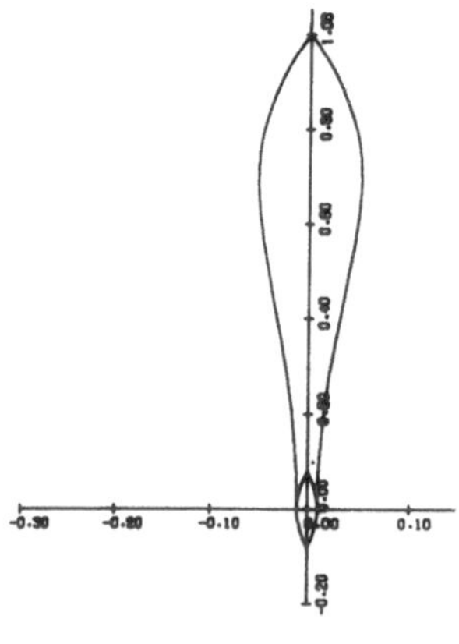

FIGURE 17 BARRIER CROSS SECTION AT x_3 = 180°

TORPEDO AVOIDANCE BY SHIPS

In order to approximate a "ship shaped" target we will take the "right" side of the ship to be given by

$$\theta_1 = (x_1 + a)^2 + x_2^2 - (a^2 + b^2) = 0 \tag{18}$$

and the "left" side to be given by

$$\theta_2 = (x_1 - a)^2 + x_2^2 - (a^2 + b^2) = 0 \tag{19}$$

as illustrated in figure (2). For the smooth portion of the right side(+) or left side(-) we have the additional transversality conditions $\lambda_{1f} = [2\mu\, x_{1f} \pm a]$, $\lambda_{2f} = 2\mu\, x_{2f}$, and $\lambda_{3f} = 0$ where μ is a positive constant. At the cusp points the transversality condition is given by the requirement that the final value of the adjoint vector must lie in the gradient cone to the surface, see [Peng (1974)] . From (4) we may now write the requirement $H_f = 0$ as

$$(x_1 \pm a)\ \alpha \sin x_{3f} + x_{2f}(\alpha \cos x_{3f} - 1) \pm a\, u_{1f}\, x_{2f} = 0\ . \tag{20}$$

Thus for a given value of x_{3f} this condition in conjunction with (18) or (19) will yield terminal points on the smooth portion of the terminal set provided u_{1f} is known.

Terminal Control

In order to determine the proper terminal control we must consider additional conditions at the cusp points. At the upper(+) and lower(-) cusp points, we have the transversality conditions

$$\lambda_{1f} = 2a(\mu_1 - \mu_2) \tag{21}$$

$$\lambda_{2f} = \pm\, 2b(\mu_1 + \mu_2) \tag{22}$$

where μ_1 and μ_2 are positive constants. If we set $\lambda_{2f} = \pm 1$ we can determine λ_{1f} from the condition $H_f = 0$. At the cusp points we have

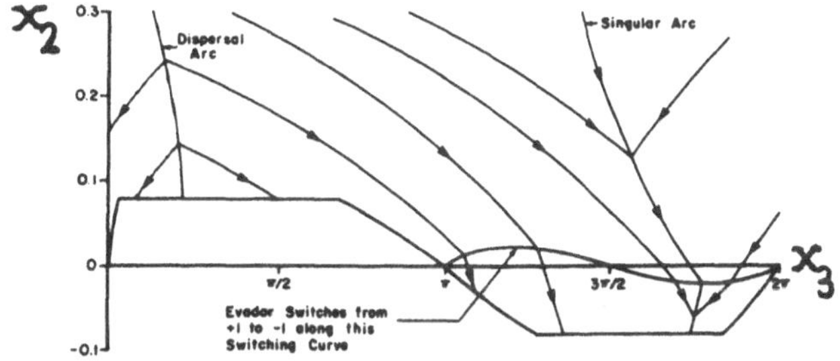

FIGURE 18. A QUALITATIVE SKETCH OF TRAJECTORES ON THE RIGHT BARRIER

$$\lambda_{1f}(\alpha \sin x_{3f} \pm u_{1f} b) + (\alpha \cos x_{3f} - 1) = 0 \tag{23}$$

Thus

$$\lambda_{1f} = (1 - \alpha \cos x_{3f})/(\pm u_{1f}\, b + \alpha \sin x_{3f}) \;. \tag{24}$$

For $\alpha = 1/2$, $b = .08$, the numerator is positive. Let $\varepsilon = \sin^{-1} b/\alpha = 9.2^{\circ}$, then the denominator is positive in the range $\varepsilon \leqslant x_{3f} \leqslant 180^{\circ} - \varepsilon$ at the lower cusp. As will be shown later, barrier trajectories terminat at the upper and lower cusps within these x_{3f} ranges.

At the upper and lower cusps we have $\sigma_{1f} = \pm \lambda_{1f}\, b$. Hence we conclude that at either cusp points

$$u_{1f} = \begin{cases} +1 & \varepsilon \leqslant x_{3f} \leqslant 180^{\circ} - \varepsilon \\ -1 & 180^{\circ} + \varepsilon \leqslant x_{3f} \leqslant 360^{\circ} - \varepsilon \end{cases} \tag{25}$$

The terminal control for u_1 on the smooth part of the target is obtained by evaluating (σ_{1f}) at such points. We obtain that for the right (+) and left(-) sides $\sigma_{1f} = \pm\, 2\mu a\, x_{2f}$ from which we may deduce the control unless $x_{2f} = 0$. At such points we must examine $\mathring{\sigma}_{1f}$ where $\mathring{\sigma}_{1f} = -2\mu(x_{1f} \pm a)$. Since $x_{1f} = 0$ only at the cusp point, we have that terminal control for u_1 on the smooth part of the target is given by

$$u_{1f} = \begin{cases} +1 & \text{if} \quad x_{1f}x_{2f} > 0 \quad \text{or} \quad x_{1f}x_{2f} = 0 \text{ with } x_{1f} < 0 \\ -1 & \text{if} \quad x_{1f}x_{2f} < 0 \quad \text{or} \quad x_{1f}x_{2f} = 0 \text{ with } x_{1f} > 0 \end{cases} \tag{26}$$

It remains now to determine the terminal control u_{2f} . If $\lambda_{1f} \neq 0$ we may combine the expression for $\mathring{\sigma}_{2f}$ and H_f to give

$$\mathring{\sigma}_{2f} = \frac{\alpha\lambda_{1f}}{1 - u_{1f}x_{1f} - \alpha\cos x_{3f}} \left[\cos x_{3f} - \alpha + u_{1f}(x_{1f}\cos x_{3f} - x_{2f}\sin x_{3f})\right] \tag{27}$$

For the parameters chosen, $\mathring{\sigma}_{2f} = 0$ at the upper(+) and lower(-) cusp points, i.e., when

$$\cos x_{3f} = \alpha \pm u_{1f} b \sin x_{3f} \tag{28}$$

Singular control is possible at points where (28) is satisfied. Terminal control for u_2 at all other points must be deduced from (5) in view of (27).

The terminal control for u_1 and u_2 as obtained from these conditions is illustrated in figure 3b.

Barrier cross sections

The barrier cross sections for the torpedo avoidance problem are shown in figures (11)-(17). (Cross sections from 180°-360° may be obtained as previously noted.) As previously noted in figure 3b, the barriers will emanate from the cusp points for the majority of cross sections. The essential features of the cross sections are not unlike the game cross sections of the previous example. However the cross sections do not illustrate the nature of the maneuvers just prior to contact.

In Figure (18) a qualitative sketch of the trajectories on the right barrier in the vicinity of the target is illustrated. We see that for values of x_{3f} such that $\Pi \leq x_{3f} \leq 2\Pi$ trajectories intersect on the lower half of the target set, however just prior to termination the evader control changes sign. The effect is to swing the vulnerable rear portion of the ship away from the torpedo at the last moment.

REFERENCES

1. A. Blaquiere, F. Gerard and G. Leitmann, Quantitative and Qualitative Games, Academic Press, New York, 1969.
2. A.E. Bryson and Y.C. Ho, Applied Optimal Control, Blaisdell, Waltham, Massachusetts, 1969.
3. R.G. Cottrell, Optimal Intercept Guidance for Short-Range Tactical Missiles, AIAA Journal, Vol. 9, No.7, 1414-1415, 1971.
4. W.J. Grantham and T.L. Vincent, A Controllability Minimum Principle, Journal of Optimization Theory and Applications, 1974, (In print).
5. Rufus Isaacs, Differential Games, John Wiley and Sons, Inc., New York, 1965.
6. G.J. Olsder and J.V. Breakwell, Role Determination in an Aerial Dogfight, Navy Workshop in Differential Games, Annapolis, Maryland, 1973.
7. W.Y. Peng, Controllability and Qualitative Game Transversality Conditions for Non-Smooth Targets, Ph.D. Dissertation, University of Arizona, Tucson, 1973.
8. S. Ramo and A. Pucket, Guided Missile Engineering, McGraw Hill, New York, pp 176-180, 1959.
9. T.L. Vincent, E.M. Cliff, W.J. Grantham and W.Y. Peng, A Problem of Collision Avoidance, University of Arizona Engineering Experiment Station Report No. 39, 1972.
10. T.L. Vincent and W.Y. Peng, Ship Collision Avoidance, Navy Workshop in Differential Games, Annapolis, Maryland, 1973.

PURSUIT GAMES: A SURVEY

Otomar Hájek

Department of Mathematics and Statistics, and
Systems Research Centre, Case Western Reserve University
Cleveland, Ohio 44106 U.S.A.*

ABSTRACT. A survey of results on pursuit games in finite-dimensional spaces is presented, with emphasis on computable methods of winning. Two proto-principles are treated: neutralisation of effect of opponent's control, and maintanance of constant observation vector. The appendices illustrate their application.

1. INTRODUCTION

The broad category of differential games comprises cost-optimal games, with the two-person zero-sum as central but by no means sole situation; games of pursuit and of evasion (these are not dual to each other); and a fascinating collection of "others". While it is possible to bring to bear general definitions covering the first two groups, the process is formal and rather unsatisfactory; the problem, the type of question to which a meaningful answer is forthcoming, the methods and apparatus used, and the results seem to be specific to each subgroup. Adopting this admittedly vague classification, the purpose of this paper is to present a survey of pursuit games in finite dimensions. It is suggested that the subject has passed out of the prehistoric stage (analogous to the Bernoullis' period in differential equations) in which the study of individual problems is of prime importance; that we are already reaching out for more general methods, with only technical difficlties remaining in each specific case; and that we may have already sighted something just as

---

* Present mailing address.

J. D. Grote (ed.), The Theory and Application of Differential Games. 281-291.

all-pervasive at Pontrjagin's maximum principle.

A pursuit game, in Euclidean n-space R^n, is described in part by its dynamical equation

$$\dot{x} = f(x,p,q) , \tag{1}$$

the pursuer and quarry constraints $p(t) \in P$, $q(t) \in Q$, and a target set Ω. (The description is completed, implicitly, by studying only winning positions and non-anticipatory strategies as described below.) It will be assumed throughout that f is continuous, P and Q are non-void and compact (subsets of, say, R^n again), and Ω is a non-void closed set in R^n.

A pursuer control is a measurable mapping $p:R^1 \to P$ (analogously for quarry, $q:R^1 \to Q$). Given these and a point $x_0 \in R^n$, a corresponding state response is generalised solution $x(\cdot)$ of the initial value problem of (1),

$$x(t) = x_0 + \int_0^t f(x(s), p(s), q(s))ds \tag{2}$$

for all t in a neighborhood of 0.

A non-anticipatory pursuer strategy is a mapping σ, from the set of quarry controls to that of the pursuer controls (notation such as $\sigma(q)(t)$), which is non-anticipatory (past-determined, hereditary, deterministic) in the following sense: whenever two quarry controls satisfy $q_1 = q_2$ almost everywhere on an interval $(-\infty,\alpha)$, then also $\sigma(q_1) = \sigma(q_2)$ a.e. $(-\infty,\alpha)$. We will say that σ forces an initial position x_0 to target within a time-interval I if each state response (2) with $p = \sigma(q)$ satisfies the termination condition

$$x(t) \in \Omega \text{ for some } t \in I$$

(t may depend on $q:R^1 \to Q$; it may be necessary to extend the domain of $x(\cdot)$). The set of all $x_0 \in R^n$ for which such a strategy σ exists is called the winning set W_I. In case $I = [0,t]$ we will abbreviate to W_{0t}, and to W_t for $I = [t,t]$ (isochronous capture).

A special case of non-anticipatory strategy is the case that, at each time t, $\sigma(q)(t)$ depends on $q(t)$ and t only: $\sigma(q)(t) = \sigma(q)(t) = \tau(q(t),t)$ with $\tau:Q \times R^1 \to P$; these are the stroboscopic strategies. Even more special are the indifferent strategies, with $\sigma(q)(t)$ independent of q.

These terms will be used rather loosely but, it is hoped, suggestively. The standard passage from R^n to R^{n+1} extends our results to the allonomous case $\dot{x} = f(t,x,p,q)$ as long as the player constraint sets can be kept autonomous (e.g. as in $\dot{x} = a(t,x) + B(t,x)p + C(t,x)q$); in particular this covers time-dependent termination conditions. There is little difficulty in treating the important modification that the state space R^n is replaced by a differential n-manifold. We will not treat external constraints of the state variable x (Man and lion, Lady in the lake) nor games with a life-line. Possibly this might be a further development, having manifolds-with-boundary as state spaces. We will not treat infinite dimensional state spaces. Finally, we will naively identify the game with its state-space description, ignoring the natural and fruitful concepts of equivalent games, reduced state space and minimal representation.

The results that are reported on here mostly appear in [3] and [5].

2. LINEAR GAMES: ISOCHRONOUS AND STROBOSCOPIC CAPTURE

Here (1) is specialised to the n-dimensional linear equation

$$\dot{x} = Ax - p + q \; ; \tag{3}$$

the player constraint sets P, Q are, in addition, taken convex and symmetric about 0 (the latter condition is removable with slight trouble). We will treat several types of target sets.

First Duality Theorem. A point x can be forced to target 0 stroboscopically and isochronously at time t if, and only if, x can be steered to 0 at t within the dual control system

$$\dot{y} = Ay - u \quad , \quad u(t) \in U \, , \tag{4}$$

where the constraint set is

$$U = P \underline{*} Q = \cap\{q \in Q : P - q\} \, .$$

Furthermore, a correspondence between the stroboscopic strategies σ and controls u is

$$\sigma(q)(s) = u(s) + q(s) \tag{5}$$

(all $s \in [0,t]$, all quarry controls $q:R^1 \to Q$).

Note that, according to (5), in (3) the term $-p+q = -(u+q) + q =$ thus pursuer first neutralises quarry completely, and then uses left-over energy u to steer to origin within (4). The significance of our theorem is that this simple tactic is the only possible one, within the confines of our situation. A less transparent occurrence of what might be termed the Principle of Neutralisation appears in the second duality theorem below. Applications are suggested in the appendix. The difference operator $\underline{*}$ was introduced by Pontrjagin [8].

Corollary 1. In the situation above, a necessary condition

for presence of non-trivial positions from which termintion can be achieved stroboscopically and isochronously is that $Q \subset P$. If $Q \subset \text{Int } P$, or, more generally, if the system (4) is controllable, then there is a non-void open set of winning positions.

Second Duality Theorem. Let $\Omega = a + L$ with L a linear subspace of R^n. A point x can be forced to Ω stroboscopically and isochronously within time t if, and only if, $x \in e^{-At}(R(t)+\Omega)$, where R(t) is the reachable set at time t of the control system

$$\dot{y} = Ay - v \quad , \quad v(s) \in V(s) \quad , \tag{6}$$

where the constraint set is $V(s) = (P + e^{-As}L) \underline{*} Q$.

There is an analogous but somewhat more complicated correspondence between strategies and the controls in (5). If a $m \times n$ matrix M is used to describe $L = \{x: Mx = 0\}$, it follows that

$$Me^{As}Q \subset Me^{As}P \quad \text{for all} \quad s \in [0,t] \tag{7}$$

is necessary and sufficient for presence of initial points which can be forced to $\Omega = \{x: Mx = Ma\}$ at given time $t > 0$. Since this requires knowledge of the behaviour of the matrix solution e^{As}, (7) is not directly a working condition, but rather a source of these.

Corollary 2. In the above situation, $MQ \subset MP$ is necessary, and $MQ \subset \text{Int } MP$ is sufficient or presence of non-trivial positions from which termination can be forced isochronously and stroboscopically.

Preserving the notation, define pursuer control order as the least integer $k \geq 1$ such that

$$MP = 0 \ , \ MAP = 0 \ , \ \ldots \ , \ MA^{k-2}P = 0$$

(either $k \leq n$ = state space dimension, or $MA^jP = 0$ for all j, and pursuer control is "ineffective"); and analogously for quarry control order.

Corollary 3. A necessary condition for presence of non-trivial (isochronous, stroboscopic) winning positions is that

quarry control order $\geq$ pursuer control order.

Setting k = pursuer control order, a more precise necessary condition is

$$MQ = MAQ = \dots = MA^{k-2}Q = 0 \ , \ MA^{k-1}Q \subset MA^{k-1}P;$$

a sufficient condition is

$$MQ = MAQ = \dots = MA^{k-2}Q = 0 \ , \ MA^{k-1}Q \subset \text{Int } MA^{k-1}P.$$

In both corollaries, the sufficient condition ensures presence of a non-void open set of points in position to win isochronously and stroboscopically at small times $t > 0$.

Still retaining the linear dynamical equation, let us treat targets Ω which are countable unions of analytic manifolds Ω_k : each Ω_k is an intersection of hypersurfaces $\{x:F(s) = 0\}$ with $F:R^n \to R^1$ having $F(x)$ analytic in the coordinates of x.

Reduction Lemma. If a point can be forced to Ω stroboscopically and isochronously at some time, then it is actually forced to an affine subspace entirely contained within one of the Ω_k.

The conclusion would be vacuous if the affinesubtarget were a single point. Then, however, one is forcing to a singleton target in Ω, and this is impossible for points $x \notin \Omega$ if $Q \not\subset P$ (see Corollary 2, with $M = I$). Thus the thrust of the reduction lemma is rather negative: some targets automatically rule out the possiblity of isochronous and stroboscopic capture.

3. LINEAR GAMES: FURTHER TOPICS

The requirement that capture be isochronous turns out to be rather severe; however, the added assumption that strategies be stroboscopic seems quite minor (e.g., the necessary portion of Corollary 3 applies even to non-stroboscopic strategies):

The Necessary Condition for Isochronous Capture. If there are any positions which can be forced to target Ω at positive time, then

$$Q \subset P + R^1\Omega \quad (R^1\Omega = \{tx : t \in R^1 \ , \ x \in \Omega) \ .$$

For affine targets $\Omega = \{x: Mx = a\}$ necessarily

(quarry control order) $\geq$ (pursuer control order)

and, if the latter is denoted by k, actually

$$MQ = MAQ = \dots = MA^{k-2}Q = 0 \ , \ MA^{k-1}Q \subset MA^{k-1}P.$$

The author does not have any example to show that the

restriction of isochronous capture to stroboscopic is actually present. In some cases there is no loss at all.

Thus, if the target set is a half-space, $\Omega = \{x \in R^n: c'x \leq \alpha\}$, and there exists a non-anticipatory strategy which forces x to Ω at time t, then there also exists an indifferent (hence, stroboscopic) strategy with the same effect. In particular, if there is a time-optimal strategy (and for linear games this can be established), then there also exists a time-optimal stroboscopic strategy. In the same situation, for points in the winning set W_{0t} with small enough t there always exists an isochronous strategy. Thus one might say, somewhat vaguely, that optimal strategies are ultimately isochronous and stroboscopic.

Next, if Ω is closed and convex and if Q is a direct summand of P (in the sense that $P = Q + U$ for some $U \subset R^n$), then again isochronous capture can be effected stroboscopically. Note that the direct sum condition is satisfied automatically if P and Q are one-simensional and $Q \subset P$ holds (this inclusion is necessary if $\Omega = 0$). The following is a variation on this last result: if P is a polytope and Q a segment with $P \underline{*} Q = 0$, and if 0 is the target, then all winning sets $W_t = 0$.

After being forced to 0 a point can be held there thereafter if $Q \subset P$; thus capture with holding implies isochronous capture. More generally, we have the

<u>Third Duality Theorem</u>. An initial position x can be forced to an affine target $\Omega = a + L$ stroboscopically within time t and held in Ω thereafter if, and only if, x can be steered to Ω and held in Ω thereafter within the control system

$$\dot{y} = Ay - v , \quad v(t) \in V$$

where $V = (P + L_A) \underline{*} Q$ and L_A is the largest subspace of L invariant under A,

$$L_A = \bigcap_{k=0}^{n-1} \{x : A^k x \in L\}.$$

One further remark. On occasion capture can be effected by isochronous and stroboscopic forcing to one or several intermediate target sets in sequence [4]; the combination is usually not isochronous.

4. NONLINEAR GAMES

Of course these are far more difficult, and the results less ambitious. I wish to touch upon one concept only, which is amenable to direct computation even in the nonlinear case, and which provides non-trivial information.

Retaining the notation from Section 1, consider a non-trivial winning position $x \notin \Omega$ and some non-anticipatory strategy σ which forces x to Ω within a bounded time interval. The corresponding jet J consists of all state responses $x(t)$ to various quarry controls, with t varying between 0 and the first intersection time (also take closures). The portion of target set boundary used by J is then $J \cap \partial\Omega \neq \emptyset$; letting x and σ vary, we obtain the used part of target, closure $(\bigcup J \cap \partial\Omega)$. Now consider also all possible closed sets S which intersect every non-trivial jet J (e.g., $\partial\Omega$, or the used part). Curiously enough, among these there is a unique minimal one, to be called the essential part of target, ess Ω; this can be proved from an alternate characterisation,

$$\text{ess}\ \Omega = \lim_{t\to 0+} (W_{0t} \setminus \Omega) = \bigcap_{t>0} \overline{W_{0t} \setminus \Omega}$$

under moderate assumptions on f in (1). Thus the essential points are the limits of non-trivial winning positions; in particular, the latter exist if, and only if, the relatively small part ess Ω of the target set boundary is non-void (and the concept may be so used with considerable effect, [5]). Below we collect several conditions for essentiality, of a direct computational character.

<u>Necessary Conditions</u>. Assume that $\Omega = \{x : F(x) = 0\}$ with $F: R^n \to R^\nu$ of class C^1. If $x \in$ ess Ω, then there exists a half-ray H_1 through the origin in R^ν with the following property: for each value $q \in Q$ there is a $p \in P$ such that

$$DF(x) \cdot f(x,p,q) \in H_1 \quad . \tag{8}$$

If (1) is replaced by $\ddot{x} = g(x,\dot{x},p,q)$ and the termination condition is $Mx = a$ with $(n\times\nu)$ - matrix M, then a necessary condition for $(x,\dot{x}) \in$ ess Ω is that there exist a half-plane H_2 in R^ν, with both 0 and $M\dot{x}$ on its boundary line, such that

$$M \cdot g(x,\dot{x},p,q) \in H_2 \quad . \tag{9}$$

for each $q \in Q$ and some $p \in P$.

From the other side, we also have sufficient conditions for essentiality, analogous but of lesser generality:

<u>Proposition</u>. Assume that $\Omega = \{x : F(x) \leq 0\}$ with $F: R^n \to R^1$ of class C^1, and $x \in \Omega$. Then a point $x \in \overline{\Omega}$ is essential if

$$\min_{p\in P} \max_{q\in Q} DF(x) \cdot f(x,p,q) < 0 \quad .$$

<u>Proposition</u>. In the linear game (3) let $\Omega = \{x : Mx = a\}$ with an $(n\times\nu)$-matrix M. Then $x \in \Omega$ is essential if

$$MQ \subset \text{Int}\ M(P - Ax + [\varepsilon,+\infty)c)$$

for some $\epsilon > 0$ and $c \in R^n$, interior taken in R^ν; in the positive case nearby points are forced to Ω by indifferent strategies.

The geometric nature of these conditions is apparent. If (1) has the player dynamics separated in the sense that $f(x,p,q) = a(x,p) + b(x,q)$, then (8) becomes

$$Mb(x,Q) \subset (-M)a(x,P) + S_1 \text{ for } (M = DF(x); \text{ in particular, } \dim Mb(x,Q) \leq 1 + \dim Ma(x,P).$$

Similarly for (9), resulting in

$$\dim Mb(x,\dot{x},Q) \leq 2 + \dim Mb(x,\dot{x},P).$$

There is a pleasing apposition with the condition $\dim MQ \leq \dim MP$, necessary for isochronous capture (see Section 3).

There also appears to be a connection with the swerve manoeuvre in the homicidal chauffeur game , [6] and with the tactic, in prusuit with curvature constraints, of maintaining a constant direction of the line-of-sight vector [1], [9]: the necessary condition states that, in the end-game, and in the limit, the tactic is the only one possible. Conceivably this might be the manifestation of something more general, which might tentatively be called a Principle of Constant Oservation.

In [2, p. 81] a min-max condition analogous to that of the proposition is required of all points on the boundary of the target.

Appendix 1: Applications of first duality theorem.

A good justification of a mathematical theory is in its ability to provide results outside of itself. Here we propose to apply pursuit game theory to control problems.

The first is not too surprising: control of a system under the effect of unpredictable observable disturbances (games against nature). A good instance is Zermelo's Navigation Problem [11], [10] of piloting an airplane between two given points against the effect of a changing wind. The dynamical equation in R^2 is $\dot{x} = p-q$, with $x = 0$ for termination (after an affine translation in R^2), and pilot control bounds $|p(t)| \leq \alpha$, wind velocity bound $|q(t)| \leq \beta$. The proof of the First Duality Theorem is an immediate generalisation of the intuitively obvious "solution".

In the second we treat a non-linear control problems, with n-dimensional dynamical equation

$$\dot{x} = f(x) - u$$

a constraint on the control u and a target set. With this we associate the linear pursuit game

$$\dot{x} = Ax - p + q \qquad (p = u,\ g = f(x) - Ax)$$

with judiciously chosen A, and suitable quarry constraint set containing all values of the control q of a notional opponent. The method may be called unorthodox linearisation: if the game can be won, the resulting strategy steers the original control system to actual (rather than approximate) target, non-optimally but perhaps suboptimally.

As an example, consider the torque control of a frictionless mathematical pendulum, $\ddot{x} + \sin x = u$, $|u(t)| \leq 1$; and suppose we wish to steer the initial position $x = \frac{1}{2}$, $\dot{x} = 0$ to rest $x = 0 = \dot{x}$. After about three decades of development of control theory, the solution of the prolem is simple in principle (see [7 , 446-455] even with damping allowed); but computationally it remains messy. Thus a four step integration technique, using interval arithmetic, yields the following estmate for the optimal time t:

$$1.260 < t < 1.442 \quad . \tag{10}$$

We carry out the unorthodox linearisation,

$$\ddot{x} + \alpha x = u + (\alpha x - \sin x)$$

(α to be chosen subsequently); apply the neutralisation principle, and solve the dual linear control system $\ddot{y} + \alpha y = v$. Optimal times for the latter provide upper estimates on t. Summary:

Choice of α	Optimal time
$\alpha = 0$	1.960
$\alpha = 1$	1.498
best linear in $[-\frac{1}{2}, \frac{1}{2}]$	1.324
least squares in -"-	1.323

The last two are better upper estimates, and are obtained with far less effort, that that in (10). Information was lost by treating $\alpha(x(t) - \sin x(t)$ as unpredictable; the results suggest that not too much was actually lost.

It would be interesting to find out whether a similar technique, "unorthodox reduction" to non-lag systems, could be applied to

$$\ddot{x}(t) + x(t-1) = u(t) \ ,$$

possibly via $\ddot{x} + x = u + (x(t) - x(t-1))$.

Appendix 2: Application of the necessary conditions.

We treat pursuit with curvature constraints, [1], [9]; two players move in Euclidean n-space, each with constant speed, and with an upper bound on the curvature; termination occurs at coincidence of positions.

The curvature $\varkappa$ of a piecewise C^2-curve $x:R^1 \to R^n$ is

$$\varkappa = \left|\frac{dT}{ds}\right| = \left|\frac{dt}{ds} \cdot \frac{d}{dt}\frac{\dot{x}}{|\dot{x}|}\right| = \frac{|\ddot{x}|}{|\dot{x}|^2}\,|\sin \dot{x}, \ddot{x}| \;;$$

constant speed is maintained if $\dot{x}$ is perpendicular to $\ddot{x}$. Thus we may reformulate the conditions on a player's motion as

$$|\dot{x}(0)| = s \;, \; \dot{x} \quad \ddot{x} \;, \; |\ddot{x}| \le s^2/\rho \;;$$

or, in terms of controls u varying over the unit ball B^{n-1} of R^{n-1},

$$\ddot{x} = \alpha\, D(\dot{x})u \;,$$

Here $\alpha = s^2/\rho$, and $D(\dot{x})$ is an $n \times (n-1)$ matrix whose columns are orthonormal and perpendicular to $\dot{x}$ (for $n > 2$, $D(\dot{x})$ is only available locally, and $\dot{x} \to D(\dot{x})$ is analytic). Finally, the game has a dynamical equation in R^{4n} induced by the n-dimensional equations

$$\ddot{x} = \alpha\, D(\dot{x}): \;, \quad \ddot{y} = \beta\, D(\dot{y})q \;,$$

coinciding player constraint sets B^{n-1}, and $x = y$ for termination condition. Condition (9) yields

$$D(\dot{y})B^{n-1} \subset \frac{\alpha}{\beta}\, D(\dot{x})B^{n-1} + R^1\alpha \qquad (11)$$

for some $d \in R^n$. To eliminate d, choose an matrix E (type: $(n-1) \times n$) with rows orthornormal and perpendicular to d; premultiply (11) by E, noting that the matrices ED are orthogonal, to obtain $B^{n-1} \subset \alpha/\beta\, B^{n-1}$, i.e. $\beta \le \alpha$, as necessary condition for presence of non-trivial winning positions. Thus, if (quarry s^2/ρ) > (pursuer s^2/ρ), then quarry can always escape unless he was initially in contact with pursuer.

REFERENCES

1. E. Cockayne, Plane pursuit with curvature constraints, SIAM J. Appl. Math. 15(1967) 1511-1516.
2. A. Friedman, Differential Games, Wiley-Interscience, 1971.
3. O. Hájek, Lectures on Linear Pursuit Games, Case Western Reserve University, 1973 (unpublished).
4. O. Hájek, Strategy design in pursuit games; Proceedings of the 1973 conference on differential games and control theory, University of Rhode Island (to appear).
5. O. Hájek, The essential part of target in pursuit (to appear).
6. R. Isaacs, Differential Games, Wiley, 1967.
7. E. B. Lee and L. Markus, Foundations of Optimal Control Theory, Wiley, 1967.
8. L. S. Pontrjagin, On linear differential games 2, Doklady AN

SSSR 4(175) 1967.
9. G. T. Rublein, On pursuit with curvature constraints, SIAM J. Control 10(1972) 37-39.
10. R. von Mises, Ueber Fluggeschwindigkeit, Windstäarke und Eigengeschwindigkeit des Flugzeuges, Zeitschr. f. Flugtechnik u. Motorluftschiffart 19-20 (1917) 145-151.
11. E. Zermelo, Ueber das Navigationsproblem bei ruhender oder veraendlicher Windverteilung, Zeitschr. f. angew. Math. u. Mech. 11(1931) 114-124.

VALUES OF POSITIONAL GAMES

N.J. Kalton

Department of Pure Mathematics, University College of Swansea, Singleton Park, Swansea SA2 8PP.

1. INTRODUCTION

This paper is a summary of results to be published in detail elsewhere ([6]).

Many situations in the theory of two person games can be described in broad terms as follows. Two players play a game possibly involving random moves whose result is a point s of a topological space S. The pay-off, which one player seeks to maximize and the other to minimize, is given by $f(s)$ where f is a continuous real-valued function on S. Thus the game has two distinct phases; the first phase, which includes the actual mechanics of the game, being independent of the second phase which is the evaluation of the pay-off. Of course, the particular function f will influence the behaviour of the players in the first phase. Such a situation we will call a *positional game*.

Our main motivation for the study of positional games is the particular example of a differential game of survival with purely terminal pay-off. In fact any differential game of survival can be reduced to one with purely terminal pay-off. In this case S can be taken as ∂F the boundary of the terminal set F. If we admit the possibility of random moves we can also treat stochastic differential games governed by a stochastic differential equation of the type

$$dx = f(t,x,y,z)dt + \sigma(t,x)dw$$

where w is an n-dimensional Brownian motion and $\sigma(t,x)$ is an $n \times n$-matrix; see Friedman [5] for a discussion of the problems involved in such games.

J. D. Grote (ed.), The Theory and Application of Differential Games. 293-300.

Even a two-person matrix game can be described as a positional game with S a finite set with mn elements, where m and n are the numbers of pure strategies available to each player.

Suppose S is compact and Hausdorff and $f \in C(S)$. Let $V(f)$ be any notion of value associated with the game with pay-off f. It is clear that $V : C(S) \to R$ must have certain properties to be reasonable as a definition of value

(i) $V(f) \geq V(g)$ whenever $f \geq g$,

(ii) $V(\alpha f) = \alpha V(f)$ whenever $\alpha \in R$, with $\alpha \geq 0$,

(iii) $V(f+\alpha) = V(f) + \alpha$ whenever $\alpha \in R$.

Such a functional on $C(S)$ we shall call a *gamonic functional.*

For example, consider a differential game of fixed duration; the terminal set may be taken to be compact (since the set of points attainable at time T is relatively compact). Then both the upper and lower values (see [3]) considered as functions of the pay-off are gamonic. Also the Danskin σ-value for $0 \leq \sigma \leq 1$ (see [1] and [4]) and the value with relaxed controls ([2]) are gamonic.

Consider a positional game from the point of view of the minimizer. If he adopts a given strategy Σ then according to his opponent's choice of strategy the result of the game (taking into account random moves) may be any point in a set $A(\Sigma)$ of regular probability measures on S. Thus the value or upper value to the minimizer takes the form

$$V(f) = \inf_{\Sigma} \sup_{A(\Sigma)} \int f\, d\mu .$$

The main result of this note is that any gamonic functional on $C(S)$ can be represented in this form; precisely

$$V(f) = \min_{C \in \mathcal{C}} \max_{\mu \in C} \int f\, d\mu$$

where $\mathcal{C}$ is a collection of weak*-closed convex subsets of $P(S)$ the regular probability measures on S. Thus any gamonic functional on $C(S)$ can be realized in a certain sense as the upper value of a positional game. Of course it follows there is equally a representation in the form

$$V(f) = \max_{D \in \mathcal{D}} \min_{\mu \in D} \int f\, d\mu .$$

Similar results can be obtained for functionals on $C_0(S)$ where S is locally compact. Here condition (iii) must be replaced by $V(f) - V(g) \leq \|f - g\|$, and a further condition is required to ensure that V does not depend too much on points near infinity (see [6]).

Of course if a positional game is purely deterministic (i.e. involves no random moves) then we may expect a representation in the form

$$V(f) = \min_{E \in \mathscr{E}} \max_{s \in E} f(s)$$

where $\mathscr{E}$ is a collection of closed subsets of S. We obtain an external characterization of such functionals in §4. Let Φ be a 'utility transformation', i.e. a continuous increasing map $\Phi : R \to R$. Then if V is deterministic (i.e. represents a deterministic game)

$$\Phi[V(f)] = V(\Phi \circ f).$$

Thus a change in utility does not affect the optimum strategy of the minimizer. We show in §4 that if V is gamonic and satisfies this equation for some non-linear utility transformation then V has a representation

$$V(f) = \min_{E \in \mathscr{E}} \max_{s \in E} f(s).$$

As noted at the beginning of this section, these results are proved in detail in a forthcoming note [6]. One of the aims of this line of research is to study an abstract theory of differential games, akin to modern potential theory [7]

2. REPRESENTATION THEOREMS

We begin by considering an abstract situation. Suppose X is a real normed space and $V : X \to R$ is a positively homogeneous functional. Suppose V is uniformly continuous; then it follows that V satisfies a Lipschitz condition

$$|V(x) - V(y)| \leq K\|x-y\|. \qquad x, y \in X$$

A sublinear functional p is a positively-homogeneous functional, satisfying in addition

$$p(x) + p(y) \geq p(x+y) \qquad x, y \in X.$$

It follows from the Hahn-Banach theorem that if p is a continuous sublinear functional then

$$p(x) = \max_{x^* \in C(p)} x^*(x)$$

where $C(p) = \{x^* \in X^* : x^*(x) \leq p(x),\ x \in X\}$ is a weak*-compact convex subset of X^*. Conversely if C is a weak*-compact convex subset of X^* then $\max_{x^* \in C} x^*(x)$ is a continuous sublinear functional.

If V is a positively homogeneous functional then we define $p_V : X \to R$ by

$$p_V(x) = \sup_{y \in X} [V(x+y) - V(y)].$$

Lemma 2.1 If V is uniformly continuous then p_V is a continuous sublinear functional.

The corresponding convex subset $C(p_V)$ of X^* is called the *support* of V or $\mathrm{Supp}V$.

A weak*-compact convex subset C of X^* is *V-admissible* if $\max_{x^* \in C} x^*(x) \geq V(x)$ for $x \in X$. It is clear that $\mathrm{Supp}V$ is V-admissible. By an argument based on Zorn's Lemma it is easy to verify that

Lemma 2.2 Every V-admissible subset of X^* contains a minimal V-admissible set.

We shall denote by $\mathcal{M}(V)$ the class of minimal V-admissible subsets of X^*; from Lemma 2.2 it follows that $\mathcal{M}(V) \neq \emptyset$. We now come to our main representation theorem.

Theorem 2.3 If V is a uniformly continuous positively homogeneous functional on X then

$$V(x) = \min_{M \in \mathcal{M}(V)} \max_{x^* \in M} x^*(x).$$

Proof (Sketch). For given $y \in X$ we define

$$C(y) = \{x^* \in \mathrm{Supp}V : x^*(y) \leq V(y)\}.$$

The main part of the proof consists of showing that $C(y)$ is V-admissible (although possibly not minimal). This is done by proving by the Hahn-Banach theorem that for any $z \in X$ there exists $x^* \in \mathrm{Supp}V$ such that

$$x^*(z) \geq V(z)$$

$$x^*(y) \leq V(y).$$

This demonstrates that

$$\max_{x^* \in C(y)} x^*(z) \geq V(z) \qquad z \in X$$

and hence that $C(y)$ is V-admissible.

Now suppose $C \subset C(y)$ is minimal and V-admissible. Then clearly $V(y) = \max_{x^* \in C} x^*(y)$ so that the theorem will follow.

Corollary 2.4 $\overline{\text{co}}[\cup(C : C \in \mathcal{M}(V))] = \text{Supp} V$.

It is now an easy matter to establish our main application, which enables us to identify certain functionals as resulting from a positional game.

Theorem 2.5 Suppose S is a compact Hausdorff space and $V : C(S) \to \mathbb{R}$ is positively homogeneous and in addition satisfies

(i) $V(f) \geq V(g)$ whenever $f \geq g$

(ii) $V(f+\alpha) = V(f) + \alpha$ whenever $\alpha \in \mathbb{R}$.

Then

$$V(f) = \min_{C \in \mathcal{M}} \max_{\mu \in C} \int_S f \, d\mu$$

where $\mathcal{M}$ is the collection of weak*-closed convex subsets C of the set $P(S)$ of regular probability measures on S, which are minimal with respect to the condition

$$\max_{\mu \in C} \int_S f \, d\mu \geq V(f).$$

Theorem 2.5 is proved by observing that under conditions (i) and (ii), $p_V(f) \leq \max_{s \in S} f(s)$ for any $f \in C(S)$.

We shall call a functional on $C(S)$ satisfying (i) and (ii) *gamonic*. Note that if V is gamonic then so is V^* where $V^*(f) = -V(-f)$. From this we obtain

Corollary 2.6 If $V : C(S) \to \mathbb{R}$ is gamonic then

$$V(f) = \max_{c \in \mathcal{N}} \min_{\mu \in C} \int_S f \, d\mu$$

where $\mathcal{N}$ is the collection of weak*-closed convex subsets of $P(S)$ minimal subject to the condition

$$\min_{\mu \in C} \int_S f \, d\mu \leq V(f) \qquad f \in C(S).$$

3. THE INDICATOR FUNCTION

It is natural to consider the extension of V to certain discontinuous functions. One might hope to obtain a natural extension to all bounded Borel functions as for the case of linear functionals on $C(S)$. This seems to be impossible, but it is possible to define an extension to (upper or lower) semi-continuous functions. Our main result in this direction is:

Theorem 3.1 Suppose $\{f_\lambda\}$ is a decreasing net of continuous functions on S and $u = \lim_\lambda f_\lambda$ pointwise is a bounded function. Then

$$\lim_\lambda V(f_\lambda) = \inf_{C\in \mathcal{M}} \sup_{\mu\in C} \int u\, d\mu.$$

The same result is true for increasing nets. Thus if u is a bounded upper-semi-continuous function we may define

$$\tilde{V}(u) = \inf_{C\in \mathcal{M}} \sup_{\mu\in C} \int u\, d\mu$$

and then expect to obtain decent results. In particular if E is a closed subset of S we define

$$\delta_V(E) = \tilde{V}(\chi_E) = \inf_{C\in \mathcal{M}} \sup_{\mu\in C} \mu(E)$$

and then δ_V is called the *indicator function* of V. The indicator function does not determine V except in one special case (see the next section).

One result we can obtain is a 'value-existence' result that

$$\delta_V(E) = \inf_{C\in \mathcal{M}} \sup_{\mu\in C} \mu(E) = \sup_{C\in \mathcal{N}} \inf_{\mu\in C} \mu(E)$$

for any closed subset E of S.

4. DETERMINISTIC FUNCTIONALS

A gamonic functional V is called *deterministic* if V has a representation

$$V(f) = \min_{E\in \mathcal{E}} \max_{s\in E} f(s)$$

where $\mathcal{E}$ is a collection of closed subsets of S. In this section we obtain some external characterizations of deterministic functionals.

Proposition 4.1 If V is a deterministic gamonic functional, then for any continuous increasing map $\Phi : R \to R$

$$\Phi(V(f)) = V(\Phi \circ f) \qquad \text{for } f \in C(S).$$

Our main result will be a strong converse of Proposition 4.1; first we need to relate deterministic functionals with properties of the indicator functional.

Proposition 4.2 Suppose for every closed subset E of S $\delta_V(E) = 0$ or $\delta_V(E) = 1$. Then V is deterministic.

To prove 4.2 we define $\mathcal{E}$ to be the set of closed E such that $E \cap F \neq \emptyset$ whenever $\delta_V(F) = 1$. It is then possible to show that if $V(f) = \alpha$ then the set $G = \{s : f(s) \leq \alpha\}$ belongs to $\mathcal{E}$.

Theorem 4.3 Suppose $\Phi : R \to R$ is any non-linear continuous increasing map, and suppose that

$$\Phi(V(f)) = V(\Phi \circ f) \qquad \text{for } f \in C(S).$$

Then V is deterministic.

Proof (Sketch) Φ is almost everywhere differentiable; if c is a point of differentiability then

$$\Phi[\tilde{V}(c + t\chi_E)] = \tilde{V}(\Phi(c + t\chi_E))$$

for $t > 0$ so that

$$\Phi(c + t\delta) = \tilde{V}[\Phi(c) + (\Phi(c+t) - \Phi(c))\chi_E]$$

$$= \Phi(c) + \delta(\Phi(c+t) - \Phi(c))$$

where $\delta \equiv \delta_V(E)$. Hence

$$\Phi(c+t\delta) - \Phi(c) = \delta[\Phi(c+t) - \Phi(c)].$$

Hence for any natural number n,

$$\Phi(c+t\delta^n) - \Phi(c) = \delta^n[\Phi(c+t) - \Phi(c)].$$

Suppose $0 < \delta < 1$; then letting $n \to \infty$

$$t\Phi'(c) = \Phi(c+t) - \Phi(c)$$

and hence Φ is linear on $\{\xi : \xi \geq c\}$. It quickly follows that Φ is everywhere linear.

Hence $\delta = 0$ or 1 and we use 4.2.

We note that under the hypotheses of Theorem 4.3, each $C \in M$ is a face of $P(S)$, i.e. if $C \in M$ there exists a closed subset E of S such that

$$C = \{\mu \in P(S) : \mu(E) = 1\}.$$

REFERENCES

1. J.M. DANSKIN, Differential games with perfect information, to appear.

2. R.J. ELLIOTT and N.J. KALTON, The existence of value in differential games, Mem. Amer. Math. Soc. No.126(1972).

3. ——— ———, Upper values of differential games, J. Diff. Eqns. 14(1973) 89-100.

4. R.J. ELLIOTT, A. FRIEDMAN and N.J. KALTON, Alternate play in differential games, J. Diff. Eqns. 15(1974) 560-588.

5. A. FRIEDMAN, Stochastic differential games, J. Diff. Eqns. 11(1972) 79-108.

6. N.J. KALTON, On a class of non-linear functionals on spaces of continuous functions, in preparation.

7. ———, The minimum principle and an abstract approach to differential games, in preparation.

CATASTROPHE THEORY IN DIFFERENTIABLE GAMES

L. Markus

Department of Mathematics
Universities of Minnesota, USA, and Warwick, UK

ABSTRACT. A family of two-person zero-sum games with payoffs $P_t(x, y)$ is analysed qualitatively. In particular the dependence of the local equilibria or game saddles on the parameter t is studied. The theory of catastrophe is used to classify the resulting discontinuities and singularities.

1. INTRODUCTION TO THE PROBLEM OF VARIABLE GAMES

Consider a two-person zero-sum game with pure strategies x and y each ranging over the real number line R. The payoff $P(x, y)$, from the y-player to the x-player, will be assumed to be a differentiable (this shall mean of class C^∞) real function

$$P : R \times R \to R.$$

Since the x-player seeks to maximise P, the y-player to minimise P, the game value is defined by

$$\max_x \min_y P(x, y) = \min_y \max_x P(x, y)$$

provided these extrema are assumed in the (x, y) plane R^2. Pure strategies (x_o, y_o) which provide a solution of the game, satisfy the equilibrium or saddle condition

$$P(x, y_o) \leq P(x_o, y_o) \leq P(x_o, y)$$

J. D. Grote (ed.), The Theory and Application of Differential Games. 301-310.

so
$$\frac{\partial P}{\partial x} = \frac{\partial P}{\partial y} = 0 \text{ and } \frac{\partial^2 P}{\partial x^2} \leq 0, \frac{\partial^2 P}{\partial y^2} \geq 0 \text{ at } (x_o, y_o) .$$

The prototype of such games is described by the payoff

$$P(x, y) = y^2 - x^2 + 1 ,$$

with value 1 achieved at the unique equilibrium $x_i = 0$, $y_o = 0$.

We shall be concerned with the qualitative description of local equilibria or game saddles, as distinct from the global equilibrium of the above prototype $(y^2 - x^2 + 1)$. Moreover we shall study the dependence of such local equilibria on various parameters t entering the payoff function $P_t(x, y)$. Almost entirely we restrict the parameter t to be a real variable in R. Such parametrized families of payoff functions are analysed from the viewpoint of catastrophe theory. Moreover new aspects of the theory of catastrophe singularities arise because of the game-theoretic nature of our problem..

Our results are incomplete, hardly more than a beginning. Some of the statements made here must be accepted only guardedly, as they are little more than considered judgements.

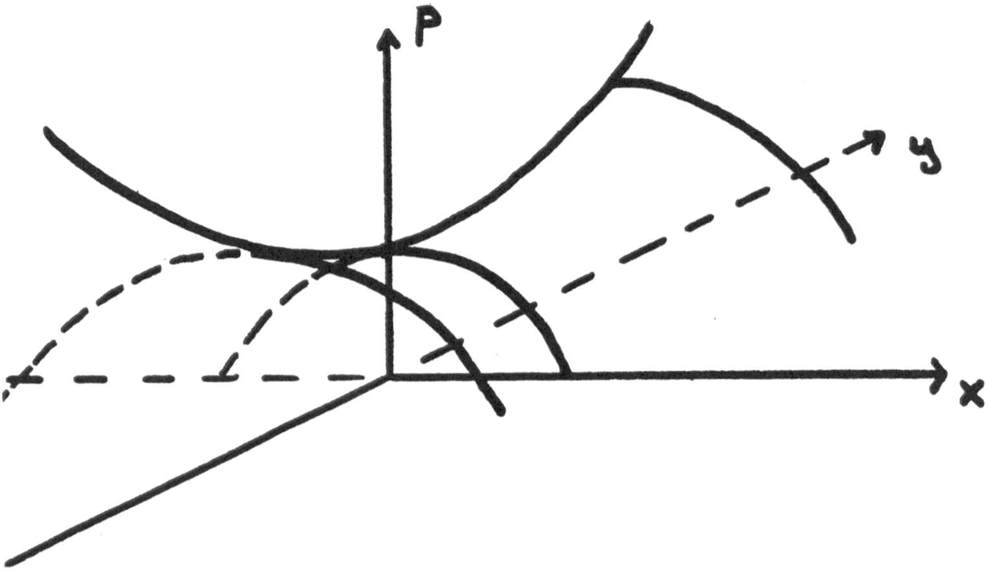

Graph of P(x, y) as hyperbolic-paraboloid or saddle surface over the (x, y) -plane.

2. LOCAL SOLUTIONS OF DIFFERENTIAL GAMES

Consider a two-person zero-sum game with differentiable payoff $P(x, y)$ for pure strategies along the axes of the (x, y)-plane R^2.

<u>Definition</u>. A local equilibrium or game-solution is a pair of pure strategies (x_o, y_o) that lie interior to some compact rectangle, R_o, with sides parallel to the axes, in which

$$\max_x \min_y P(x, y) = \min_y \max_x P(x, y),$$

$$\text{for } (x, y) \in R_o ,$$

and this local game-value is attained at (x_o, y_o) .

<u>Theorem</u>. At a local equilibrium (x_o, y_o) the payoff $P(x, y)$ has a critical point, that is,

$$\frac{\partial P}{\partial x} = \frac{\partial P}{\partial y} = 0 \quad \text{at } (x_o, y_o)$$

Furthermore the Taylor series yields

$$P(x,y) = P(x_o,y_o)+\tfrac{1}{2}(A(x-x_o)^2+2B(x-x_o)(y-y_o)+C(y-y_o)^2)+ \ldots$$

where

$$B^2 - AC \geq 0 \text{ and } A = \frac{\partial^2 P}{\partial x^2} \leq 0, \quad C = \frac{\partial^2 P}{\partial y^2} \geq 0 .$$

On the other hand if (x_o,y_o) is any critical point of $P(x,y)$ where $A < 0$ and $C > 0$, then (x_o,y_o) is a local equilibrium.

<u>Remark</u>. Disregarding special coincidences or degeneracies corresponding to the strict vanishing of A or C, we can recognise a local game-solution as a critical point of $P(x,y)$ where $A < 0$ and $C > 0$. Each such critical point is a nondegenerate saddle point of $P(x,y)$ in R^2 .

3. REMARKS ON CRITICAL POINT THEORY AND CATASTROPHE SINGULARITIES

Let F_n be the collection of differentiable real functions $f(x^1, \ldots, x^n)$ on R^n, so each

$$f : R^n \to R$$

is of class C^∞. A critical point q_o of $f \in F_n$ is a point of R^n where $df = 0$, so

$$\frac{\partial f}{\partial x^1} = \frac{\partial f}{\partial x^2} = \ldots = \frac{\partial f}{\partial x^n} = 0 \text{ at } q_o \,.$$

The critical point q_o is nondegenerate in case the Hessian symmetric matrix H $({}^{o}\partial^2 f/\partial x^i\, \partial x^j)$ is nonsingular. In this case the signature of H is an invariant, under differentiable changes of local coordinates about q_o in R^n. Moreover, as proved by M. Morse, there then exist local coordinates $(\xi^1, \ldots, \xi^n)$ in which $f - f(q_o)$ is a quadratic sum or difference of squares

$$f(\xi) - f(q_o) = \pm (\xi^1)^2 \ldots \quad \pm (\xi^n)^2 \,.$$

Thus nondegenerate critical points of a function $f \in F_n$ can be classified, up to local diffeomorphism, by the signature of H.

Also a nondegenerate critical point q_o of $f \in F_n$ is structurally stable in the sense that any approximation $\tilde{f}$ of f must have a unique critical point $\tilde{q}_o$ near q_o, and $\tilde{f}$ at $\tilde{q}_o$ is nondegenerate of the same type as f at q_o. Such structural stability fails for every degenerate critical point. For instance, take $f = x^3$ in F_1 at $q_o = 0$ and consider the close approximation $f = x^3 - \varepsilon x$ which has a maximum and a minimum approaching the origin as $\varepsilon \to 0+$.

Clearly it is often useful to approximate a function $f \in F_n$ by some other $f \in F_n$ which has only nondegenerate critical points. We require a precise concept of approximation which will permit higher accuracy in various delicate localities, say on some sequence of points which tend to infinity in R^n. The appropriate mode of approximation and perturbation is described in terms of the Whitney C^∞-topology on F_n. Here a neighbourhood N of $f \in F_n$ is defined by choosing some order $k \geq 0$ of differentiation and a real positive tolerance $\varepsilon(x) \in F_n$. Then $f \in F_n$ belongs to N in case

$$|D^{\alpha_1 + \alpha_2 + \ldots + \alpha_n} (f - \tilde{f}) \quad | < \varepsilon\,(x)$$

for all partial derivatives D^α, with $\alpha = \alpha_1 + \ldots + \alpha_n$, of total order $0 \leq \alpha \leq k$ and for all $x \in R^n$. Then F_n becomes a topological space (nonmetrizable) with the Baire property that any countable intersection of open-dense subsets of F_n is still dense in F_n. For this reason we often call a subset $\mathcal{J} \subset F_n$ (or the property specifying $\mathcal{J}$) generic in case $\mathcal{J}$ is open-dense in

F_n, or merely that $\mathfrak{J}$ contains a countable intersection of open-dense subsets of F_n.

We summarize our considerations with the known generic property of the Morse functions $\mathfrak{M}_n$: The subset $\mathfrak{M}_n \subset F_n$ of all functions $f \in F_n$ having only nondegenerate critical points is open-dense in F_n.

It would thus appear that for practical studies of applied mathematics where small uncertainties or perturbations might arise we could restrict our attention to the Morse functions $\mathfrak{M}_n$. But this conclusion is not warranted if we wish to study a family of functions $f_t(x)$ (here $f_t(x) = f(x,t) \in F_{n+1}$) where t ranges over the real line R (or perhaps over the r-vector space R^r). Then it is possible that the function $f_t(x) \in F_n$, for each fixed $t \neq 0$, has only nondegenerate critical points, but these become degenerate and change type at $t = 0$. Moreover it could happen that any perturbation of the whole family to f(x) in F_{n+1} would still display such degenerate critical points for some t_o near $t_o = 0$. Such a critical point of the family $f_t(x)$ would define a catastrophe singularity at t_o, as defined later, and it must be considered as a significant physical or geometric reality that cannot be avoided by perturbing the family $f_t(x)$.

Example Consider $P_t(x,y) = y^2 - (x^3 - tx) + 1$ in F_3. For each $t \to 0+$ there is a local saddle point and a local maximum very near (0,0) in the (x,y)-plane. At $t = 0$ these two nondegenerate critical points coalesce into the degenerate critical point of $(y^2 - x^3 + 1)$. For $t < 0$ there are no critical points of $P_t(x,y)$ in R^2.

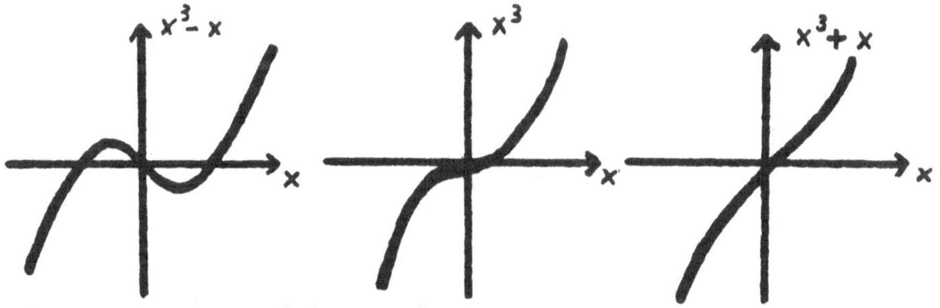

Coalescence of critical points as $t \to 0+$.

Moreover it seems clear that any perturbation $\tilde{P}_t(x,y)$ of $P_t(x,y)$ in F_3 must also display a degenerate critical point near the origin in R^2, for some fixed $\tilde{t}_o$ near 0. This is apparent if we recall that the local saddle of $\tilde{P}_t$, for $t > 0$, can always be continued as t decreases unless it becomes degenerate. Also for $t < 0$ the function $\tilde{P}_t$ has no critical points in R^2 since it closely approximates P_t which has none. Thus for some $\tilde{t}_o$ near 0 the function $\tilde{P}_{\tilde{t}_o}$ has a degenerate critical point.

A further analysis of the nature of this degenerate critical point is possible by considering the Hessian matrix. The Hessian of the saddle point P_t is

$$H(t) = \begin{pmatrix} A(t) & B(t) \\ B(t) & C(t) \end{pmatrix}$$

with $A(t)$, $B(t)$, $C(t)$ defining a differentiable curve in the (A,B,C)-space R^3. In this (A,B,C)-space we note this curve $H(t)$ as it leaves the saddle region $B^2 - AC > 0$ in R^3 to meet the boundary 2-surface $S : B^2 - AC = 0$. We have already noted that for any perturbed family $\tilde{P}_t$ the corresponding curve $\tilde{H}(t)$ also meets S in R^3. This is reasonable since a curve in R^3 can be expected to meet a surface $S \subset R^3$ in a transversal manner. However we would not expect such a perturbed curve $\tilde{H}(t)$ to meet the 1-dimensional subset of R^3 where $B = 0$ and $AC = 0$, and a suitable perturbation does miss this locus of co-dimension 2 in R^3. However if t were a 2-vector parameter, then some such higher degeneracy of the critical point might prove unavoidable.

This example, and the rambling discussion, gives a taste of the theory of catastrophes, and we next state this famous theorem of Thom [1] following the development of Zeeman [2].

Let F_{n+r} denote the space of real C^∞-functions on R^{n+r}, with the Whitney C^∞-topology. Let $f : R^n \times R^r \to R$ be a differentiable function in F_{n+r} and define $M_f \subset R^{n+r}$ to be given by

$$\frac{\partial f}{\partial x^1}, \ldots, \frac{\partial f}{\partial x^n} = \text{grad}_x f = 0 .$$

Here $(x^1, \ldots, x^n)$ are coordinates in R^n and $(t^1, \ldots, t^r)$ are coordinates in R^r for fixed $n \geq 1$ and $r \geq 1$. Generically M_f is a differentiable r-manifold (smooth r-surface) of codimension n because it is given by n equations in R^{n+r}. Let $\chi_f : M^r_f \to R^r$ be the map induced by the projection $R^{n+r} \to R^r$. We call χ_f the catastrophe map of f and a point of the critical manifold M_f at which χ_f is singular (Jacobian has rank $< r$) is called a catastrophe singularity. We can now state Thom's theorem.

<u>Theorem</u>. If $r \leq 5$ there is an open-dense set $F^*_{n+r} \subset F_{n+r}$ which we call generic functions, with the following properties: For $f \in F^*_{n+r}$

1) M_f is a differentiable r-manifold imbedded in R^{n+r}

2) Any singularity of χ_f in $M_f^r \to R^r$ is equivalent to one of a finite number of types called <u>elementary catastrophes</u>

3) χ_f is locally (structurally) stable at all points of M_f, with respect to small perturbations of f in F^*_{n+r} .

The number of elementary catastrophes depends on r as follows

r	1	2	3	4	5	6
elem.cats.	1	2	5	7	11	∞

We now comment on the nature of the equivalence and stability mentioned in the theorem.

Here equivalence means the following: two maps $\chi : M \to N$ and $\chi' : M' \to N'$ are equivalent if there exist diffeomorphisms h and k such that the diagram commutes

$$\begin{array}{ccc} M & \xrightarrow{\chi} & N \\ h\downarrow & & \downarrow k \\ M' & \xrightarrow{\chi'} & N' \end{array}, \text{ so } k^{-1}\ \chi' h = \chi : M \to N.$$

Suppose the maps χ and χ' have singularities at points x and x', respectively. Then the singularities are equivalent if the above definition holds locally near the singularities with hx = x'.

Next let $\chi_f : M_f \to R^r$ be the catastrophe map for a given $f \in F^*_{n+r}$.

<u>Definition</u>. The catastrophe map χ_f is locally stable at $(x_o,t_o) \in M_f$ if: given a neighbourhood N of (x_o,t_o) in R^{n+r} , there exists a neighbourhood V of f in F^*_{n+r} such that -

for each $g \in V \ \exists\ (x_1,t_1)$ in $N \cap M_g$
and χ_f at (x_o,t_o) is locally equivalent to χ_g at (x_1,t_1).

Of course a critical point x_o of $f \in F^*_{n+r}$, for a fixed t_o, is nondegenerate if and only if χ_f is a local diffeomorphism of M_f into R^r near (x_o,t_o). Also the nondegenerate critical points of f, for a specified fixed signature, fill an open subset of the r-manifold M_f with a boundary in M_f consisting entirely of singular points of the catastrophe map χ_f.

For the case r = 1 where the parameter t is on the real line R, there is just one qualitative type of elementary catastrophe. We refer back to our example $P_t(x,y) = y^2 - (x^3 - tx) + 1$ to illustrate this catastrophe. We compute

$$\frac{\partial P}{\partial x} = -3x^2 + t = 0, \quad \frac{\partial P}{\partial y} = 2y = 0$$

to define the curve $y = 0$, $t = 3x^2$ in (x,y,t)-space. Hence, generically, a saddle point varies smoothly with the real parameter t until it coalesces with an extremum (here a minimum) to form a degenerate saddle, and then both disappear. No other behaviour for any other payoff function, which varies with time t, is possible -- at least in the generic case

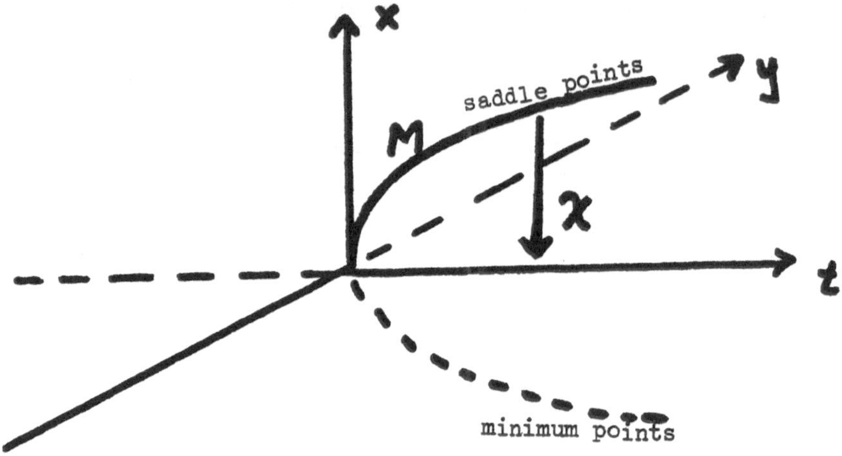

Elementary catastrophe for r = 1.

4. GAME CATASTROPHES

In the above analysis of the critical points of a family $P_t(x,y)$ of payoff functions we classified the nondegenerate critical points in R^2 for each fixed t, and later the catastrophe singularities, up to local diffeomorphism. However such an equivalence relation does not respect the game structure which distinguishes the x and y axes as pure strategies for the two players. Hence we should allow only local coordinate transformations of the form

$$\bar{x} = q(x), \quad \bar{y} = \psi\,(y)$$

for C^∞ functions q and ψ with q' and ψ' nowhere zero. Critical points equivalent under such a transformation will be called game-equivalent.

<u>Definition</u>. Let $P(x,y) \varepsilon F_2$ have a critical point $q_o=(x_o,y_o)$ so

$$P(x,y)=P(x_o,y_o)+\tfrac{1}{2}(A(x-x_o)^2+2B(x-x_o)(y-y_o)+C(y-y_o)^2)+\ldots \quad .$$

Then q_o is a nondegenerate saddle in case $B^2 - AC>0$. Further q_o is then a

1) game saddle in case $A < 0$ and $C > 0$
2) reverse game saddle in case $A > 0$ and $C < 0$
3) non-game saddle in case $0 < AC < B^2$

4) ambiguous game saddle in case $A = 0$ or $C = 0$

Also we call a degenerate saddle, $B^2 - AC = 0$ but not all A, B, C vanishing, a collapsed saddle.

Remark. The above classes of saddle points are clearly invariant under a game-equivalence transformation of local coordinates about q_o. In fact, using game-equivalent local coordinates we can reduce the quadratic terms to the standard forms, in the respective cases:

1) $A = -1$, $C = +1$

2) $A = +1$, $C = -1$

3) $|A| = |C| = 1$ and $AC = +1$

4) $A = 0$ and $C = \pm 1$ (say $A = 0$, $C \neq 0$), $B = 1$.

The first two cases arise when the eigenvectors of H (asymptotic lines) are separated by the x and y-axes in R^2. Case 3 occurs when the eigenvectors lie in the same quadrant, and case 4 when one of the eigenvectors lies along an axis. The collapsed saddle corresponds to the case when the eigenvectors coincide.

The first goal of this detailed classification would be to prove the following expected theorem.

Theorem. A generic payoff family $P_t(x,y)$ in F_3 has a critical manifold M in R^{2+1} whereon the game saddles fill an open set GS of M. Moreover each point q of the boundary of GS in M corresponds to an ambiguous game saddle which is also a boundary point for the non-game saddles. However q cannot be a boundary of the reverse game saddles. Note in particular that q_o is a nondegenerate saddle.

It is easy to see that GS is open in the 1-manifold M and that each boundary point q must be of type 2, 3, 4 or even a collapsed saddle. But at q we have either $A = 0$ or $B = 0$ so only case 4 or the collapsed saddle are possibilities. But in the generic case we can expect the eigenvectors of H to swing across an x or y-axis as $P_t(x,y)$ reaches the point corresponding to q. If one of these eigenvectors swung past an axis at q, then q would correspond to an ambiguous game saddle and it would bound a set of non-game saddles.

It would not be generic for one of the eigenvectors to swing up to coincidence with an axis without thereafter crossing past the axis. Thus we can expect q to bound a set of non-game saddles.

Also it seems unlikely that both eigenvectors of H would coalesce precisely on the x or y-axis, since this is not a generic behaviour.

In order to make this argument into a rigorous proof we need to define the appropriate open-dense subset $F_3^{*G} \subset F_3^* \subset F_3$, and then deduce the conclusions presented above. We do not follow this path to completion here.

Further classification of catastrophe singularities under the relation of game-equivalence remains untouched, especially for the case where the parameter t runs over R^r for some $r \geq 2$, or x and $y \in R^k$ for $2k = n \geq 4$.

REFERENCES

1. R. Thom, Stabilite structurelle et morphogenese, Benjamin, New York, 1972.

2. E. C. Zeeman, The classification of elementary catastrophes of codimension ≤ 5, University of Warwick Report, 1974.